Hydrologic and Geochemical Characterization of the Santa Rosa Plain Watershed, Sonoma County, California

Edited by Tracy Nishikawa

Chapter A
Introduction to the Study Area
By Tracy Nishikawa, Joseph A. Hevesi, Donald S. Sweetkind, and Linda R. Woolfenden

Chapter B
Hydrology of the Santa Rosa Plain Watershed, Sonoma County, California
By Donald S. Sweetkind, Joseph A. Hevesi, Tracy Nishikawa, Peter Martin, and Christopher D. Farrar

Chapter C
Groundwater Quality and Source and Age of Groundwater in the Santa Rosa Plain Watershed, Sonoma County, California
By Peter Martin, Loren F. Metzger, Jill N. Densmore, and Roy A. Schroeder

Chapter D
Conceptual Model of Santa Rosa Plain Watershed Hydrologic System
By Tracy Nishikawa, Joseph A. Hevesi, Donald S. Sweetkind, and Peter Martin

Prepared in cooperation with the Sonoma County Water Agency

Scientific Investigations Report 2013–5118

U.S. Department of the Interior
U.S. Geological Survey

U.S. Department of the Interior
SALLY JEWELL, Secretary

U.S. Geological Survey
Suzette M. Kimball, Acting Director

U.S. Geological Survey, Reston, Virginia: 2013

For more information on the USGS—the Federal source for science about the Earth, its natural and living resources, natural hazards, and the environment, visit http://www.usgs.gov or call 1–888–ASK–USGS.

For an overview of USGS information products, including maps, imagery, and publications, visit http://www.usgs.gov/pubprod

To order this and other USGS information products, visit http://store.usgs.gov

Suggested citation:
Nishikawa, Tracy, ed., 2013, Hydrologic and geochemical characterization of the Santa Rosa Plain watershed, Sonoma County, California: U.S. Geological Survey Scientific Investigations Report 2013–5118, 178 p.

Contents

Chapter A

Introduction to the Study Area

Contents—Continued

Chapter B

Hydrology of the Santa Rosa Plain Watershed, Sonoma County, California

Contents—Continued

Contents—Continued

Chapter C

Groundwater Quality and Source and Age of Groundwater in the Santa Rosa Plain Watershed, Sonoma County, California

Contents—Continued

Chapter D
Conceptual Model of Santa Rosa Plain Watershed Hydrologic System

Figures

Continued

Figures—Continued

Figures—Continued

Chapter C

Chapter D

Tables

Chapter A

Chapter B

Chapter C

Conversion Factors

Inch/Pound to SI

Multiply	By	To obtain
Length		
inch (in.)	25.4	millimeter (mm)
foot (ft)	0.3048	meter (m)
mile (mi)	1.609	kilometer (km)
mile, (mi)	1.609	kilometer (km)
Area		
acre	4,047	square meter (m^2)
acre	0.4047	hectare (ha)
acre	0.4047	square hectometer (hm^2)
acre	0.004047	square kilometer (km^2)
square mile (mi^2)	259.0	hectare (ha)
square mile (mi^2)	2.590	square kilometer (km^2)
Volume		
gallon (gal)	3.785	liter (L)
acre-foot (acre-ft)	1,233	cubic meter (m^3)
acre-foot (acre-ft)	0.001233	cubic hectometer (hm^3)
Flow rate		
acre-foot per day (acre-ft/d)	0.01427	cubic meter per second (m^3/s)
acre-foot per year (acre-ft/yr)	1,233	cubic meter per year (m^3/yr)
acre-foot per year (acre-ft/yr)	0.001233	cubic hectometer per year (hm^3/yr)
cubic foot per second (ft^3/s)	0.02832	cubic meter per second (m^3/s)
inch per year (in/yr)	25.4	millimeter per year (mm/yr)
Hydraulic conductivity		
foot per day (ft/d)	0.3048	meter per day (m/d)
Leakance		
foot per day per foot [(ft/d)/ft]	1	meter per day per meter
foot per second (ft/s)	0.3048	meter per second (m/s)
foot per day (ft/d)	0.3048	meter per day (m/d)
cubic foot per second (ft^3/s)	0.02832	cubic meter per second (m^3/s)
cubic foot per day (ft^3/d)	0.02832	cubic meter per day (m^3/d)
inch per year per foot [(in/yr)/ft]	83.33	millimeter per year per meter [(mm/yr)/m]
Radioactivity		
picocurie per liter (pCi/L)	0.037	becquerel per liter (Bq/L)
Transmissivity*		
foot squared per day (ft^2/d)	0.09290	meter squared per day (m^2/d)

Temperature in degrees Celsius (°C) may be converted to degrees Fahrenheit (°F) as follows:

°F=(1.8×°C)+32

Temperature in degrees Fahrenheit (°F) may be converted to degrees Celsius (°C) as follows:

°C=(°F-32)/1.8

Vertical coordinate information is referenced to the North American Vertical Datum of 1988 (NAVD 88) and National Geodetic Vertical Datum of 1929 (NGVD29).

Horizontal coordinate information is referenced to the North American Datum of 1983 (NAD 83).

Altitude, as used in this report, refers to distance above the vertical datum.

*Transmissivity: The standard unit for transmissivity is cubic foot per day per square foot times foot of aquifer thickness [(ft^3/d)/ft^2]ft. In this report, the mathematically reduced form, foot squared per day (ft^2/d), is used for convenience.

Specific conductance is given in microsiemens per centimeter at 25 degrees Celsius (μS/cm at 25°C).

Concentrations of chemical constituents in water are given either in milligrams per liter (mg/L) or micrograms per liter (μg/L)

Abbreviations

APHA	American Public Health Association
asl	above sea level
bls	below land surface
CB	Cotati Basin
CDPH	California Department of Public Health
CDWR	California Department of Water Resources
ECON	The Environment, Community, and Opportunity Network
ELAP	Enviromental Laboratory Accreditation Program (California)
GAMA	Groundwater Ambient Monitoring and Assessment
ET	evapotranspiration
GIS	geographic information system
GMWL	global meteoric water line
LLNL	Lawrence Livermore National Laboratory
LMWL	local meteoric water line
MCL	maximum contaminant level (USEPA or California)
NL	notification level (California)
NOSAMS	National Ocean Sciences Accelerator Mass Spectrometry Facility
NSF	North San Francisco
PES	PES Environmental, Inc.
SMCL	secondary maximum contaminant level
SPR	spring site
SRP	Santa Rosa Plain
SRPW	Santa Rosa Plain Watershed
SWRCB	State Water Resources Control Board (California)
SW	surface-water site
T/R/S	township/range/section
UPL	upland
USEPA	U.S. Environmental Protection Agency
USGS	U.S. Geological Survey
VAL	valley
VSMOW	Vienna Standard Mean Ocean Water
WB	Windsor Basin
WG	Wilson Grove
W	well site (groundwater)
As	arsenic
^{12}C	carbon-12
^{13}C	carbon-13
^{14}C	carbon-14
DIC	dissolved inorganic carbon
DO	dissolved oxygen

Abbreviations—Continued

Fe	iron
^1H	hydrogen isotope
^2H or D	deuterium
^3H	tritium
^3He	helium-3
N	nitrogen
NO_3	nitrate
^{16}O	oxygen-16
^{18}O	oxygen-18
pmc	percent modern carbon
t	time
TDS	total dissolved solids
TU	tritium unit
VOCs	volatile organic compounds
$\delta_i E$	delta notation; the ratio of the heavier isotope (i) to the more common lighter isotope of an element (E), relative to a standard reference material, expressed as per mil
‰	per mil (parts per thousand)

Well-Numbering System

Wells are identified and numbered according to their location in the rectangular system for the subdivision of public lands. Identification consists of the township number, north or south; the range number, east or west; and the section number. Each section is divided into sixteen 40-acre tracts lettered consecutively (except I and O), beginning with "A" in the northeast corner of the section and progressing in a sinusoidal manner to "R" in the southeast corner. Within the 40-acre tract, wells are sequentially numbered in the order they are inventoried. The final letter refers to the base line and meridian. In California, there are three base lines and meridians; Humboldt (H), Mount Diablo (M), and San Bernardino (S). All wells in the study area are referenced to the San Bernardino base line and meridian (S) Well numbers consist of 15 characters and follow the format 011N001E24Q008S. In this report, well numbers are abbreviated and written 11N/1E-24Q8. Wells in the same township and range are referred to only by their section designation, 24Q8. The following diagram shows how the number for well 11N/1E-24Q8 is derived.

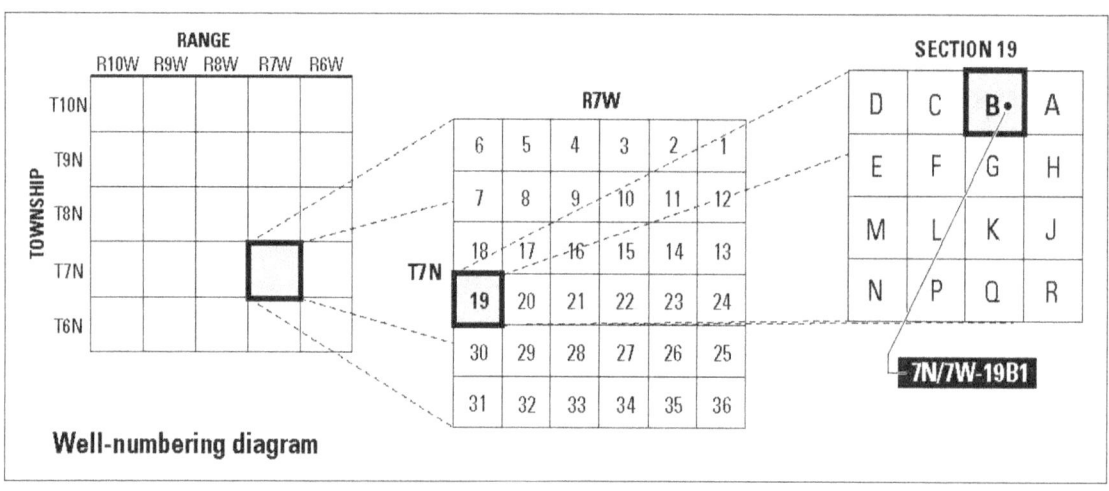

Well-numbering diagram

Acknowledgements

The authors acknowledge the support of Sonoma County Water Agency. The authors thank the city of Santa Rosa, the city of Sebastopol, the city of Rohnert Park, the city of Cotati, the town of Windsor, California American Water, and Sonoma County and their respective personnel. The authors also acknowledge the data contributions by the California Department of Water Resources and the California Department of Public Health.

The authors thank the two USGS colleague reviewers; their comments greatly improved this report.

Hydrologic and Geochemical Characterization of the Santa Rosa Plain Watershed, Sonoma County, California

By Tracy Nishikawa

Executive Summary

The Santa Rosa Plain is home to approximately half of the population of Sonoma County, California, and faces growth in population and demand for water. Water managers are confronted with the challenge of meeting the increasing water demand with a combination of water sources, including local groundwater, whose future availability could be uncertain. To meet this challenge, water managers are seeking to acquire the knowledge and tools needed to understand the likely effects of future groundwater development in the Santa Rosa Plain and to identify efficient strategies for surface- and groundwater management that will ensure the long-term viability of the water supply. The U.S. Geological Survey, in cooperation with the Sonoma County Water Agency and other stakeholders in the area (cities of Cotati, Rohnert Park, Santa Rosa, and Sebastopol, town of Windsor, Cal-American Water Company, and the County of Sonoma), undertook this study to characterize the hydrology of the Santa Rosa Plain and to develop tools to better understand and manage the groundwater system.

The objectives of the study are: (1) to develop an updated assessment of the hydrogeology and geochemistry of the Santa Rosa Plain; (2) to develop a fully coupled surface-water and groundwater-flow model for the Santa Rosa Plain watershed; and (3) to evaluate the potential hydrologic effects of alternative groundwater-management strategies for the basin. The purpose of this report is to describe the surface-water and groundwater hydrology, hydrogeology, and water-quality characteristics of the Santa Rosa Plain watershed and to develop a conceptual model of the hydrologic system in support of the first objective. The results from completing the second and third objectives will be described in a separate report.

Study Area

The area of interest of this study is referred to herein as the Santa Rosa Plain watershed.

- The population centers are the cities of Santa Rosa, Rohnert Park, Cotati, Sebastopol, and the town of Windsor.

- The 167,410-acre Santa Rosa Plain watershed area includes all of the Mark West Creek watershed

(161,410 acres), with areas to the northwest and south of the Mark West Creek watershed boundary added to include most of the Santa Rosa Plain groundwater subbasin as defined the California Department of Water Resources. The Mark West Creek watershed includes three drainage basins: Mark West Creek, Santa Rosa Creek, and Laguna de Santa Rosa.

- The Santa Rosa Plain watershed includes the Santa Rosa Plain and Rincon Valley groundwater subbasins, as well as parts of the Wilson Grove Formation Highlands groundwater basin, Kenwood Valley groundwater basin, Healdsburg area groundwater subbasin, and Alexander Valley groundwater subbasin.

Geohydrology

The geohydrology of the study area was refined through the incorporation of recently published studies of subsurface geology, geophysics, and surface geologic mapping. The specific goals of this study were to identify the surface and subsurface configuration of the water-bearing units, the textural variations in these units, the major structures in the study area, and the three-dimensional shape of the basins.

Major findings regarding the geohydrology from this study are the following:

- Four principal aquifer units were defined for the study area: the Glen Ellen Formation (including Quaternary alluvial deposits), Wilson Grove Formation, the Petaluma Formation, and the Sonoma Volcanics. The units were identified through borings, geophysics, and surface exposures.

- Reported hydraulic-conductivity values for the Glen Ellen Formation range from 13 to 23 feet per day. Reported specific-yield values for the formation range from 3 to 7 percent.

- Reported hydraulic-conductivity values for the Wilson Grove Formation range from 2 to 65 feet per day. Estimated and reported specific yield and storativity values for the formation range from 10 to 20 percent and from 0.00095 to 0.08, respectively.

- Reported transmissivity values for the Petaluma Formation range from 130 to 1,600 square feet per day.

- Reported transmissivity values for the Sonoma Volcanics range from 0.8 to 5,300 square feet per day. Reported specific-yield values for the formation range from 0 to 15 percent.

- Geophysical data indicate that two main groundwater sub-basins exist beneath the Santa Rosa Plain; the basins are as deep as 4,500 ft below the town of Windsor and as deep as 10,000 ft beneath Rohnert Park. The two basins are separated by a buried bedrock ridge, the Trenton Ridge, which has a minimum depth of about 1,000 ft below land surface.

- Faults could be barriers to groundwater flow in the Santa Rosa Plain watershed. Recent surface geologic mapping and geophysical studies have refined the locations of the major faults within the Santa Rosa Plain watershed.

- Three-dimensional subsurface models of lithologic variations indicate a west-to-east transition from dominantly fine-grained marine sands to heterogeneous continental sediments interbedded with Sonoma Volcanics. In contrast to previous studies, the three-dimensional models of the Santa Rosa Plain watershed indicate that the Petaluma Formation extends throughout the deeper parts of the basins beneath the Santa Rosa Plain.

- The most consistently productive wells in the study area extract water from the Wilson Grove Formation. The Glen Ellen Formation is heterogeneous and contains substantial amounts of clay such that well production from this unit is highly variable. Previous reports have reported that the Glen Ellen Formation is as thick as 3,000 ft; however, recent work, incorporated in this report, indicates that the thickness of the formation is variable, but typically hundreds of feet. The Petaluma Formation is the deepest and thickest aquifer; it is generally less permeable than others, but water is produced from sandy horizons within the formation.

- For the purposes of this study, five groundwater storage units were defined on the basis of previous work, hydrogeology, and fault locations.

  - The Windsor Basin (WB) storage unit is located north of the Trenton Ridge fault, west of the Mayacmas Mountain foothills, and east of the Sebastopol fault.

  - The Cotati Basin (CB) storage unit is located south of the Trenton Ridge fault, west of the Sonoma Mountain foothills, and east of the Sebastopol fault.

  - The Wilson Grove (WG) storage unit is located between the Mendocino Range and the Sebastopol fault.

  - The Valley (VAL) storage unit includes the alluvial fill of the Rincon, Bennett, and Kenwood valleys.

- The Uplands (UPL) storage unit includes the Mayacmas and Sonoma mountains east of the Rodgers Creek fault zone but excludes the Valley storage unit.

Surface-Water Hydrology

The surface-water hydrology of the study area was evaluated on the basis of streamflow records from 15 U.S. Geological Survey stream gages. Seven gages within the Santa Rosa Plain watershed had records less than 10 years, and the remaining gages had records of about 12 years. Records from three additional gages in watersheds adjacent to the Santa Rosa Plain watershed were also evaluated to extend the record of the lower Mark West Creek gage (gage number 11466800, Mark West Creek near Mirabel Heights). The length of record for the three gages outside of the Santa Rosa Plain watershed varied from about 36 to 69 years. The data from the Napa River near Napa gage (gage number 11458000) were selected for use in extending the Mark West Creek near Mirabel Heights record.

Major findings regarding the surface-water hydrology from this study are the following:

- Highly variable streamflow—Streamflow within the Santa Rosa Plain watershed is variable with high flows (7,200 cubic feet per second at Mark West Creek near Mirabel Heights) during winter and very low to zero flow during summer.

- Rapid response to precipitation—Streamflow within the Santa Rosa Plain watershed is characterized as having a rapid response relative to precipitation events.

- Very low summer flows—Baseflow is a minor component of total mean flow, but it constitutes the only flow during the dry, summer months. Most of the stream reaches within the Santa Rosa Plain watershed are intermittent.

- Flow-duration curves—Flow-duration curves were used to summarize flow conditions at stream gages in and next to the Santa Rosa Plain watershed. Daily mean discharges of 600 cubic feet per second and higher occurred 10 percent of the time at only one gage within the Santa Rosa Plain watershed that had continuous records (11466800, Mark West Creek near Mirabel Heights). Daily mean discharges of 10 cubic feet per second and less occurred 35 percent of the time at all gages with continuous records within the Santa Rosa Plain watershed.

- Record extension—Correlations between the Mark West Creek near Mirabel Heights gage (11466800) and the Napa River near Napa gage (expressed as coefficient of determination, r^2) were found to be good ($0.59 \leq r^2 \leq 0.91$). The maintenance of variance-extension, type 1 (MOVE.1) method was used to extend (water-years 1960–2005) the record at the Mark West Creek near Mirabel Heights gage. Considering the combined record of flood flowrates between 1959 and 2010 and assuming flood flowrates of 6,000 or 8,000 cubic feet per second, Mark West Creek flooded 62 and 25 days, respectively.

- Flooding in the Laguna de Santa Rosa—In the Laguna de Santa Rosa floodplain, streamflow can be affected by backwater from the Russian River. Flow can reverse during periods of flooding on the Russian River.

Groundwater, Groundwater Movement, and Groundwater Levels

The groundwater reservoir in the study area includes the saturated sedimentary rocks and sediments underlying the floor of Santa Rosa Valley and neighboring lowlands, as well as the volcanic rocks underlying the mountains to the east of the Santa Rosa Valley where such rocks are sufficiently permeable to yield water. Beneath the floor of Santa Rosa Valley, the principal groundwater reservoir is lithologically heterogeneous but consists of one continuous body of saturated material. This reservoir is typically in hydraulic communication laterally with permeable consolidated rocks that underlie the uplands surrounding the Santa Rosa Valley and interfinger with the basin-fill deposits. Groundwater in the principal reservoir is contained in the pore spaces of the Quaternary alluvial materials and Tertiary sedimentary material, including the Glen Ellen Formation, the Wilson Grove Formation, and the Petaluma Formation. Groundwater is also contained in locally permeable areas within the Sonoma Volcanics. Major findings regarding groundwater movement and levels from this study are the following:

- Significant sources of groundwater recharge in the Santa Rosa Plain watershed are infiltration of precipitation, infiltration from streams, and irrigation-return flow.

- Significant groundwater sinks are groundwater pumpage, evapotranspiration from phreatophytes, and baseflow in streams.

- The water-level contour maps show that, on a larger scale, groundwater flows from the Mayacmas and Sonoma Mountains in the Uplands storage unit westward, toward the Laguna de Santa Rosa on the western edge of the Santa Rosa Plain, and eastward, from the highlands in the Wilson Grove storage unit toward the Laguna de Santa Rosa.

- In general, the water-level contours indicate that Santa Rosa Creek gains water east of the Rodgers Creek fault zone, and Mark West and Santa Rosa creeks gain water in the western part of the Santa Rosa Plain.

- Water-level contours, based on available data, indicate cones of depression in the Cotati/Rohnert Park area and north of Sebastopol from Spring 1974 to Fall 2001; however, water levels partially recovered in the Cotati/Rohnert Park area by 2007 in response to decreased pumping.

- Groundwater levels in monitoring wells showed response to pumping, with variable response with depth.

- Pumping tests published in 1987 to determine if the Sebastopol fault is a barrier to groundwater flow were inconclusive.

Groundwater Quality

Groundwater quality was characterized for the Santa Rosa Plain watershed by using analyses for selected physical properties and inorganic constituents compiled from previous investigations and databases maintained by the U.S. Geological Survey, California Department of Public Health, California Department of Water Resources, and public-supply purveyors, for the period 1947–2010. These data were used to characterize the areal, vertical, and temporal variations in groundwater quality and to identify constituents of potential concern. Stable and radioactive isotopes analyzed from groundwater samples collected in 2004 as part of the U.S. Geological Survey Groundwater Ambient Monitoring and Assessment Program or analyzed from surface-water and groundwater samples collected from 2006–10 as part of this study, or concurrently by private consultants, were used to help identify the recharge source and age of groundwater in the study area.

Major findings on water quality from this study are the following:

- Major ions.

  - Samples from springs in the Mayacmas Mountains were a mixed cation-bicarbonate type water, with dissolved-solids concentrations less than about 100 milligrams per liter—the lowest dissolved-solids concentrations in the Santa Rosa Plain watershed. The median dissolved-solids concentrations of well samples in the Uplands and Valley storage units were 330 and 392 milligrams per liter, respectively. The higher dissolved-solids values compared with springs in hard rock settings reflect increasing sediment and water interaction and, perhaps, anthropogenic effects.

  - Most of the well samples from the Windsor Basin and Cotati Basin storage units were mixed cation-bicarbonate and sodium-bicarbonate type waters. Data indicated limited vertical mixing between upper and lower aquifers.

  - Most well samples in the Wilson Grove storage unit were calcium-bicarbonate type or mixed cation-bicarbonate type waters with dissolved-solids concentrations less than 300 milligrams per liter.

  - Data indicated that the Rodgers Creek fault zone restricts the lateral movement of water from the Uplands and Valley storage units to the Windsor Basin and Cotati Basin storage units.

- Data indicated little mixing across the Sebastopol fault.

- Temporal variation of specific conductance and chloride.

  - The specific conductance of water from almost 75 percent of the wells evaluated increased over time, with about half increasing by more than 10 percent during the period of record.

  - Chloride concentrations increased in about 67 percent of the wells evaluated, with just over half increasing by more than 10 percent during the period of record.

  - The largest increases in specific conductance, chloride concentration, or both were in wells located in the vicinity of the cities of Rohnert Park and Cotati.

- Concentrations of manganese, iron, and arsenic in 73, 43, and 12 percent, respectively, of all samples from wells in the study area were greater than State and federal drinking water standards prior to any treatment to achieve drinking water standards.

- Stable isotopes.

  - In general, the isotopic values of samples from the Uplands and Valley storage units grouped together in the lighter range of all isotopic values, which is consistent with groundwater in these storage units being recharged by precipitation falling on higher elevations.

  - The isotopic values for well samples from the Wilson Grove storage unit also grouped together, but were within the heavier range of isotopic values from all wells in the Santa Rosa Plain watershed, which is consistent with groundwater in this storage unit being recharged by precipitation falling on lower elevations.

  - The isotopic values from wells in the Windsor Basin and Cotati Basin storage units spanned the entire range of isotopic values in the Santa Rosa Plain watershed, indicating a mixture of water sources. Furthermore, data indicated that some of the recharge to the Windsor Basin and Cotati Basin storage units originates as precipitation directly falling on the lower elevations of the Santa Rosa Plain.

  - The isotopic values of samples collected from a borehole in the Petaluma Formation within the Rodgers Creek fault zone, on the east side of the Santa Rosa Plain, and from a well perforated in the Petaluma Formation near the Laguna de Santa Rosa, on the west side of the Santa Rosa Plain, were significantly lighter (more negative) than the other surface-water or well samples. These results are consistent with groundwater derived from recharge at higher altitudes or under cooler conditions than have prevailed in the post-glacial period.

- Age dating.

- Tritium (^{3}H) concentrations were analyzed in 35 samples collected from 30 wells and ranged from less than 0.3 (detection limit of samples collected in 2004) to 2.4 tritium units.

- About 43 percent of the samples had detectable ^{3}H concentrations, indicating that these samples contain some modern water (water recharged since 1952).

- Modern recharge was more prevalent in the shallow-well samples.

- The vertical migration of recharge in the Santa Rosa Plain probably is restricted by the presence of low-permeability clay deposits in the Glen Ellen and Petaluma Formations.

- Carbon-14 (^{14}C) activities for the 16 well samples analyzed ranged from 1.3 to 84.7 percent modern carbon.

  - ^{14}C activities represent uncorrected ages of 1,000 to 34,000 years before present.

  - The deep well samples had uncorrected ^{14}C ages of 4,000 years or older, and five of the deep well samples were 10,000 years or older. These ages are consistent with other chemical and isotopic data indicating limited vertical mixing.

  - The two oldest samples (27,000 and 34,000 years before present) were collected from a borehole in the Petaluma Formation within the Rodgers Creek fault zone on the east side of the Santa Rosa Plain and from a well perforated in the Petaluma Formation near the Laguna de Santa Rosa on the west side of the Santa Rosa Plain. The old ages of these samples, coupled with major-ion data, indicated a long groundwater-flowpath passing through the Cotati Basin storage unit.

Conceptual Model

A conceptual model of the hydrologic system for the Santa Rosa Plain watershed was developed (1) to better understand the movement and storage of water in the Santa Rosa Plain watershed; (2) to aid in interpretations of field data, such as water quality, well hydrographs, and stream-gage records; and (3) to provide a framework for the development of a numerical flow model to help understand the response of the Santa Rosa Plain watershed to current and historic water uses and to predict the response of the hydrologic system to potential future conditions. The conceptual model of the Santa Rosa Plain watershed is based on known and estimated physical and hydrologic characteristics of the surface-water and groundwater systems and how these characteristics influence the flow and storage of water in the Santa Rosa Plain watershed. Following Markstrom and others (2008), the hydrologic system of the Santa Rosa Plain watershed is conceptualized as having three regions: region 1 consists of the plant canopy, the land surface, and the soil zone; region 2 consists of streams, lakes, and wetlands; and region 3 is the subsurface zone that consists of an unsaturated zone and an underlying saturated zone.

The characteristics of the conceptual model for this study include the following:

- Boundary conditions—75 percent of the Santa Rosa Plain watershed boundary is no flow, with the remaining parts allowing hydraulic communication between the Santa Rosa Plain watershed and the Kenwood Valley and Sonoma Valley, the Cotati-Rohnert Park area and Petaluma Valley, the area west of the Sebastopol Fault and Wilson Grove Formation Highlands, and the Windsor area and the Russian River Basin. The lower boundary of the aquifer system is the contact with the low-permeability basement rock, which was assumed to be no flow.

- Region 1—Region 1 of the Santa Rosa Plain watershed includes the plant canopy, the land surface, and the soil zone. Region 1 provides the primary link between climatic factors and the Santa Rosa Plain watershed hydrologic system. The plant canopy includes natural vegetation, crops (for example, row crops, pastures, vineyards, and orchards) and landscaped urbanized areas (for example, residential areas, parks, and golf courses). The land surface includes areas covered by soil; areas with naturally exposed bedrock areas free of soil cover; and areas covered by anthropogenic features such as buildings, roads, and parking lots. The soil zone represents the upper unsaturated zone and is conceptually defined as the layer extending from the ground surface to the base of the root zone (Markstrom and others, 2008). The soil zone stores and transmits water between the atmosphere and the underlying unsaturated and saturated zones.

  - Inflows to region 1 include precipitation, irrigation-return flow, surface water, and groundwater discharge.

  - Outflows from region 1 include evapotranspiration, surface-water runoff, and infiltration to the unsaturated zone.

  - Storage components in region 1 include plant canopy, depressions in impervious surfaces, depressions in pervious surfaces, and soil-zone storage.

- Region 2—Region 2 of the Santa Rosa Plain watershed consists of surface-water bodies such as streams, lakes, and wetlands. Three major streams drain the Santa Rosa Plain watershed: (1) Mark West Creek, (2) Santa Rosa Creek, and (3) the Laguna de Santa Rosa. The surface-water bodies, primarily streams, are the primary storage components and flow conduits for surface water.

  - Inflows to region 2 include overland flow and interflow from region 1, groundwater discharge (gaining streams

include reaches of Mark West and Santa Rosa creeks), precipitation, and reclaimed municipal wastewater.

  - Outflows include surface-water discharge, streambed losses (infiltration), evaporation, and diversions (for example, irrigation).

    - The 5-year average surface-water outflow from the Santa Rosa Plain watershed was 210,000 acre-feet per year.

    - Losing streams include reaches of Mark West and Santa Rosa creeks to the west of the Rodgers Creek fault zone and in some upstream reaches within the uplands.

- Region 3—Region 3 of the Santa Rosa Plain watershed consists of the subsurface zone, which comprises an unsaturated zone and a saturated zone.

- The areally extensive unsaturated zone connects the soil zone to the saturated part of the groundwater system. The unsaturated zone is beneath stream channels in parts of the Santa Rosa Plain watershed. Percolation through the soil zone in region 1 and infiltration through streambeds and lakebeds in region 2 becomes infiltration to the top of the unsaturated zone and is a function of the physical characteristics of the unsaturated zone (vertical hydraulic conductivity, thickness, effective porosity, and vertical saturation profile).

  - Inflows to the unsaturated zone of region 3 include infiltration from the soil zone, streams, and water bodies and capillary inflow from the water table.

    - More infiltration is expected in areas with Wilson Grove and Glen Ellen formations because of higher permeability.

    - Less infiltration is expected in mountains because of low-permeability Sonoma Volcanics and the Franciscan formation.

    - Infiltration from stream channels could be greatest in the transition from the mountains to the Santa Rosa Plain.

  - Outflows from the unsaturated zone include flows into the saturated zone (recharge) and transpiration.

    - Outflow is expected in the Wilson Grove storage unit because of permeable unsaturated/saturated zone interface.

    - Less outflow is expected in the Windsor Basin and Cotati Basin storage units because of less permeable unsaturated/saturated zone interface (the clay-rich Petaluma formation underlies the Glen Ellen or Quaternary formations).

- Important hydrologic characteristics of the Santa Rosa Plain watershed groundwater-flow system include those that determine the ability of the groundwater system to transmit water, to store and release water, and to allow for vertical passage of water between layers, as well as those that control the flow of water across geologic or hydrologic boundaries. The movement of infiltrated recharge through the saturated groundwater system is controlled by topography, aquifer and aquitard properties, the magnitude of natural discharge, and pumping. Aquifer and aquitard properties depend on the type of sediments and rocks composing the hydrogeologic system.

  - Groundwater is contained in the pore spaces of the Quaternary alluvial materials and Tertiary sedimentary rocks, including the Glen Ellen Formation, the Wilson Grove Formation, the Petaluma Formation, and Sonoma Volcanics.

    - These units, although lithologically heterogeneous, generally form a continuous body of saturated material.

    - The most permeable formations are the Glen Ellen and Wilson Grove formations.

    - The Petaluma Formation has low productivity but is widely distributed.

  - Water-quality data indicated the following:

    - The source of natural recharge is precipitation falling on the Santa Rosa Plain and on the surrounding mountains.

    - There is limited hydraulic communication between shallow and deeper aquifer systems

    - There is limited groundwater underflow across the Rodgers Creek fault zone from the uplands to the Santa Rosa Plain.

    - The Sebastopol fault could be a partial barrier to groundwater flow.

  - Outflows from the saturated zone include pumpage, evapotranspiration, discharge to streams and lakes, discharge to the soil zone, and underflow to surrounding basins.

    - Pumping for public supply, agricultural, and domestic uses is the largest component of groundwater discharge.

      - Reported public-supply pumpage ranged from 3,900 to 10,100 acre-feet per year.

      - Estimated agricultural pumpage ranged from 8,900 to 46,600 acre-feet per year.

      - Estimated domestic pumpage ranged from 11,000 to 23,200 acre-feet per year.

      - Estimated total actual evapotranspiration from the soil, unsaturated, and saturated zones (regions 1 and 3) by a preliminary watershed model was about 265,700 acre-feet per year.

      - Water-level contours indicated that Santa Rosa Creek consistently gains streamflow east of Rodgers Creek fault zone.

      - Groundwater discharged from springs is a source of baseflow for streams or is lost to evapotranspiration.

      - Underflow to neighboring basins is insignificant.

  - Faults can be barriers to groundwater flow.

    - Geochemical data indicated that Rodgers Creek fault is a barrier to groundwater flow.

    - It is unclear whether the Sebastopol fault is a barrier; water quality and reported results are conflicting.

Data Gaps

Analyses of hydrologic, hydrogeologic, and geochemical data collected for this study indicated gaps in the data available for the SRPW.

- Lack of depth-dependent water-level and water-quality data make it difficult to calibrate a groundwater-flow model by aquifer or layer.

- Additional water-quality data are needed to help explain variability in observed water-quality data in the Cotati Basin storage unit.

- Better estimates, or direct measurements, of agricultural and domestic pumpage are required to improve calibration of a numerical groundwater and surface-water flow model. The locations of these wells are often unknown or unreported; therefore, the locations of these wells should be determined.

Chapter A. Introduction to the Study Area

By Tracy Nishikawa, Joseph A. Hevesi, Donald S. Sweetkind, and Linda R. Woolfenden

Introduction

The Santa Rosa Plain (SRP) is home to approximately half of the population of Sonoma County, Calif., which has over 249,000 residents (California Department of Finance, 2012b), and is experiencing growth in population and demand for water. Water managers face the challenge of meeting the increasing water demand with a combination of Russian River water—which has uncertainties in its future availability—local groundwater resources, and extending recycled water and other water conservation programs (Sonoma County Water Agency, 2012). To meet this challenge, water managers are seeking to acquire the knowledge and tools needed to understand the likely effects of future groundwater development in the SRP and to identify efficient strategies for surface- and groundwater management that will ensure the long-term viability of the water supply. The U.S. Geological Survey (USGS), in cooperation with the Sonoma County Water Agency (SCWA) and other stakeholders in the area, undertook this study to characterize the hydrology of the SRP and to develop tools to better understand and manage the groundwater system.

The SRP is located about 50 miles (mi) north of San Francisco, Calif. (fig. 1). The population centers are the cities of Santa Rosa, Rohnert Park, Cotati, Sebastopol, and the town of Windsor (fig. 2). The study area is a modified form of the Mark West Creek watershed (also referred to in previous studies as the Laguna de Santa Rosa watershed), which includes three surface-water drainage basins: Mark West, Santa Rosa, and the Laguna de Santa Rosa (Cal-Atlas, 2007; fig. 2). The study-area boundary extends beyond the Mark West Creek watershed, along the northwestern and southern sections of the boundary, to better represent the complete area of the SRP groundwater subbasin as defined by the California Department of Water Resources (2003; fig. 2). In addition, a small area on the west side of the boundary is included to account for the entire drainage area upstream of the tributary junction with the Russian River. This western addition is not included in some published versions of the Mark West Creek watershed (Cal-Atlas, 2007). For the purpose of this study, the modified form of the Mark West Creek watershed is referred to as the Santa Rosa Plain watershed (SRPW; fig. 2).

The SRP groundwater subbasin is part of the larger Santa Rosa Valley groundwater basin and is the largest groundwater subbasin in Sonoma County (California Department of Water Resources, 1980; Sonoma County Water Agency, 2006). The 262 square mile (mi²) SRPW contains the SRP groundwater subbasin, Rincon Valley groundwater subbasin, the northern half of the Kenwood Valley groundwater basin, eastern portions of the Wilson Grove Formation Highlands groundwater basin, southern portion of the Healdsburg area groundwater subbasin, and the southern portion of the Alexander Valley groundwater subbasin (California Department of Water Resources, 1980; Sonoma County Water Agency, 2006; City of Rohnert Park, 2007; fig. 3).

Previous Work

Cardwell (1958) characterized the hydrogeology of the Santa Rosa and Petaluma Valleys, but did not consider the hydrogeology of the surrounding mountains. He described the geology of the Jurassic and Cretaceous consolidated rocks, the Tertiary sedimentary and volcanic rocks, and the late Tertiary to Quaternary unconsolidated sediments. The late Tertiary to Quaternary unconsolidated sediments (Merced, now Wilson Grove, and Glen Ellen Formations), as defined by Cardwell (1958), were described as the most important water-bearing formations; the Tertiary rocks were of secondary importance.

In addition to the hydrogeology, Cardwell (1958) described the groundwater sources and sinks, the water quality, and the potential storage capacity of the groundwater subbasins. The primary sources of groundwater recharge were rainfall and seepage from streams. The primary groundwater sinks were natural discharge (groundwater flow to streams, springs, evapotranspiration, and adjacent basins) and pumping. In 1949, domestic and agricultural pumping accounted for 23 and 73 percent, respectively, of the total pumping in the Santa Rosa Valley area (Cardwell, 1958, table 7). In general, the groundwater quality met health-based standards for potable use at the time of publication. Cardwell (1958) assumed that the zone for available groundwater storage extended from 10 to 200 ft below land surface (bls). The total estimated available storage in the Santa Rosa Valley area was about 1 million acre-feet (acre-ft; Cardwell, 1958, table 12).

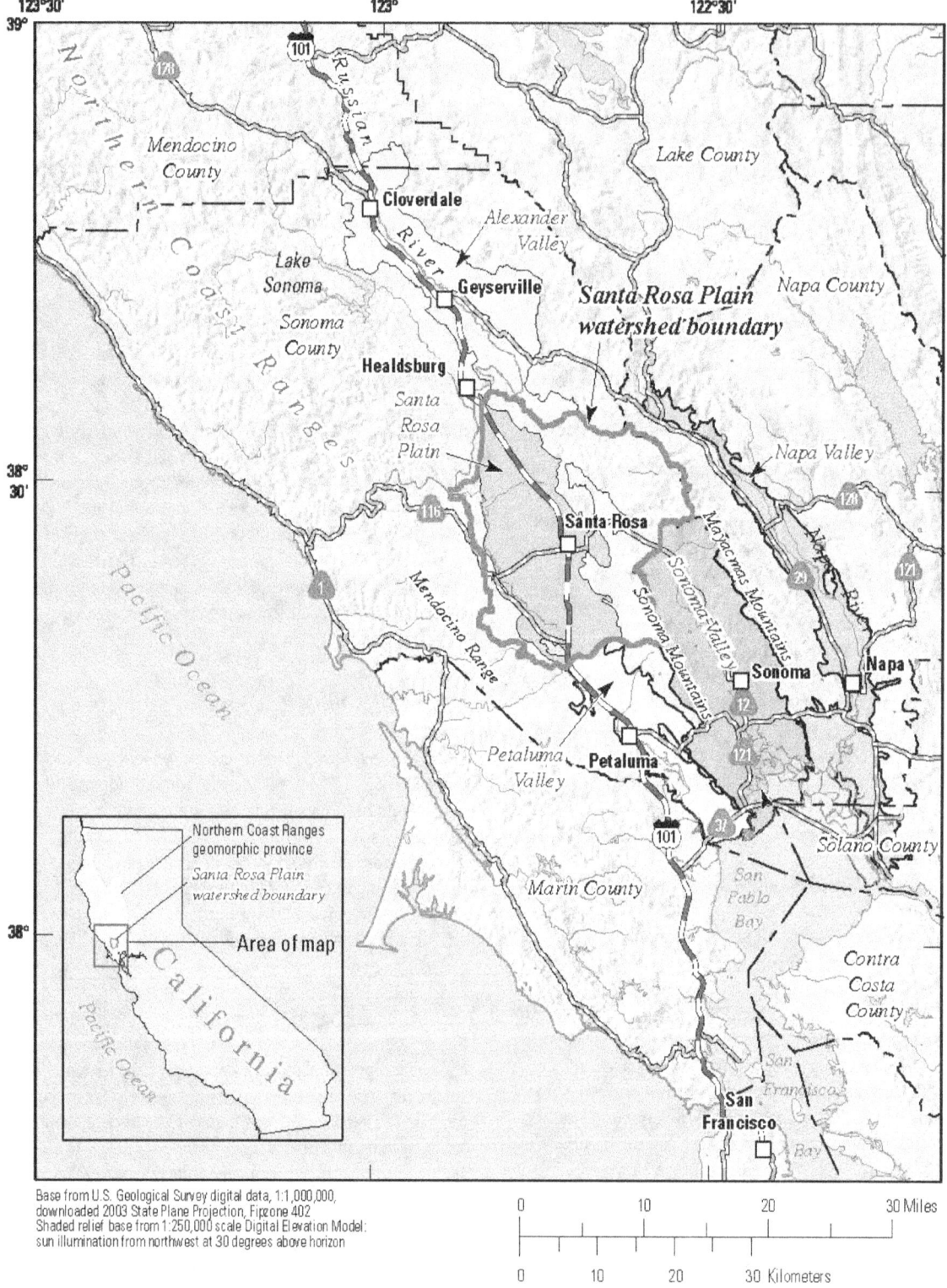

Figure 1.   Location of the Santa Rosa Plain watershed study area within Sonoma County, California.

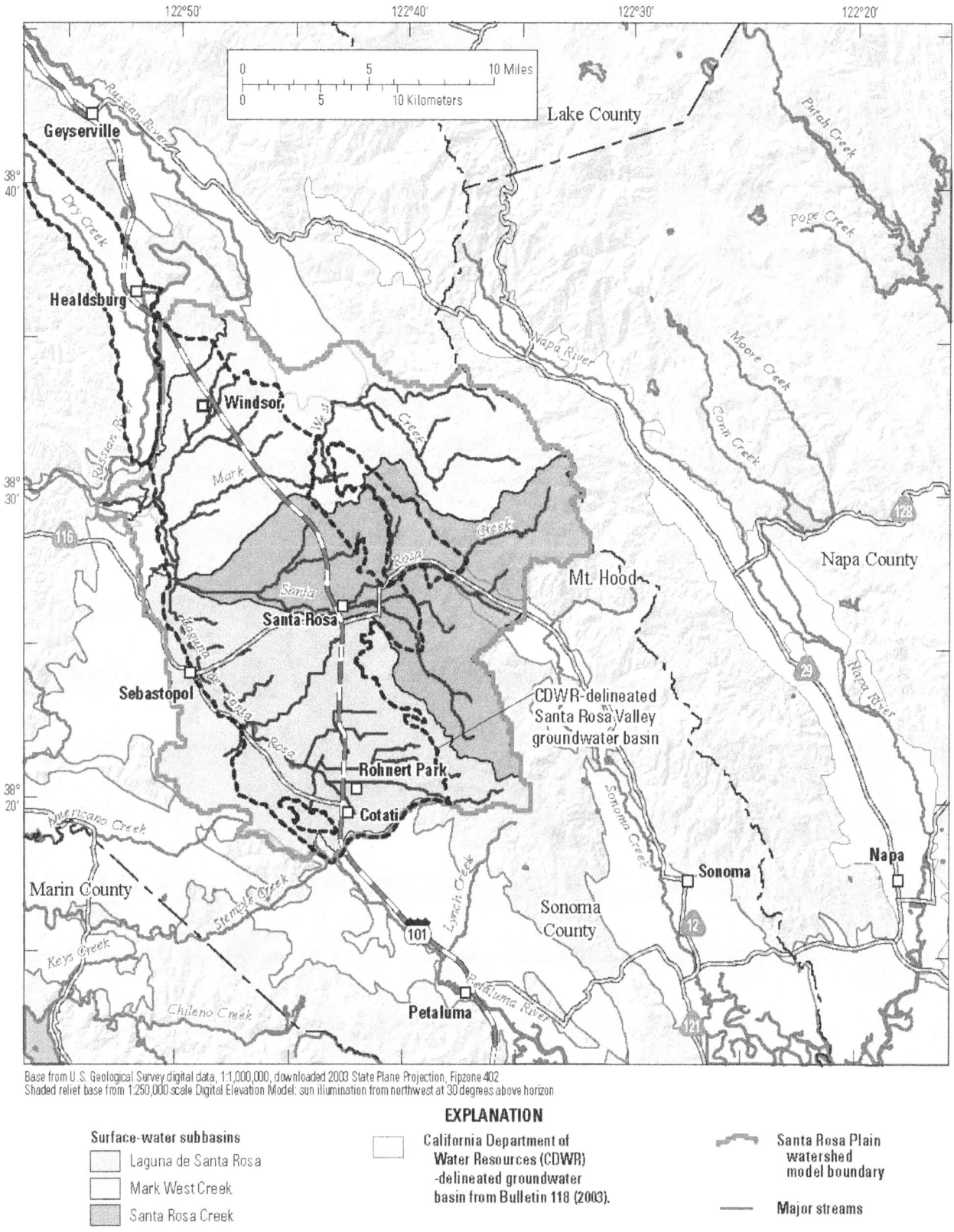

Figure 2. Santa Rosa Plain watershed boundary with the Mark West Creek, Santa Rosa Creek, and Laguna de Santa Rosa drainage basins, Sonoma County, California.

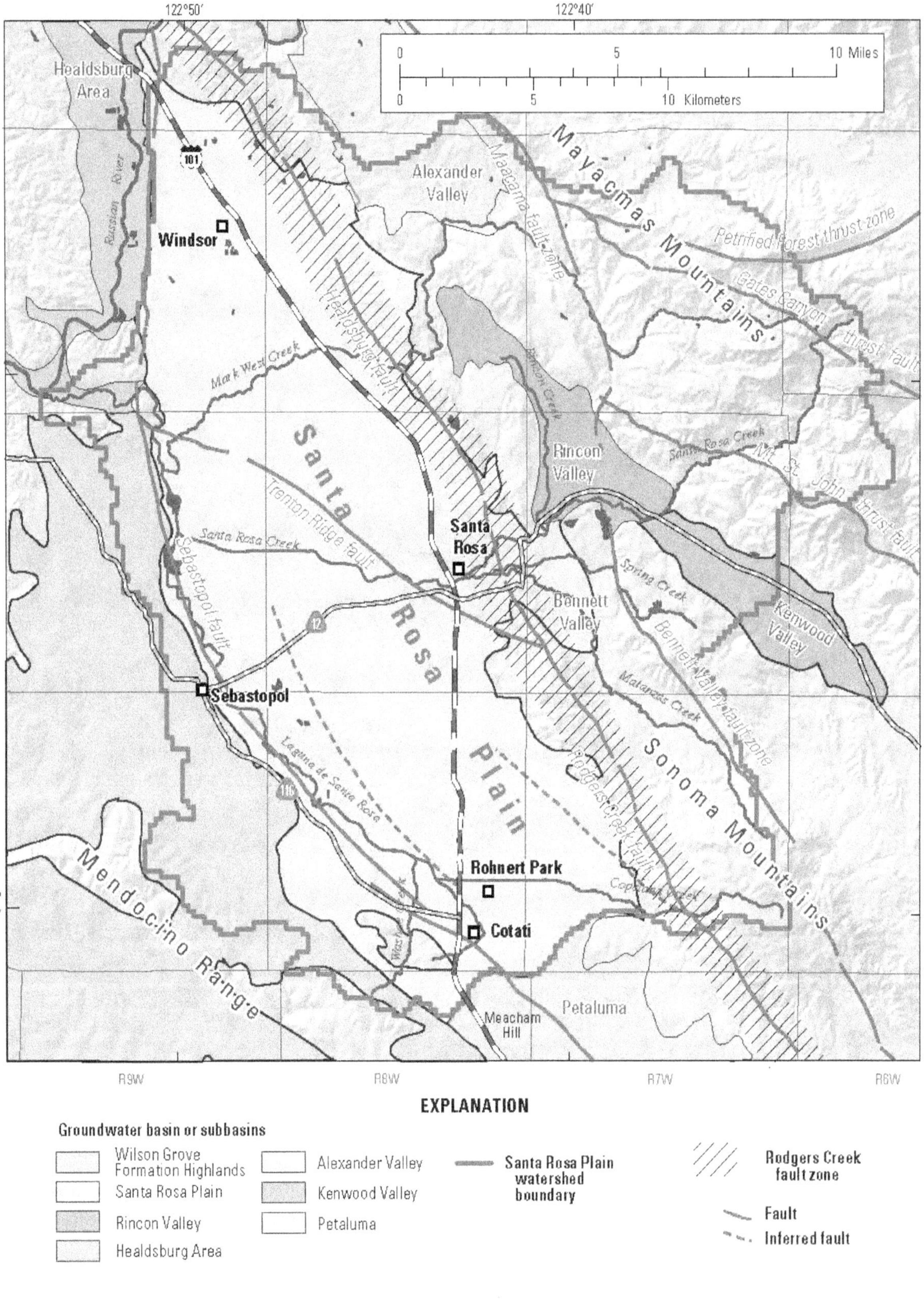

Figure 3.   Santa Rosa Plain watershed boundary with groundwater subbasins, Sonoma County, California.

The California Department of Water Resources (CDWR) published a series of reports describing the hydrogeology of the SRP (Ford, 1975; Herbst and others, 1982; Kadir and McGuire, 1987). Ford (1975) presented a general overview of the geology, hydrology, water-supply systems, soil types associated with septic tanks, available groundwater, and water quality of groundwater basins in Sonoma County. Ford reported a gross groundwater storage capacity of almost 10 million acre-ft for the Santa Rosa groundwater basin, which was almost an order of magnitude greater than the value estimated by Cardwell (1958). However, the saturated thickness used by Ford ranged from 630 to 1,160 feet (ft), compared to the 190 ft used by Cardwell (1958). Ford (1975) suggested that the Santa Rosa groundwater basin be given a high priority for further study.

Herbst and others (1982) carried out a similar study as Ford (1975); however, their study was focused on the SRP and a link-node type of groundwater-flow model (TRANSCAP; Miyazaki, 1980) of the subbasin was developed. The total available storage capacity of the SRP groundwater subbasin was estimated at about 4.3 million acre-ft, assuming an average saturated thickness of 400 ft. Note that the study area used by Herbst and others (1982) was smaller than the one used by Ford (1975). Total estimated natural recharge from 1960 to 1975 was 439,200 acre-ft, or about 29,300 acre-ft per year (acre-ft/yr), and total estimated groundwater pumpage for the period was 445,300 acre-ft, or about 29,700 acre-ft/yr. The authors determined that there were insufficient data to calibrate and verify the model and made suggestions for additional data collection.

Kadir and McGuire (1987) updated the model developed by Herbst and others (1982) by using an updated version of the link-node type model. The model domain used by Kadir and McGuire (1987) is shown in figure 4 overlain on the current study area. Kadir and McGuire (1987) treated the Sebastopol fault (fig. 3) as a no-flow boundary and the area west of the fault was not simulated. There were limited data to calibrate the model; therefore, the model was calibrated using 1978–81 water-level data and was verified using 1981–83 data. Calibration hydrographs were presented; however, neither a simulated hydrologic budget nor results from simulating water-management scenarios were presented.

Kulongoski and others (2010) presented an assessment of the groundwater quality of multiple groundwater basins in the north San Francisco Bay area, of which one was the SRP groundwater subbasin, as part of the USGS component of the state Groundwater Monitoring and Assessment (GAMA) program. The USGS collected water-quality samples from 89 wells in 2004, of which 28 were located in the study area, and analyzed water-quality data for 2001–04 from the California Department of Public Health database. The water-quality samples were analyzed for a wide range of constituents, including volatile organic compounds, pesticides, and naturally occurring inorganic constituents. A complete review and analysis of the GAMA dataset as applied to the SRPW is presented in *chapter C* of this report.

Purpose and Scope

The USGS, in cooperation with the Sonoma County Water Agency, cities of Cotati, Rohnert Park, Santa Rosa, and Sebastopol, town of Windsor, Cal-American Water Company, and the County of Sonoma, undertook this study to evaluate the groundwater resources of the SRP and to develop tools to better understand and manage the groundwater system. The objectives of the study are (1) to develop an updated assessment of the hydrogeology and geochemistry of the SRP, (2) to develop a fully-coupled surface- and groundwater-flow model for the SRPW, and (3) to evaluate the potential hydrologic effects of alternative groundwater-management strategies for the basin. The purpose of this report is to describe the surface- and groundwater hydrology, hydrogeology, and water-quality characteristics of the SRPW and to develop a conceptual model of the hydrologic system in support of the first objective. The results from completing the second and third objectives will be described in a separate report.

This report comprises four chapters. *Chapter A* summarizes the purpose and scope of the study, provides a description of the study area, and presents an overview of previous work. *Chapter B* provides a more detailed description of the geology and the surface- and groundwater hydrology of the study area. The geology section provides a description of the geologic setting, stratigraphic units, major faults, and basin depth and geometry. The surface-water hydrology section provides a detailed description of the hydrography of the study area, the characteristics of the main stream channels, the processes controlling streamflow in the study area, and the known characteristics of streamflow in the study area based on the available records. The groundwater hydrology section describes the groundwater subbasins and storage units, aquifer system, groundwater recharge and discharge, and groundwater flow. *Chapter C* focuses on groundwater-quality conditions and the sources and ages of groundwater in the basin. *Chapter D* presents a conceptual model of the SRPW based on the data presented in the previous chapters.

The data and interpretive results presented in this report form the foundation for the fully-coupled surface- and groundwater-flow model developed for the study area (described in a separate report), which provides further insights into the overall water budget, potential effects of groundwater development, and distribution of groundwater recharge and discharge.

Study Area Description

The following description of the study area provides an overview of the physiography, climate, hydrography, soils, vegetation and land cover, and land uses of the SRPW. The description of physiography includes an overview of geologic controls on landforms within and surrounding the study area. The description of land use provides a brief history

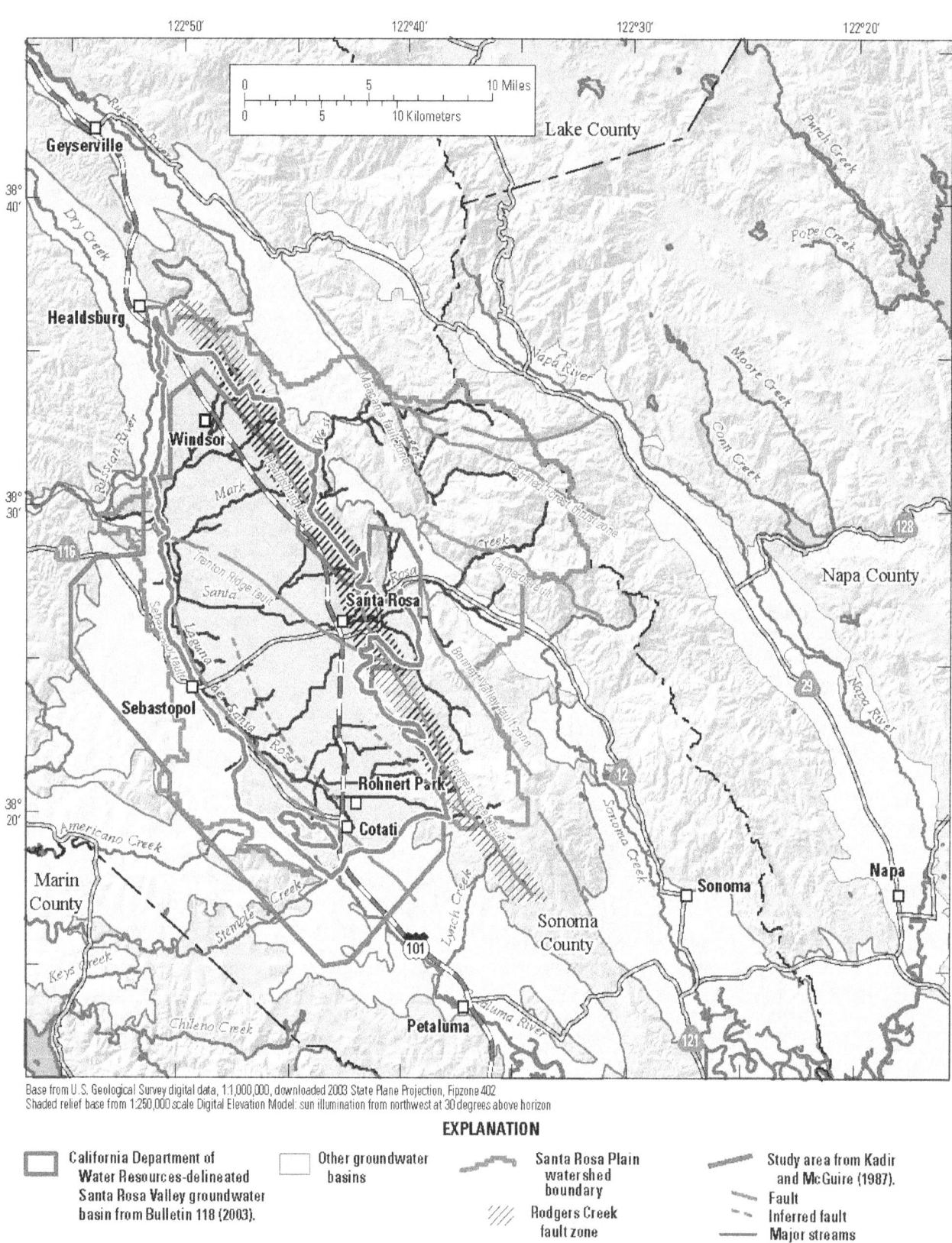

Base from U.S. Geological Survey digital data, 1:1,000,000, downloaded 2003 State Plane Projection, Fipzone 402
Shaded relief base from 1:250,000 scale Digital Elevation Model: sun illumination from northwest at 30 degrees above horizon

EXPLANATION

California Department of Water Resources-delineated Santa Rosa Valley groundwater basin from Bulletin 118 (2003).	Other groundwater basins	Santa Rosa Plain watershed boundary
		Rodgers Creek fault zone

Study area from Kadir and McGuire (1987).
Fault
Inferred fault
Major streams

Figure 4.   Kadir and McGuire (1987) model domain in relation to study area, Santa Rosa Plain watershed, Sonoma County, California.

of land-use changes and includes an overview of the effect of land use change on the hydrology of the SRPW, as documented in previous studies. More detail on the study area is provided in later chapters.

Physiography

The SRPW lies within the Coast Range geomorphic province (fig. 1) that occupies most of the northwestern edge of California and consists of a series of small mountain ranges and ridges that trend generally northwest-southeast, subparallel to the Pacific coast line (Jenkins, 1938; California Geological Survey, 2002). The northern portion of the Coast Range geomorphic province extend from the San Francisco Bay northward to the California-Oregon border.

The principal mountain ranges within and adjoining the study area, from the Pacific Ocean coastline to approximately 30 mi inland, include the Mendocino Range, the Sonoma Mountains, and the Mayacmas Mountains (fig. 1). All of these highlands are of modest relief, generally less than 2,500 feet above sea level (ft asl), and most ridge lines are between 500 and 1,500 ft asl. The Mendocino Range is a large, primarily rugged and heavily wooded mountain block that extends over 200 mi from the San Francisco Bay north to Humboldt Bay. Within the study area, however, the Mendocino Range is made up of low, rounded hills that range in altitude between 600 and 1,200 ft asl. The Mendocino Range commonly exhibits a knobby, irregular topography that results from erosion of the tectonically mixed basement rocks and overlying poorly cemented rock units (Graymer and others, 2007). Large and small landslides are common to this area (Wentworth and others, 1997). The Sonoma Mountains rise from near sea level to altitudes of 1,000–2,500 ft asl southeast of Santa Rosa. Within the SRPW, the maximum altitude of the Sonoma Mountains is 2,452 ft asl and is found along the southeastern-most extent of the study-area boundary. The Mayacmas Mountains are less steep and generally range between 500 and 2,500 ft asl. The maximum altitude within the SRPW is 2,730 ft asl at the summit of Mt. Hood in the Mayacmas Mountains.

The flat-floored, elongate valleys that lie between many of the ranges of north coastal California have provided land for urban and agricultural development. Santa Rosa Valley and its eastern neighbors, Sonoma Valley and Napa Valley, are three of the largest centers of development (fig. 1). These valleys have substantial urban population centers surrounded by large tracts of agricultural land and rural populations. Each of the valleys forms the lowlands of well-delineated and separate watersheds with separate underlying groundwater-flow systems (California Department of Water Resources, 2003). Although the groundwater systems within Sonoma Valley are separated from those in the Santa Rosa Valley by a groundwater-flow divide, the two valleys are topographically connected through the narrow, northwest-to-west arcing Kenwood Valley, which extends from the upper end of Sonoma Valley into the minor lowlands of Rincon Valley and Bennett Valley (fig. 3). Rincon and Bennett valleys lie 1 to 2 mi east of Santa Rosa Valley and are mostly separated from it by the Sonoma Mountains and a narrow ridge in the Mayacmas Mountains. The two valleys occupy an approximately 7-mi long northwest-trending structural trough that parallels the eastern side of Santa Rosa Valley, but connect to it through a narrow gap in the mountains at the eastern side of the city of Santa Rosa (fig. 3).

The SRP is a lowland area of about 90 mi² in a north-west trending structural depression that separates the Mendocino Range to the west from the Sonoma Mountains and Mayacmas Mountains to the east (fig. 3). The valley floor lies mostly between altitudes of about 50 and 150 ft asl. The north-northwest to south-southeast axis of the valley extends for about 20 mi, from near the Russian River on the north to Meacham Hill on the south; the valley width ranges mostly from 4 to 7 mi. The floor of the valley is relatively flat compared to the surrounding mountains, but is not without internal topographic features. Most of the valley floor consists of a low, uneven topography developed on poorly cemented sedimentary rocks and weakly compacted sediments underlying the alluvial flood plains, terraces, and fans that have been deposited by west-flowing intermittent streams (Sowers and others, 1998).

Climate

The climate for the study area is generally Mediterranean, with cool, wet winters, warm, dry summers, and a strong coastal influence on climate that significantly moderates temperature extremes (Sloop and others, 2009).

Precipitation

The spatial distribution of mean annual precipitation is strongly affected by topography and varies considerably within the SRPW. Temporal variability in precipitation is primarily controlled by the seasonal pattern of cool, wet winters and warm, dry summers. Mean annual precipitation calculated for the California Data Exchange Center (CDEC) climate station in the city of Santa Rosa (SRO) (California Data Exchange Center, 2011), for the period 1906–2010, was 30 in. (fig. 5). For the period 1990–2005, about 98 percent of the annual precipitation falls from October through May (Sonoma County Water Agency, 2006). At the Santa Rosa climate station, the wettest month has been January, with an average precipitation of 6.4 in. for the period 1990–2005 (Sonoma County Water Agency, 2006). February and December are the next wettest months, with averages of 5.3 and 5.2 in., respectively. The months of May through September all have less than 1 in. average monthly precipitation; July is the driest month, at 0.03 in., for the period 1990–2005.

In addition to seasonal variation, there is also significant year-to-year variability in precipitation due to natural cycles and trends in global circulation patterns that strongly affect climate for a given year. Wetter-than-normal periods occur

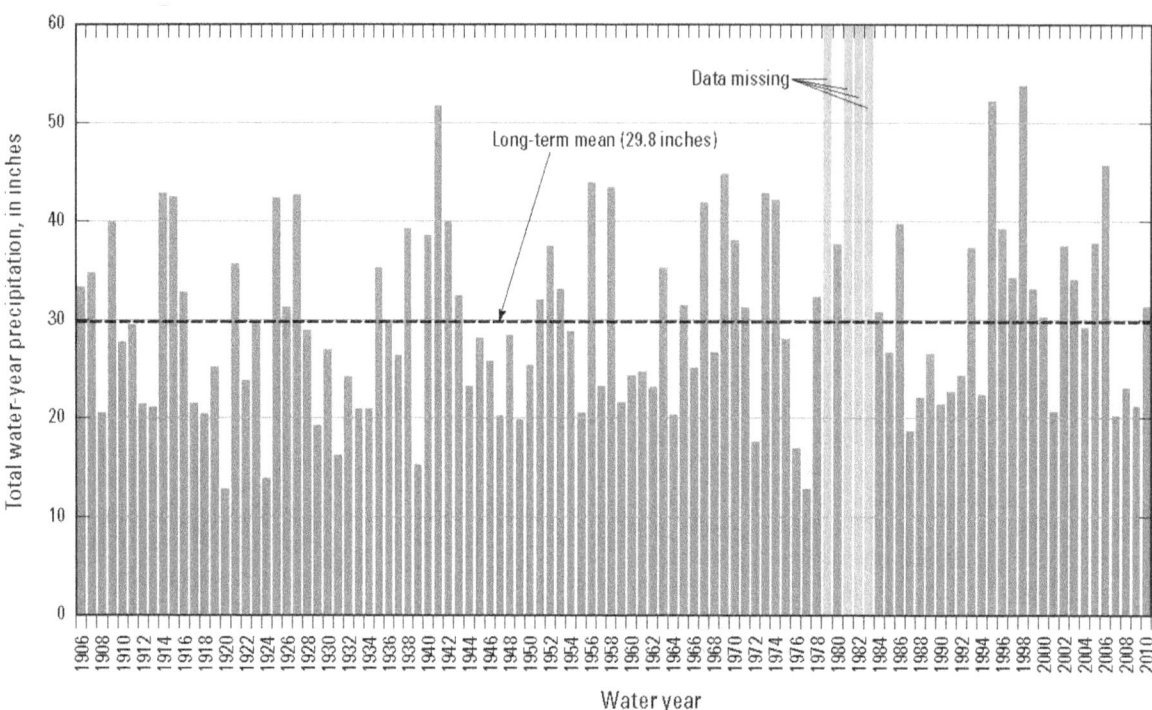

Figure 5.   Total annual precipitation for water-years 1906–2010, California Data Exchange Center SRO (City of Santa Rosa) climate station, Sonoma County, California (California Data Exchange Center, 2011).

when conditions are favorable for establishing a persistent storm track centered on the SRPW. These conditions generally result in an increase in the frequency and duration of storms. For example, wetter than normal conditions prevailed during the winter of 2005–06, when a relatively high frequency of storms, followed by a relatively large storm event, resulted in significant flooding in the SRP on December 21, 2005, through January 1, 2006. In contrast, circulation patterns can also cause a decrease in the frequency of winter storms, resulting in drier than normal conditions. Such were the conditions during water year 1977, and annual precipitation was only 12.3 in. at the CDEC SRO climate station (fig. 5). When there are drier-than-normal winters, the main stream channels will often go dry during the summer, although there still can be some flow downstream of the developed areas as a result of irrigation runoff.

Estimates of mean annual precipitation for the period 1971–2000, obtained using Parameter-elevation Regressions on Independent Slopes Model (PRISM; Daly and others, 2004), were used to indicate the spatial and temporal distribution of precipitation in the SRPW. The PRISM model provides an estimate of spatial and temporal variability in precipitation in response to (1) distance from moisture sources, (2) average storm track, (3) aspect of land surface in relation to storm track, and (4) effect of altitude on adiabatic cooling of moisture-laden air masses (Daly and others, 2004). The PRISM estimates of average annual precipitation (PRISM Climate Group, 2012) for the SRPW include minimum values of approximately 30 in. in the central part of the SRP, which

is in agreement with the measured data, and maximum values of more than 50 in. in the Mayacmas and Sonoma Mountains (fig. 6).

Air Temperature

Similar to precipitation, the spatial and temporal distributions of air temperature are also strongly affected by topography and season. Mean air temperature between 1990 and 2005 for a California Irrigation Management Information System (CIMIS) climate station near the city of Santa Rosa (CIMIS station 83) varied from a minimum of 47°F during January to a maximum of 70°F during July, with a mean temperature of 59°F (Sonoma County Water Agency, 2006).

Spatial variation of mean monthly maximum and minimum air temperatures for the SRPW is complex. For example, the mean minimum air temperature for January is 35°F for the lower altitudes in the central and western part of the SRPW compared to 40 and 44°F for higher locations in the Mayacmas and Sonoma Mountains, respectively (PRISM Climate Group, 2012). In contrast, the mean maximum air temperature in January is 52°F for the summit areas in the mountains compared to 56°F in the lowlands throughout the western part of the watershed. The mean difference between monthly maximum and minimum air temperatures is approximately 25°F, with the largest difference between maximum and minimum air temperatures of more than 30°F usually occurring during the summer.

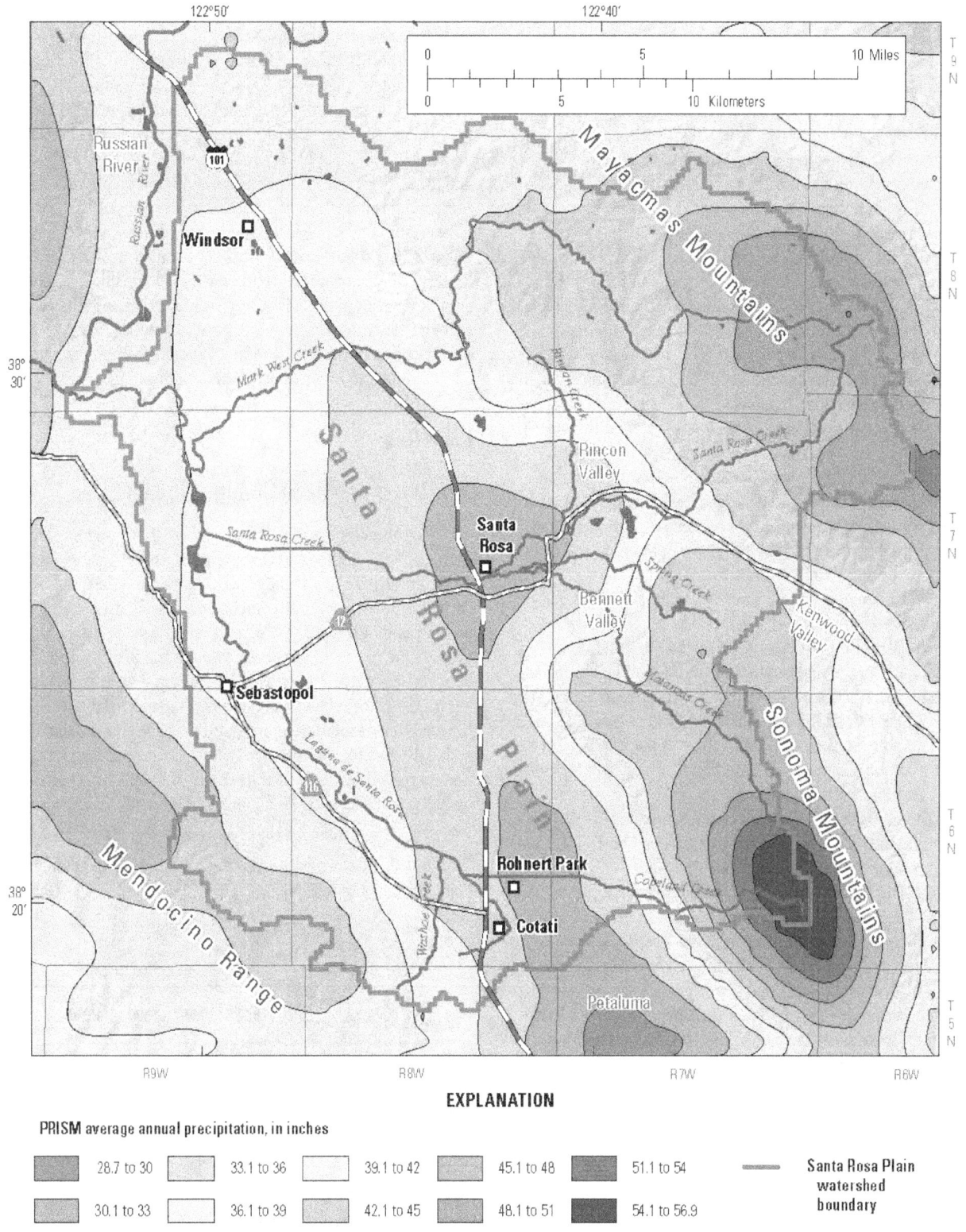

Figure 6.   Contours of average annual precipitation for 1971–2000, Santa Rosa Plain watershed, Sonoma County, California.

Soils

The U.S. Department of Agriculture (2007a) developed a spatial database of soils for the entire United States (Soil Survey Geographic, or (SSURGO). The SSURGO database defines 2,165 separate soil map units within the SRPW. The soil map units define the spatial distribution of soil types (soil components) within the study area. According to the SSURGO database, the thickness of soils is variable within the SRPW, with thinner soils located in the highlands and thicker soils located in the basins and valleys (fig. 7). Soil is absent in a few isolated locations in the more rugged terrain of the Mayacmas Mountains; these locations are dominated by rock outcrops. The average soil thickness in the Mayacmas and Sonoma Mountains is approximately 1.8 ft. The average soil thickness throughout the lowlands of the SRPW is approximately 5 ft. The thickest soils, approximately 6 ft and greater, are found within the Laguna de Santa Rosa floodplain. Soil thickness is indicated as zero for areas identified as surface water.

The SSURGO database includes basic soil properties, such as soil texture (the percent fraction of sand, silt, and clay), porosity, and permeability. In general, soil texture is highly variable throughout the SRPW. The percentage of clay in soils within the SRPW varies from 0 to 50 percent, with a mean of 29 percent (fig. 8). The soils with highest percentage of clay (47–50 percent clay) tend to be located in the southern part of the SRP, south of the city of Santa Rosa, and throughout the lowest altitudes of the Laguna de Santa Rosa floodplain. The soils with the lowest percentage of clay (0–18 percent) tend to be located along the more rugged areas of the uplands, in the northeastern part of the SRPW, and along many of the major stream channels, such as Mark West Creek and Santa Rosa Creek.

The percentage of sand in soils within the SRPW varies from 0 to 98 percent, with a mean of 33 percent (fig. 9). Soils with the highest percentage of sand (greater than 53 percent sand) tend to be located along the western part of the SRPW, west of the Laguna de Santa Rosa channel. They also are found along the lower reaches of Santa Rosa Creek and in some higher altitude locations in the Sonoma Mountains. Soils with a low percentage of sand (less than 25 percent) are found throughout the SRPW, including the Laguna de Santa Rosa lowlands, areas close to the city of Santa Rosa, and throughout a large area in the southern part of the SRPW.

The average porosity of soils is 0.34, with higher porosities mostly in the central parts of the SRP lowlands and lower porosity in sandy soils at many locations in the highlands throughout the eastern parts of the watershed. The average soil-field capacity, a measure of the volume fraction of water retained by a given soil in response to capillary forces, is 0.12 for the SRPW. The SSURGO database also includes integrated

soil properties that indicate how each soil is likely to behave under various moisture conditions. An example of such a property is the soil hydrologic group, a standard soil classification ranging from A to D (U.S. Department of Agriculture, 2007b). When thoroughly wet, soil-type A has a low runoff potential, type B has a moderately low runoff potential, type C has a moderate to high runoff potential, and type D has a high runoff potential. The runoff potential is based on a combination of soil properties, such as soil thickness, the presence or absence of restrictive layers, soil texture, permeability, and the depth to the water table. The distribution of soil hydrologic groups in the SRPW shows soils with relatively low runoff potential along the western boundary (types A and B) and soils with high to moderately high runoff potential (types C and D) in the southern part of the SRP and throughout the upland areas of the Sonoma and Mayacmas Mountains (fig. 10).

Land Use

Modern (post-1950) Changes in Land Use

Historically, the population in the SRPW was mostly rural, and agriculture was the main developed land use. In 1950, the population of the city of Santa Rosa was 17,902. At that time, the only other incorporated city was Sebastopol, with a population of 2,601 (Cardwell, 1958). The populations of the primary cities and towns in the SRPW increased between 1950 and 2010 (table 1), and there were corresponding increases in urban and residential land use (table 2). The greatest rate of increase in reported population between 1970 and 2010 was for the town of Windsor, where the population increased by more than an order of magnitude (table 1; California Department of Finance, 2012a and 2012b).

Table 1.   Population for the cities and township in the Santa Rosa Plain, Sonoma County, California, 1950, 1960, 1970, 1980, 1990, 2000, and 2010.

[A□□re□□□i□□: NR, not reported]

City or township	1950[1]	1960[1]	1970[1]	1980[1]	1990[1]	2000[1]	2010[2]
Santa Rosa	17,902	31,027	50,006	82,658	113,313	147,595	167,302
Rohnert Park	NR	NR	6,133	22,965	36,326	42,236	40,952
Cotati	NR	1,852	1,368	3,346	5,714	6,471	7,258
Sebastopol	2,601	2,694	3,993	5,595	7,004	7,774	7,380
Windsor	NR	NR	2,359	NR	13,371	22,744	26,751

[1]California Department of Finance (2012a).

[2] Estimated population on January 1, 2010, California Department of Finance (2012b).

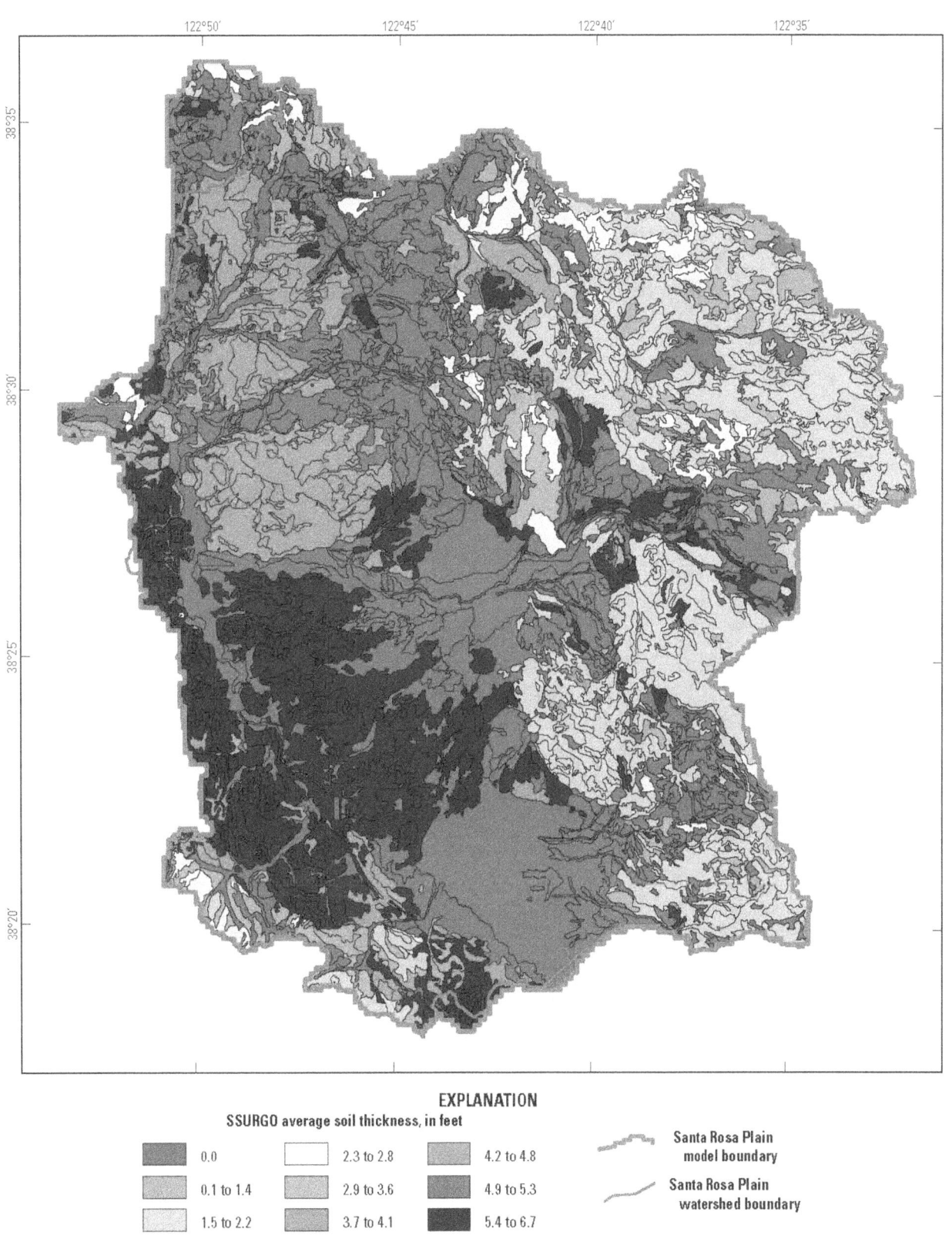

EXPLANATION

SSURGO average soil thickness, in feet

■ 0.0	□ 2.3 to 2.8	■ 4.2 to 4.8
■ 0.1 to 1.4	■ 2.9 to 3.6	■ 4.9 to 5.3
■ 1.5 to 2.2	■ 3.7 to 4.1	■ 5.4 to 6.7

Santa Rosa Plain
model boundary

Santa Rosa Plain
watershed boundary

Figure 7.   Soil Survey Geographic (SSURGO) soil thickness, Santa Rosa Plain watershed, Sonoma County, California (U.S. Department of Agriculture, 2007a). Soil thickness is indicated as zero for areas identified as water.

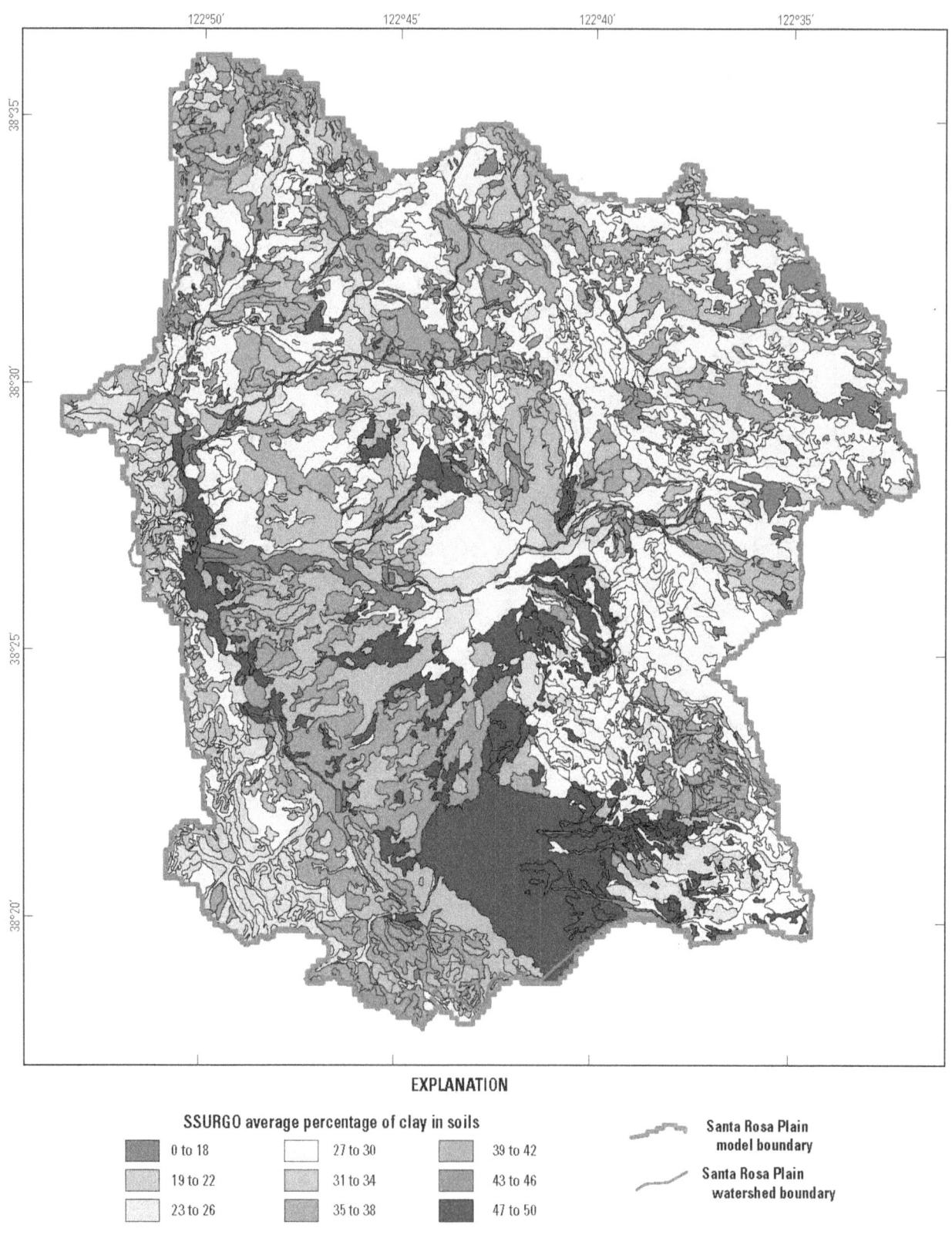

EXPLANATION

SSURGO average percentage of clay in soils

- 0 to 18
- 19 to 22
- 23 to 26
- 27 to 30
- 31 to 34
- 35 to 38
- 39 to 42
- 43 to 46
- 47 to 50

Santa Rosa Plain
model boundary

Santa Rosa Plain
watershed boundary

Figure 8.   Soil Survey Geographic (SSURGO) average percentage of clay in soils, Santa Rosa Plain watershed, Sonoma County, California (U.S. Department of Agriculture, 2007a).

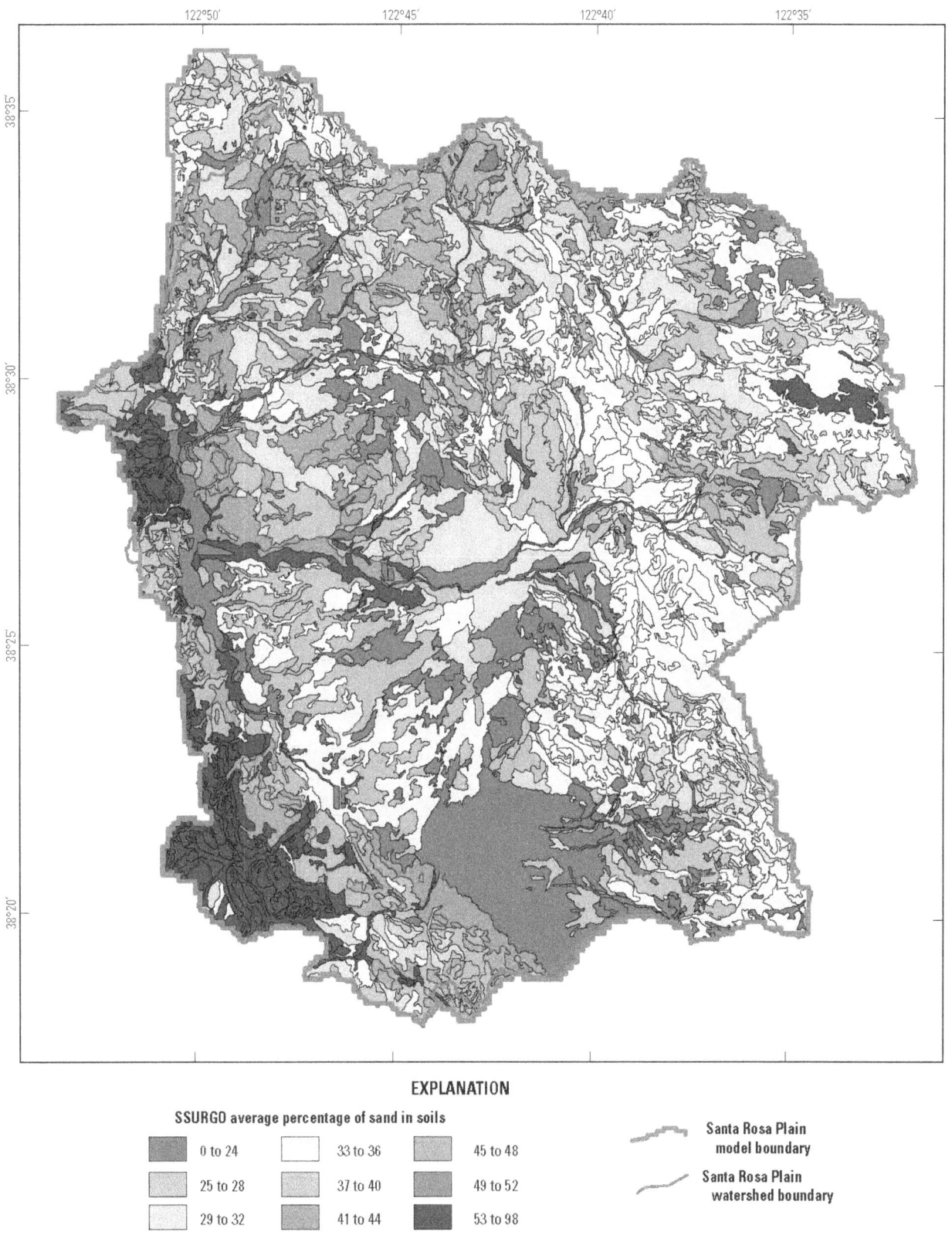

EXPLANATION

SSURGO average percentage of sand in soils

0 to 24

25 to 28

29 to 32

33 to 36

37 to 40

41 to 44

45 to 48

49 to 52

53 to 98

Santa Rosa Plain
model boundary

Santa Rosa Plain
watershed boundary

Figure 9. Soil Survey Geographic (SSURGO) average percentage of sand in soils, Santa Rosa Plain watershed, Sonoma County, California (U.S. Department of Agriculture, 2007a).

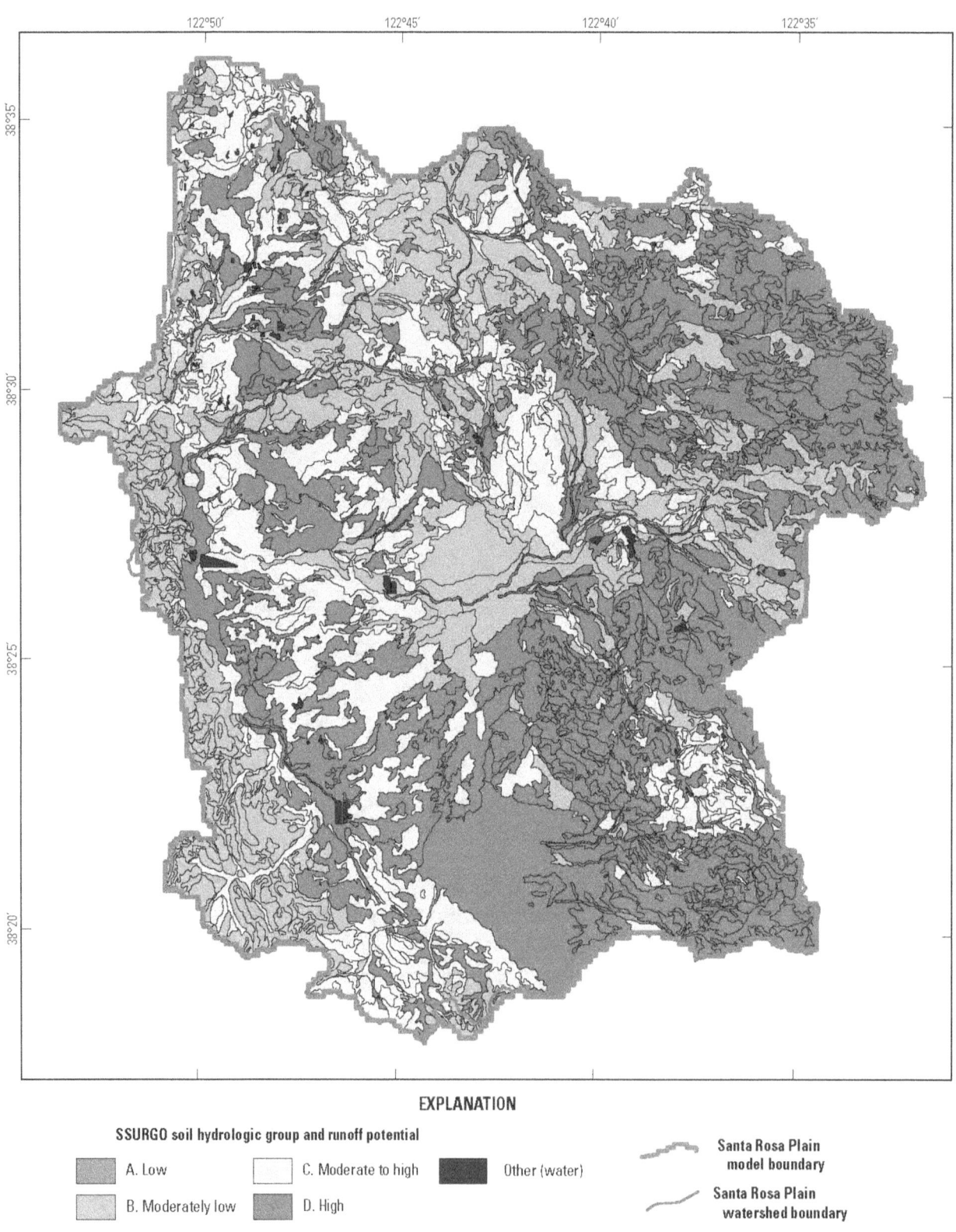

EXPLANATION

SSURGO soil hydrologic group and runoff potential

A. Low

B. Moderately low

C. Moderate to high

D. High

Other (water)

Santa Rosa Plain
model boundary

Santa Rosa Plain
watershed boundary

Figure 10.   Soil Survey Geographic (SSURGO) soil hydrologic groups, Santa Rosa Plain watershed, Sonoma County, California (U.S. Department of Agriculture, 2007a).

Table 2.   Land use in Santa Rosa Plain Watershed, Sonoma County, California, 1974, 1979, 1986, 1999, and 2008.

[Bracketed, italicized numbers are a subset of total agriculture and are not included in total area. A□□re□ □□i□□□ Mi², square miles; —, no data]

| Land-use type | Land-use surveys[1] | | | | | | | | | |
| | 1974 | | 1979 | | 1986 | | 1999 | | 2008 | |
Single use	Acres	Mi²	Acres	Mi²	Acres	Mi²	Acres	Mi²	Acres	Mi²
Urban and residential	28,154	44.0	30,919	48.3	31,827	49.7	32,265	50.4	—	—
Commercial and industrial	751	1.2	2,075	3.2	2,703	4.2	2,914	4.6	—	—
Total agriculture	24,303	38.0	23,295	36.4	25,807	40.3	23,444	36.6	25,782	40.3
[Irrigated]	*6,839*	*10.7*	*10,262*	*16.0*	*11,811*	*18.5*	*18,698*	*29.2*	—	—
[Non-irrigated]	*17,464*	*27.3*	*13,033*	*20.4*	*13,996*	*21.9*	*4,746*	*7.4*	—	—
Native vegetation and riparian	112,996	176.6	103,276	161.4	94,591	147.8	101,629	158.8	—	—
Other	836	1.3	7,475	11.7	12,112	18.9	6,788	10.6	—	—
Total area	167,040	261.1	167,040	261	167,040	260.9	167,040	261.0	—	—

[1]Modified from California Department of Water Resources, 1974, 1979, 1986, and 1999; County of Sonoma Permit and Resource Management Department; unpublished crop surveys of Sonoma County, Division of Planning and Local Assistance, Sacramento.

Land-use data for the SRPW were available for 1974, 1979, 1986, 1999, and 2008 (California Department of Water Resources, 1974, 1979, 1986, and 1999; Sonoma County Water Agency, written commun., 2010). The first four land-use surveys show eight general land-use classifications, including agricultural (irrigated, non-irrigated, and idle); residential, commercial, and industrial; suburban-native vegetation or agriculture; and native vegetation (fig. 11A–D). The land-use data for 2005–08 only distinguished the agricultural (vineyard, orchard, row crop, pasture, and unknown) areas from the non-agricultural areas (fig. 11E).

The land-use data indicated minor changes in land use for most major categories since the 1970s, but indicated substantial increases in irrigated agricultural land use and decreases in non-irrigated agriculture. The land-use data showed that native vegetation and riparian areas (including water surfaces) changed little over time, constituting 68 percent, 62 percent, 57 percent, and 61 percent of the total area for 1974, 1979, 1986, and 1999, respectively (table 2). The combination of all agricultural land uses constituted an average of about 15 percent of the total area and did not change substantially over these periods. Similarly, the total agriculture acreage in 2005–08 was about 25,800, or about 15 percent of the total area (County of Sonoma Permit and Resource Management Department, 2007); however, irrigated agriculture increased steadily from 4 percent of the total area in 1974 to 11 percent in 1999 and non-irrigated land use decreased (table 2). Between 1974 and 1999, urban and residential land uses increased slightly from 17 percent (1974) to 19 percent (1999) of the study area.

Hydrologic Effects of Changing Land Use

Although changes in land use during the last few decades were modest, more substantial changes in land use during the preceding century in the SRPW has resulted in significant changes to the hydrologic system (Sloop and others, 2009). Sloop and others (2009) described the history of important anthropogenic changes with relevance to hydrologic conditions. In general, these changes were (1) the conversion of native vegetation to grassland and agriculture that started in the mid-1800s and (2) the start of rapid urbanization in the 1940s.

The conversion of land cover from native vegetation (perennial bunch grasses and annual forbs) to grassland (for ranching) and agriculture has generally resulted in a reduction in interception storage capacity, a reduction in transpiration, a decrease in permeability, and an increase in propagation of the drainage network (Sloop and others, 2009). The combined effect of these anthropogenic changes is more runoff generation (compared to unaltered landscapes), including an increase in the "flashiness" of streamflow as characterized by a steepening of the streamflow hydrograph (Sloop and others, 2009).

With the onset of rapid population increase and associated urbanization in the 1940s, which continues to present day, agricultural land has been converted to urban land uses (mostly housing and commercial). With increased urbanization, the percentage of impervious surfaces (rooftops, parking lots, roads, among others) has probably increased within the SRPW. The hypothesized increased imperviousness could have resulted in increased runoff affecting areas within and downstream of the more heavily urbanized zones (Christopher Delaney, Sonoma County Water Agency, written commun., 2008). In addition, the increase in impervious area also contributes to the "flashiness" of streamflow (Sloop and others, 2009).

Changing land use could also affect the groundwater system. The conversion of land cover from native vegetation to agriculture could reduce direct infiltration to the soil zone; however, an increase in runoff, as described above, could result in increased streambed recharge downstream from the runoff. If agriculture requires irrigation, then irrigation-return flow is another source of water to the soil zone. A change to increased impervious area could reduce direct infiltration to, and evapotranspiration from, the soil zone.

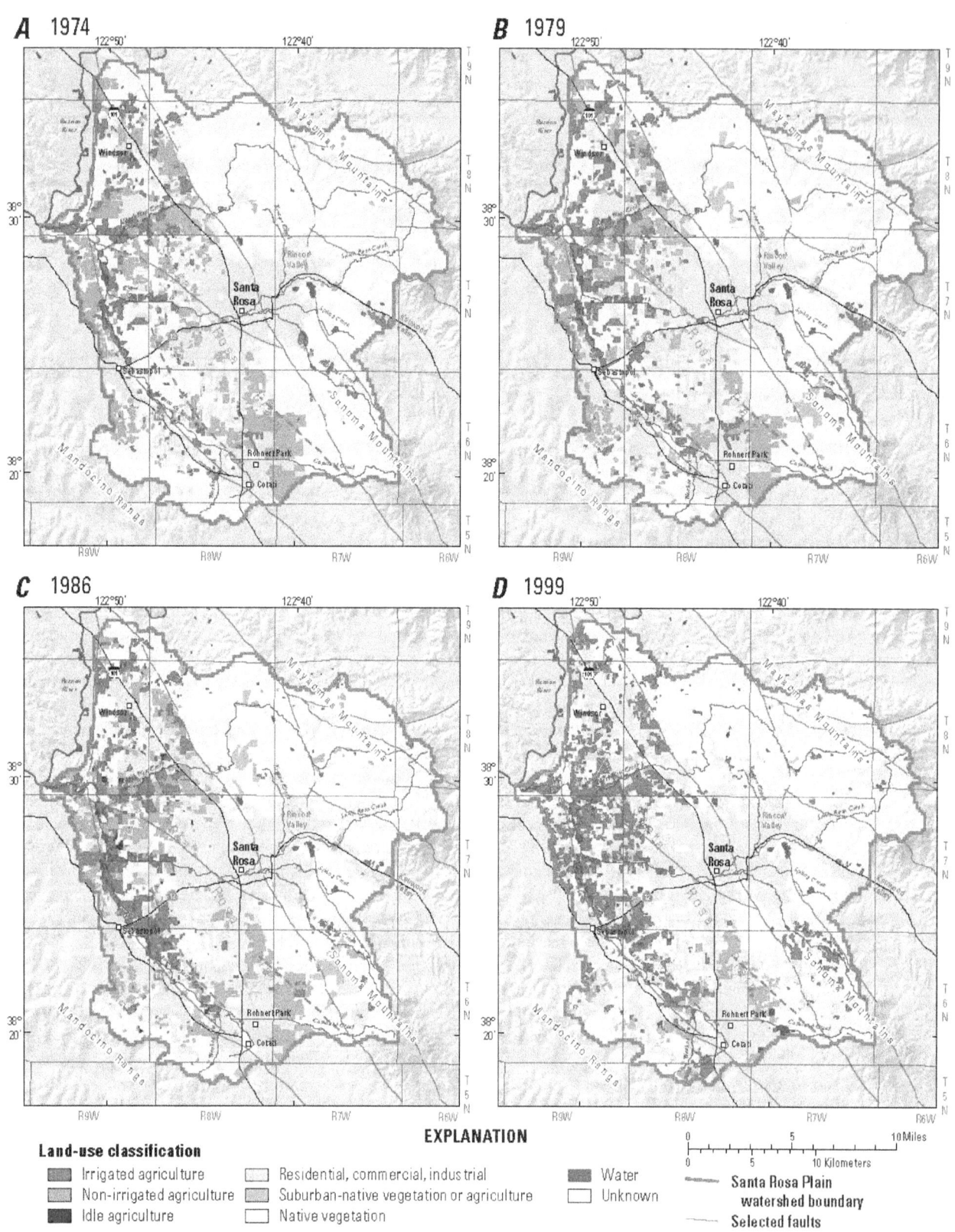

Figure 11.   Land-use maps, Santa Rosa Plain watershed, Sonoma County, California: *A*, 1974; *B*, 1979; *C*, 1986; *D*, 1999; and *E*, 2005–08.

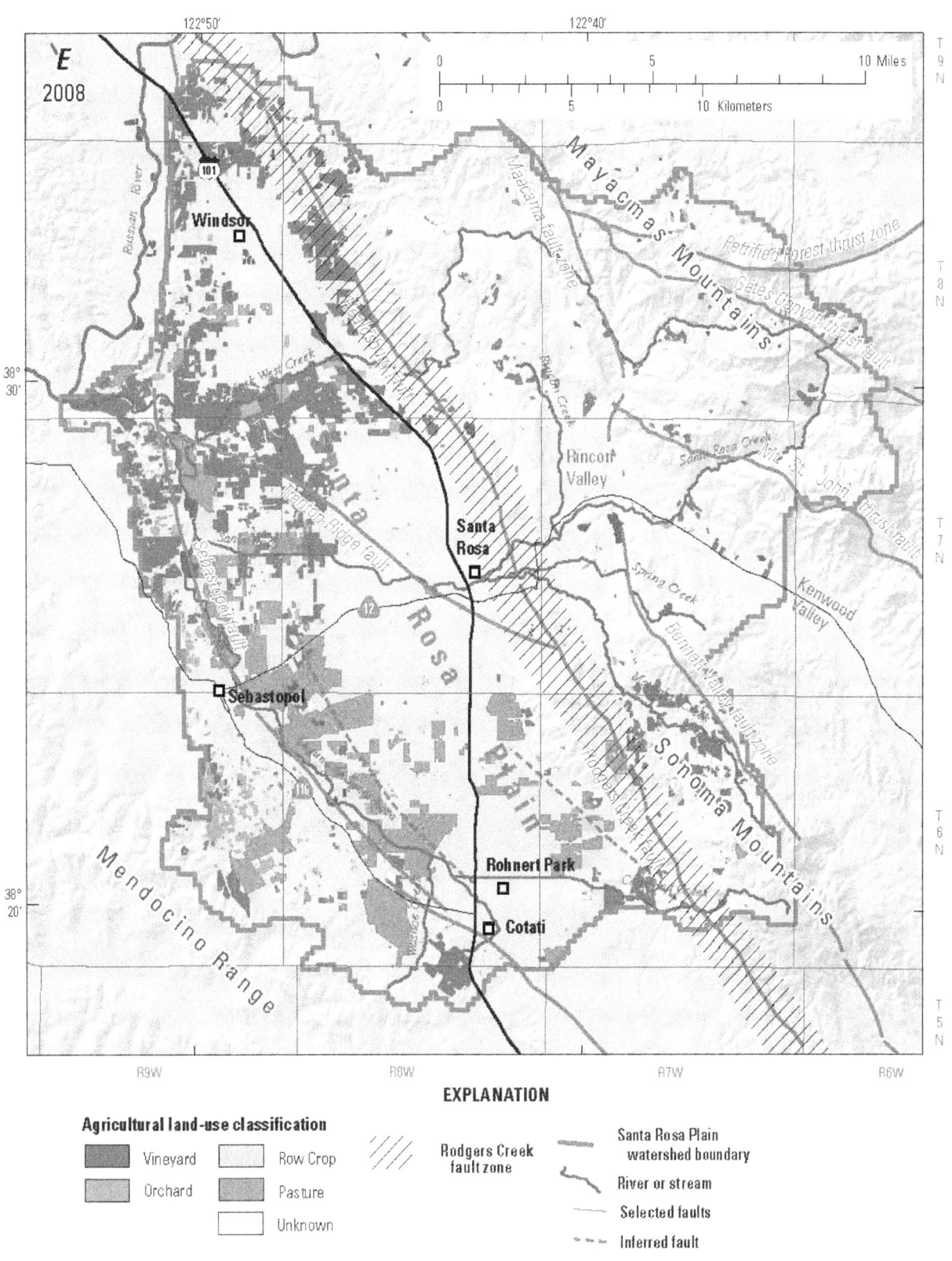

EXPLANATION

Agricultural land-use classification

■ Vineyard	Row Crop
Orchard	■ Pasture
	□ Unknown

///// Rodgers Creek fault zone

Santa Rosa Plain watershed boundary

River or stream

Selected faults

- - - Inferred fault

Figure 11. Continued

References Cited

California Data Exchange Center, 2011, Station Meta Data: Santa Rosa: Sacramento, Calif., State of California, Department of Water Resources, accessed February 28, 2011, at URL *http://cdec.water.ca.gov/cgi-progs/staMeta?station_id=SRO*

California Department of Finance, 2012a, Population totals by township and place for California Counties: 1860 to 1950: Sacramento, Calif., State of California, accessed October 18, 2012, at URL *http://www.dof.ca.gov/research/demographic/reports*

California Department of Finance, 2012b, E-4 Historical population estimates for cities, counties, and the state, 2001–2010, with 2000 and 2010 census counts: Sacramento, Calif., State of California, accessed October 18, 2012, at URL *http://www.dof.ca.gov/research/demographic/reports.*

California Department of Water Resources, 1974, 1974 Sonoma County land use survey data: unpublished data located at Division of Integrated Regional Water Management, North-Central Region, West Sacramento, California, 1:24,000 scale.

California Department of Water Resources, 1979, 1979 Sonoma County land use survey data: unpublished data located at Division of Integrated Regional Water Management, North-Central Region, West Sacramento, California, 1:24,000 scale.

California Department of Water Resources, 1980, Groundwater basins in California—a report to the Legislature in Response to Water Code Section 12924: California Department of Water Resources, Bulletin 118-80, 73 p.

California Department of Water Resources, 1986, 1986 Sonoma County land use survey data: unpublished data located at Division of Integrated Regional Water Management, North-Central Region, West Sacramento, California, 1:24,000 scale.

California Department of Water Resources, 1999, 1999 Sonoma County land use survey data: Sacramento, Calif., State of California, digital map accessed on January 28, 2011, at *http://www.water.ca.gov/landwateruse/lusrvymain.cfm*

California Department of Water Resources, 2003, California's Groundwater: Bulletin 118–Update 2003: Sacramento, Calif., State of California, 246 p., available online at URL *http://www.water.ca.gov/groundwater/bulletin118/update2003.cfm*

California Geological Survey, 2002, California Geomorphic Provinces Note 36: Sacramento, Calif., State of California, Department of Conservation, accessed November 16, 2011, at URL *http://www.consrv.ca.gov/cgs/information/publications/cgs_notes/note_36/Documents/note_36.pdf.*

California Interagency Watershed Mapping Committee, 2004, California Interagency Watershed Map of 1999 (Calwater 2.2.1): Sacramento, Calif., State of California, California Resources Agency, accessed September 7, 2007, at URL *http://atlas.ca.gov/download.html#/casil/inlandWaters*

Cardwell, G.T., 1958, Geology and ground water in the Santa Rosa and Petaluma areas, Sonoma County, California: U.S. Geological Survey Water Supply Paper 1427, 273 p., 5 plates.

City of Rohnert Park, 2007, 2005 Urban Water Management Plan: Rohnert Park, Calif., City of Rohnert Park, 108 p.

County of Sonoma Permit and Resource Management Department, 2007, LanduseByArea1: unpublished data.

Daly, Christopher, Gibson, W.P., Doggett, Matthew, Smith, Joseph, and Taylor, George, 2004, Up-to-date monthly climate maps for the conterminous United States: Proceedings of the 14th American Meteorological Society Conference on Applied Climatology, 84th AMS Annual Meeting Combined Preprints, American Meteorological Society, Seattle, Washington, January 13-16, 2004, Paper P5.1, CD-ROM.

Ford, R.S., 1975, Evaluation of ground water resources: Sonoma County, volume 1: geologic and hydrologic data: California Department of Water Resources, Bulletin 118-4, 177 p., 1 plate.

Graymer, R.W., Brabb, E.E., Jones, D.J. , J. Barnes, R.S. Nicholson, and R.E. Stamski, 2007, Geologic map and map database of eastern Sonoma and western Napa counties, California: U.S. Geological Survey Scientific Investigations Map 2956, scale 1:100,000.

Herbst, C.M., Jacinto, D.M., and McGuire, R.A., 1982, Evaluation of ground water resources, Sonoma County, volume 2: Santa Rosa Plain: California Department of Water Resources, Bulletin 118-4, 107 p., 1 plate.

Jenkins, O. P., 1938, Geomorphic Map of California: State of California Department of Natural Resources, Bulletin 158, Plate 2 .

Kadir, T.N. and McGuire, R.A., 1987, Santa Rosa Plain ground water model: California Department of Water Resources Central District, 318 p.

Kulongoski, J.T., Belitz, Kenneth, Landon, M.K., and Farrar, Christopher, 2010, Status and understanding of groundwater quality in the North San Francisco Bay groundwater basins, 2004: California GAMA Priority Basin Project: U.S. Geological Survey Scientific Investigations Report 2010-5089, 88 p.

Miyazaki, Brent, 1980, Preliminary investigations of computer program TRANSCAP and its application to ground water basin studies: California Department of Water Resources, Technical Information Record 1388-CD-1.

PRISM Climate Group, 2012, PRISM home page: accessed October 18, 2012, at *http://www.prism.oregonstate.edu/*.

Sloop, Christina, Honton, Joseph, Creager, Clayton, Chen, Limin, Andrews, E.S., Bozkurt, Setenay, 2009, Hydrology and sedimentation, Chapter 4, *of* The Altered Laguna, A Conceptual Model of Watershed Stewardship: Laguna de Santa Rosa Foundation, p. 63-110, at *http://www.lagunafoundation.org/knowledgebase/?q=node/182*.

Sonoma County Water Agency, 2006, 2005 Urban Water Management Plan, Santa Rosa, Calif., 152 p., available at URL *http://scwatercoalition.org/images/pdf/2005_uwmp_report.pdf*

Sonoma County Water Agency, 2012, Sonoma County water agency home page: accessed October 16, 2012, at URL *http://www.scwa.ca.gov/index.php*

Sowers, J.M., Noller, J.S., and Lettis, W.R., 1998, Quaternary geology and liquefaction susceptibility, Napa, California 1:100,000 quadrangle; a digital database: U.S. Geological Survey Open-File Report 98-460, 20 p., 1 sheet, scale 1:100,000.

U.S. Department of Agriculture, 2007a, Natural Resources Conservation Service, Soil Survey Geographic (SSURGO) database for Sonoma County, California: National Resources Conservation Service, accessed on May 14, 2009, at URL *http://SoilDataMart.nrcs.usda.gov*

U.S. Department of Agriculture, 2007b, Natural Resources Conservation Service, National Engineering Handbook, Part 630 Hydrology, Chapter 7 Hydrologic Soil Groups: Washington, DC, 16 p.

Wentworth, C.M., Graham, S.E., Pike, R.J., Beukelman, G.S., Ramsey, D.W., and Barron, A.D., 1997, Summary distribution of slides and earth flows in the San Francisco Bay region, California: U.S. Geological Survey Open-File Report 97-745C, 10 p.

Chapter B. Hydrology of the Santa Rosa Plain Watershed, Sonoma County, California

By Donald S. Sweetkind, Joseph A. Hevesi, Tracy Nishikawa, Peter Martin, and Christopher D. Farrar

Introduction

The Santa Rosa Plain watershed (SRPW) is geologically and hydrologically complex, with interfingering geologic units and cross-cutting structures affecting a variety of surface-water and groundwater processes (fig. 1). This chapter presents a description and analysis of the geology, surface-water hydrology, and groundwater hydrology of the SRPW for the purpose of evaluating regional groundwater availability. Aspects of the SRPW described in this chapter include the delineation of hydrogeologic units based on lithology and hydraulic properties, construction of a detailed three-dimensional hydrogeologic framework, description of the surface-water and groundwater systems, and an analysis of both predevelopment and recent groundwater recharge and discharge.

Geology

The geology of the SRPW has been previously described in groundwater resource studies of the Santa Rosa Plain (SRP; Cardwell, 1958; Ford, 1975; Herbst and others, 1982). Since these studies, geologic mapping, geophysical studies, and interpretation of borehole data have refined the understanding of the basin geometry and the identity and location of major basin-filling units (McLaughlin and others, 2008; Langenheim and others, 2010; Sweetkind and others, 2010). Three-dimensional (3D) lithologic and stratigraphic models of the SRPW constructed from borehole data, geologic mapping, and geophysical data were used to define the lithologic, stratigraphic, and structural architecture for the region (Sweetkind and others, 2010). This chapter summarizes and integrates these previous studies to develop a modern understanding of the geology of the SRPW as it relates to the groundwater resources.

Geologic Setting

The elongate ridges and valleys in the northern Coast Ranges are the geomorphic expression of folds and fault slices formed in response to compression and lateral transport of oceanic sediments and crust that collided with and accreted to the western margin of North America during the last several million years (Bailey, 1966; Norris and Webb, 1976; Dickinson, 1981). Most of the Coast Ranges are underlain by a basement of Mesozoic consolidated to weakly metamorphosed sedimentary and crystalline rocks of the Franciscan Complex (Fox, 1983), the Great Valley Group (Ingersoll, 1990), and Coast Range ophiolite (Hopson and others, 2008). Basement rocks are exposed at the surface throughout much of the Coast Ranges (Gutierrez and others, 2010) but also underlie valleys and coastal plains, where they are buried beneath variably thick accumulations of Tertiary marine and continental sediments and volcanic rocks, which, in turn, are overlain by thin Quaternary alluvial deposits.

The highlands in the eastern part of the SRPW are underlain by various types of Miocene and Pliocene volcanic rocks, interbedded in places with the largely non-marine and estuarine strata of the Petaluma Formation; both of these units unconformably overlie basement rocks (fig. 2). The rocks in the western part of the highlands are highly deformed and cut by the active, northwest-striking, right-lateral Rodgers Creek fault zone (figs. 1 and 2). The central part of the SRPW is occupied by the topographically broad, low-lying SRP (fig. 1). The southern part of the SRP is covered by Quaternary alluvial deposits (fig. 1). The northern part features low, slightly dissected exposures of late Pliocene and Quaternary (Pleistocene and Holocene) fluvial, lacustrine, and alluvial-plain deposits that have been referred to as the Glen Ellen Formation (Fox, 1983), along with younger alluvium within stream channels (Graymer and others, 2007; fig. 1). The western edge of the SRPW lies within a broad, topographically low area

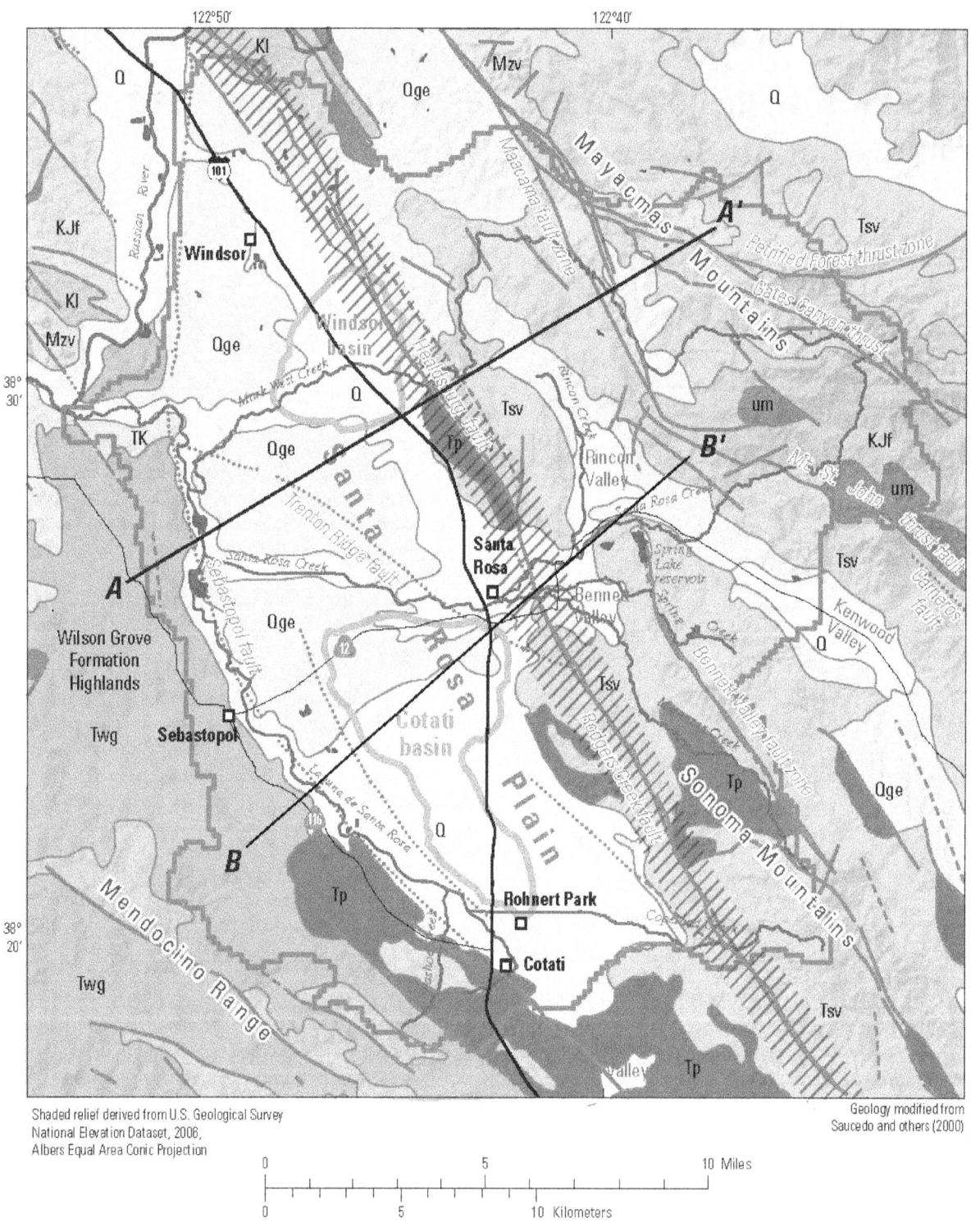

Shaded relief derived from U.S. Geological Survey
National Elevation Dataset, 2006,
Albers Equal Area Conic Projection

Geology modified from
Saucedo and others (2000)

Figure 1.   Santa Rosa Plain watershed, Sonoma County, California.

EXPLANATION

GEOLOGIC UNIT

Cenozoic

Sedimentary rocks

Quaternary

| Q | Recent alluvium, landslide, and sand dune deposits

Plio-Pleistocene and Pliocene

| Qge | Glen Ellen Formation

Pliocene and Miocene

| Twg | Wilson Grove Formation

| Tp | Petaluma Formation

Volcanic rocks
Pliocene and Miocene

| Tsv | Sonoma Volcanics

Mesozoic

Sedimentary and metasedimentary rocks

Tertiary-Cretaceous

| TK | Coastal belt rocks

Cretaceous

| Kl | Lower Cretaceous marine
| Kjf | Franciscan Complex

Plutonic, metavolcanic, and mixed rocks

| um | Ultramafic rocks
| Mzv | Volcanic and metavolcanic

Water

~~~~~ Santa Rosa Plain watershed boundary

Fault—Dashed where approximately located, dotted where concealed

/// Rodgers Creek fault zone

Location of deep basins as defined the −14 milligal isostatic gravity contour of Langenheim and others (2006)

A'
A Line of geologic section shown on figures 2, 25

Figure 1. Continued.

that is underlain by locally fossiliferous Miocene to Pliocene marine sandstone, formerly known as the "Merced" Formation (Cardwell, 1958), and now referred to as the Wilson Grove Formation (Fox, 1983). These marine strata dip gently northeastward beneath the SRPW and unconformably overlie Mesozoic rocks (figs. 1 and 2). The west and southwest sides of the SRPW are bounded by a system of poorly defined Pliocene and younger normal faults, here generalized as the Sebastopol fault (fig. 1).

Stratigraphic Units

The stratigraphic units in the SRPW include Mesozoic to Early Tertiary basement rocks, Tertiary sedimentary and volcanic rocks, and Quaternary sedimentary deposits (table 1). Mesozoic to Early Tertiary rocks of the Franciscan Complex, Great Valley Group, and Coast Range ophiolite make up the basement rocks beneath the entire study area. Tertiary sedimentary and volcanic rocks overlie the basement and are exposed in upland areas around the valleys. Quaternary sedimentary deposits underlie the valley floors and form fans along the valley margins (fig. 1).

Basement Rocks

Pre-Miocene basement includes Cretaceous and older rocks of the Franciscan Complex (Kjf; fig. 1), ultramafic rocks of the Jurassic Coast Range ophiolite (um, fig. 1), and Cretaceous marine sediments of the Great Valley Group (Kl, fig. 1; Blake and others, 1984; McLaughlin and Ohlin, 1984; Blake and others, 2002). The collective thickness of these basement units is unknown, but could be a few tens of thousands of feet (Bailey and others, 1964). These rocks are characterized by a variety of consolidated rock types, including penetratively sheared shale (melange matrix), graywacke, blocks of

blueschist, chert, greenstone, thinly interbedded shale and sandstone, and mafic to ultramafic ophiolitic rocks. Water-well drillers typically describe these rocks as either sandstone, greywacke, chert, or serpentine (table 1).

Tertiary Volcanic Rocks

The 3-8 million year old Sonoma Volcanics (Wagner and others, 2005) dominate the upland areas in the eastern part of the SRPW in the Sonoma Mountains and Mayacmas Mountains and are present within the basin fill beneath the Santa Rosa Plain (fig. 2). These volcanic rocks are well-exposed to the east of the Rodgers Creek fault zone and exist as complexly faulted slices along the southwest side of the Rodgers Creek fault zone, where they project beneath, and probably correlate with, volcanic units in the subsurface of the SRPW (McLaughlin and others, 2005; McLaughlin and others, 2008). A well drilled east of Rohnert Park (6N/8W-13R2) encountered Sonoma Volcanics at a depth of 900-ft below land surface (bls) and penetrated an additional 430 ft without reaching the underlying bedrock (California Department of Water Resources, 1979).

The Sonoma Volcanics include a thick accumulation of andesitic and basaltic tuffs containing interbedded lavas and volcaniclastic rocks (Wagner and others, 2005; McLaughlin and others, 2008). The volcanic units have a wide variety of volcanic rock types, including basaltic, andesitic, dacitic and rhyodacitic flows, flow breccias, avalanche or talus breccia, tuff, and several andesitic to rhyodacitic tephra units. Both welded and unwelded tuffs are seen in outcrops and probably reflect differences in emplacement style, accumulated thickness of individual tuff beds, and their location relative to the source vent. Many of the units have relatively limited lateral extent and appear to have been erupted from local volcanic vents. Older volcanics, such as the Tolay and Burdell

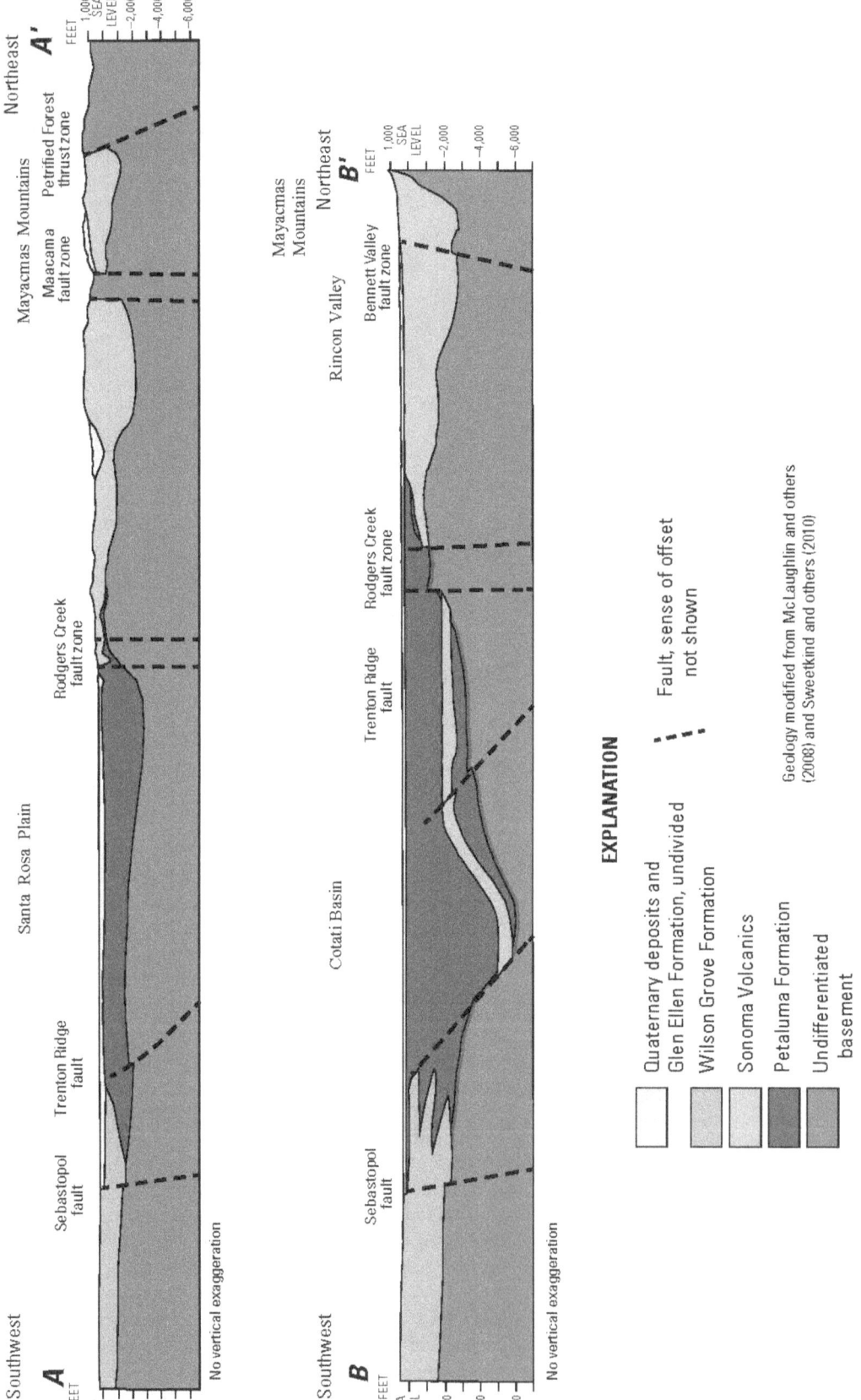

Figure 2.   Interpretive geologic cross sections of the Santa Rosa Plain watershed, Sonoma County, California.

Table 1.   Description of hydrogeologic units, Santa Rosa Plain watershed, Sonoma County, California.

[A□re□ □li□□□ >, greater than; —, no data]

Hydrogeologic unit	General age range[1]	Examples of mapped geologic units[2]	Estimated thickness, in feet	Typical depositional environment[3]	Typical lithologic description
Quaternary deposits	Quaternary	Includes units mapped as younger and older alluvium and alluvial fan and terrace deposits.	0–550	Stream-channel, flood-plain deposits, older alluvium and terrace deposits.	Gravel; sand and gravel; sand, clay, and gravel; sand and clay.
Glen Ellen Formation	Early Pleistocene (?) and Pliocene	Includes the Glen Ellen Formation, the Huichica Formation, and other unnamed Tertiary continental deposits.	0–600	Continental, piedmont, and valley alluvial fans, local lacustrine deposits.	Clay and sand, clay and gravel, sand, sand and gravel, tuff, conglomerate.
Wilson Grove Formation	Late Pliocene to late Miocene	Includes rocks formerly assigned to the Merced Formation.	0–2,700	Deep to shallow marine, locally transitional to continental environments.	Sand, sandstone, blue sandstone; clay, sand or gravel and shells; clay and sand.
Volcanic rocks	Pliocene and Miocene	Includes Sonoma, Tolay, and Burdell Mountain volcanics.	0–3,000	—	Basalt, volcanic breccia, tuff.
Petaluma Formation	Pliocene to late Miocene	Includes the Petaluma Formation.	0–3,000	Fluvial and lacustrine, with estuarine and transitional marine environment.	Clay, clay and sand, shale, sand or sandstone.
Basement rocks, undifferentiated	Pre-Miocene; predominantly Jurassic and Cretaceous	Includes the Franciscan Complex, the Great Valley Complex, including the Coast Range ophiolite.	>2,000	—	Sandstone, graywacke, chert, serpentine.

[1]General age ranges from Wagner and Bortungo (1982); Blake and others (2002); Graymer and others (2007).

[2]Mapped geologic units from Wagner and Bortungo (1982); Blake and others (2002); Graymer and others (2006 and 2007).

[3]Depositional environment listed only for Miocene and younger sedimentary rocks.

Mountain Volcanics (Wagner and others, 2011), are interbedded with the Petaluma or Wilson Grove Formations (section *B–B'* on fig. 2, table 1), whereas the younger parts of the Sonoma Volcanics overlie the Petaluma Formation and are interbedded with, or underlie, the Pliocene-Pleistocene Glen Ellen Formation (Wagner and others, 2005).

The total thickness of Sonoma Volcanics ranges up to at least a few thousand feet; however, volcanic-rock thickness is highly variable and, in general, water wells drilled in the volcanic uplands do not penetrate the entire thickness of the formation.

Tertiary Sedimentary Rocks

Three sedimentary formations of late Miocene to Early Pleistocene age dominate the basin fill beneath the SRPW: the Wilson Grove, Petaluma, and Glen Ellen Formations (table 1). In places, these three formations interfinger with each other and with the Sonoma Volcanics, forming a complex geometry. The three sedimentary units are described in the following sections in terms of their lithologic character, spatial extent, thickness, and relation to other units.

Wilson Grove Formation

The late Miocene to late Pliocene Wilson Grove Formation is mostly composed of fine- to medium-grained, thick-bedded to massively-bedded, moderate- to well-sorted, uncemented to weakly cemented marine sandstone. The cement is calcium carbonate and iron hydroxides. The formation is distinctly fossiliferous, containing stringers and lenses of shell beds (Powell and others, 2004). The unit also contains pebble and gravel stringers, some of which consist almost entirely of chert derived from the Franciscan Formation, as well as clay lenses and local thin beds of pumiceous tuff (Fox, 1983; Blake and others, 2002; Powell and others, 2004). The Wilson Grove Formation can be divided into three distinct lithologic variants that represent slightly different marine environments (Powell and others, 2004): (1) fine-grained, deep-water marine sandstones most commonly found to the west of the SRPW; (2) well-sorted, fine-to medium-grained, shallow-water marine sandstone that is represented by much of the exposed Wilson Grove Formation in the study area, especially north of Sebastopol; and (3) medium- to coarse-grained sandstone beds interbedded with pebble conglomerate beds that represent transitional marine/continental settings, commonly found south and east of Sebastopol. Along its southeastern margin, the Wilson Grove Formation interfingers with the Petaluma Formation in outcrops near the town of Cotati and at Meacham Hill immediately southwest of the SRPW (Powell and others, 2004). Wilson Grove Formation marine sandstones have been inferred to interfinger with transitional marine and non-marine deposits of the Petaluma Formation beneath the SRPW (McLaughlin and others, 2008; Sweetkind and others, 2010), but this transition zone is obscured by younger deposits (section *B–B'* on fig. 2).

In the study area, the maximum thickness of the Wilson Grove Formation has been estimated to be about 2,700 ft (Powell and others, 2004; table 1). Another study of the Wilson Grove Formation estimated that the total thickness of the formation is greater than 6,000 ft (Holland and others, 2009); however, this study aggregated measured thickness from multiple locations for portions of the unit deposited over a range of time. Exposures in the Wilson Grove Formation in the uplands to the west of the SRPW are generally 500 ft thick, whereas wells in the western part of the SRPW penetrate as much as 1,000 ft of the formation without penetrating the base of the unit. The Wilson Grove Formation is exposed over a broad area to the west of the SRPW, extending from Petaluma in the south to the Russian River on the north and westward into the Mendocino Range from the west edge of the SRPW boundary (fig. 1).

Petaluma Formation

The late Miocene to Pliocene Petaluma Formation is dominated by deposits of moderately to weakly consolidated, silty to clayey mudstone with local beds and lenses of poorly sorted sandstone and minor beds of nodular limestone and conglomerate. The Petaluma Formation can be subdivided into lower, middle, and upper members on the basis of detailed stratigraphic analysis using the coarser-grained materials and fossils (Allen, 2003). In general, the formation coarsens from claystone and siltstone at its base to sandstone and conglomerate higher up in the section (Holland and others, 2009). In outcrops near the southern part of the study area, the lower member consists of blue-gray, thick- and thin-bedded, massive-to-thinly laminated mudstone and very fine-grained sandstone (Holland and others, 2009). The lower member also includes minor interbeds of sandstone, conglomerate, limestone, dolomite, and chert. The middle and upper members are coarser-grained than the lower member and contain lenticular beds of conglomerate and thick-bedded sandstone.

The Petaluma Formation consists of transitional marine and non-marine sediments that were deposited in estuarine, lacustrine, and fluvial depositional settings (Allen, 2003; Powell and others, 2004). The continental fluvial and lacustrine facies extend westward from the Sonoma and Mayacmas mountains; they transition to marine facies beneath the SRP (McLaughlin and others, 2008; Sweetkind and others, 2010).

The Petaluma Formation is exposed in outcrops along the western slopes of the Mayacmas and Sonoma mountains from north of Santa Rosa to near Cotati, around the southern margin of SRPW, and within Bennett Valley (Clahan and others, 2003; McLaughlin and others, 2008; fig. 1). On the basis of geologic logs from wells, it is also known to be present under a large part of the valley floor of SRP (fig. 2). The formation is at least 4,000-ft thick in the Petaluma oil field, located 10-mi south of the study area, and is at least 3,000-ft thick in the study area based on outcrops and cuttings from deep, petroleum exploration wells (Morse and Bailey, 1935; Allen, 2003; Powell and others, 2004).

Glen Ellen Formation

The Pliocene to early Pleistocene (less than 3.2 million years old) Glen Ellen Formation consists of clay-rich stratified deposits of poorly sorted sand, silt, and gravel. Grain-size grades both laterally and vertically from coarse to fine, commonly over distances of a few tens to a few hundreds of feet (Cardwell, 1958). Bedding is thick to massive and often has lenticular form. Most of the clasts and probably much of the matrix were derived from the Sonoma Volcanics. Cobbles in the conglomerates are mostly subangular to rounded and range mostly between 3 and 6 in. in diameter (Weaver, 1949). The cobbles are mostly of andesitic or basaltic composition; obsidian clasts are one of the hallmark characteristics of this formation (Cardwell, 1958; McLaughlin and others, 2005). The sediments making up this formation were probably originally deposited as alluvial fans and piedmont. Some of the material beneath and next to SRP was probably deposited in lagoons or shallow bays and could grade into a marine facies.

The Glen Ellen Formation is exposed near the town of Glen Ellen in upper Sonoma Valley and to the north and west of Santa Rosa. Within the SRPW, the Glen Ellen Formation rests directly on the basement rocks of the Franciscan Complex in places, but more generally overlies the Sonoma Volcanics, Wilson Grove, or Petaluma Formations. Along the eastern margin of and beneath the SRP, the Glen Ellen Formation is overlain by alluvial units of Quaternary age (fig. 1). Cardwell (1958) reported the formation to be as thick as 3,000 ft beneath the SRP, although no drill-hole or outcrop data document such a thickness. Sweetkind and others (2010) utilized available drillers' logs in the SRPW to interpret the thickness of the Glen Ellen Formation as highly variable but, generally, to be a few hundred feet thick or less; this interpretation is used for this report.

Quaternary Deposits

Quaternary sedimentary deposits recognized and mapped within the SRPW include alluvial deposits, terrace deposits, near-shore marine and estuarine sediments, colluvium, and landslide deposits (Sowers and others, 1998; Graymer and others, 2007). Alluvial sediments of Quaternary age have been mapped as distinct deposits on the basis of the degree of consolidation, cementation, clast size and sorting, and geomorphic expression. The alluvial sediments can be divided into older (Pleistocene) and younger (Holocene) deposits, which, to some degree, are consistent with the amount of cementation and consolidation of the sediments (Cardwell, 1958; Sowers and others, 1998). Alluvial fan deposits cover the largest areal extent of any of the alluvial sediments in the study area. The fan, river terrace, and stream-channel deposits generally consist of heterogeneous mixtures of poorly- to well-sorted sand, silt, clay, gravel, cobbles, and boulders in thin to massive, interfingering beds of limited lateral extent (tens to hundreds of feet), often with lenticular form. The deposits near valley axes are mostly flat lying, but near valley margins, deposits in wedge-shaped fans dip toward the valley.

Large volumes of alluvial sediments are present in all the main valleys in the study area. The older alluvial deposits are as much as 400 ft thick; the younger alluvium is generally less than 150 ft thick. Landslide and colluvial deposits form isolated patches or thin cover near the flanks of the valleys.

Faults

The largest-offset faults in the study area are the northwest-trending, right-lateral strike-slip faults that are part of the San Andreas dextral transform system, including the Rodgers Creek fault zone, Maacama fault zone, and Bennett Valley fault zone (fig. 1); the Carneros fault zone, important farther to the southeast in Sonoma Valley, could extend into the eastern part of the study area. These faults often exhibit components of transtensional or transpressional dip-slip movement, particularly at fault bends. The major strike-slip faults in the study area typically are a complex of steeply dipping, braided strands that form zones several hundred to several thousand feet wide rather than distinct single planes of weakness. The major faults with strike-slip motion can be generally regarded as vertical for the purpose of this hydrologic study; this simplification could be less applicable to the Trenton Ridge fault, which has been interpreted as a reverse fault with a gentle dip (Fox, 1983) or steep dip (Williams and others, 2008).

Faults with large components of normal slip are interpreted to bound deep, sedimentary basins beneath the SRPW (McLaughlin and others, 2008). Normal faults have north to northeast strikes; however, many have curvilinear surface expression, such as the Sebastopol fault that bounds the west side of the SRPW and the unnamed fault that lies to the east of the Sebastopol fault (fig. 1 and section *B–B'* on fig. 2). Normal faults with dip-slip displacement are common in local areas of extension within the overall strike-slip regime. These faults often bound pull-apart basins, particularly where fault strands overlap and displacement is transferred from one fault strand to another. Where faults are at angles to the prevailing stress field, or where the principal stresses have rotated through time, compressional structures can be formed, such as the mostly buried Trenton Ridge fault and Petrified Forest thrust zone (fig. 1 and section *A–A'* on fig. 2).

The Rodgers Creek fault zone consists of two segments: a northern Healdsburg fault segment and the southern Rodgers Creek fault segment, which are separated by the Santa Rosa Creek floodplain (Gealey, 1951; Fox, 1983; McLaughlin and others, 2008). The surface traces of these two segments are offset from each other; the traces probably overlap beneath the Santa Rosa Creek floodplain (McLaughlin and others, 2008). The predominant sense of displacement on both fault segments is right lateral strike-slip and reflects the same regional pattern of displacement as on the San Andreas fault system. The fault zone is about 0.6 mi wide, and though partially obscured by landslides and urban development, the linear topography and elongate depressions, some containing wetlands or sag ponds, serve to identify the fault-zone location (McLaughlin and others, 2008). Movement along the fault zone has been complex with more than one sense of displacement during Pliocene to Holocene, dependent on the extant stress field (McLaughlin and others, 2008). The northern segment of the fault dips steeply (greater than 70 degrees) to the northeast on the basis of the hypocenters of earthquakes (Wong and Bott, 1995). North of Santa Rosa a strong subsurface gravity gradient is offset by 1.2 mi from the surface trace of the fault, indicating reverse faulting with a down-to-the-northeast sense of offset. However, Langenheim and others (2010) suggest the offset of the gravity anomaly from the mapped surface trace could represent a recent adjustment of the active fault position rather than evidence of a through-going thrust fault.

The Maacama fault zone is a northwest striking, steeply dipping, right-lateral strike-slip fault that extends across about 93 mi of northwestern California. In the study area, the Maacama fault zone is located in the Mayacmas Mountains (fig. 1), where highly disrupted slivers of Franciscan Complex basement rocks are exposed in a zone up to 1 mi wide. Although the predominant sense of displacement along the Maacama fault zone is dextral strike-slip, the fault is coincident with an elongate gravity low bounded by a strong gravity gradient, which is consistent with long-term southwest-side down offset (Langenheim and others, 2010).

The Bennett Valley fault zone is a sinuous zone of right-lateral fault strands that extends southeastward about 19 mi from the Spring Lake (or Santa Rosa Creek) reservoir into Sonoma Valley (fig. 1). Landslides and urbanization obscure much of the zone; where exposed, the fault zone is narrow (10-ft wide), steeply dipping, and north striking (McLaughlin and others, 2008). The Spring Lake reservoir was constructed over a marshy wetland area along the fault zone.

The Sebastopol fault is a curved zone of poorly exposed faults that generally parallels the contact between Pliocene Wilson Grove Formation and Quaternary alluvial deposits and the topographic break in slope between the valley floor and the low hills to the west and south of Sebastopol (figs. 1 and 2). In the vicinity of Sebastopol, the fault bends to a more northerly trend (Bezore and others, 2003; Clahan and others, 2003; Delattre and others, 2008). The Sebastopol fault probably consists of a network of subparallel, related, short fault segments; lack of strong geophysical expression indicates that total vertical and horizontal offset on these faults is small (less than 1,500 ft; fig. 2; Langenheim and others, 2010). North of Sebastopol, near Trenton, trenching for a treated-water pipeline exposed the fault and revealed that the main sense of displacement is east-side-down normal faulting. There is no evidence for any strike-slip displacement along the Sebastopol fault (R. McLaughlin, written communication, 2009).

The mountain uplands in the easternmost part of the study area are generally the result of uplift associated with young, compressional structures that largely post-date the Sonoma Volcanics (McLaughlin and others, 2005 and 2008; Graymer and others, 2006 and 2007). The Petrified Forest thrust zone and Gates Canyon thrust fault (McLaughlin and others, 2005) form an imbricate zone of south-southwest-directed, northeast-dipping reverse faults that truncate at the Maacama fault zone (fig. 1). These reverse faults trend northwest-southeast and place rocks of the Franciscan Complex above younger, ash-flow tuff and basaltic andesite of the Sonoma Volcanics (Petrified Forest thrust zone along section A–A', fig. 2) (McLaughlin and others, 2004). A similarly oriented thrust fault, the Mt. St. John thrust (Graymer and others, 2006 and 2007; Langenheim and others, 2010), juxtaposes Franciscan Complex rocks against the Sonoma Volcanics near the southern end of the Maacama fault (fig. 1).

On the basis of a prominent gravity anomaly, Langenheim and others (2010) extended the Carneros fault northwest of its mapped extent to upper Sonoma Valley and the southeast tip of the Maacama fault. Within the study area, this northern extension of the Carneros fault is buried by younger parts of the Sonoma Volcanics and by Franciscan Complex rocks that lie above the Mt. St. John thrust. Where exposed to the south, the Carneros fault juxtaposes Tertiary marine strata, including the Neroly Formation and overlying Sonoma Volcanics on the west, with Great Valley Group rocks on the east (Graymer and others, 2007; Langenheim and others, 2010). It is possible that there is a similar stratigraphic juxtaposition in the subsurface in the eastern part of the study area.

Basin Depth and Geometry

Analysis and modeling of gravity data indicate that the Tertiary sedimentary and volcanic rocks and Quaternary deposits underlying the SRPW conceal a complex basement topography developed primarily on rocks of the Franciscan Complex (Langenheim and others, 2006; McPhee and others, 2007; Langenheim and others, 2010). Taking advantage of the contrast between dense, pre-Miocene rocks (predominantly composed of Mesozoic rocks of the Franciscan Complex and the mafic Coast Range ophiolite) and less dense, Tertiary and Quaternary sedimentary rocks and Pliocene and Miocene volcanic rocks, gravity measurements detected variations in the density of rocks in the shallow and middle levels of the crust (McPhee and others, 2007). Gravity highs were measured over basement rock outcrops, areas of thin, sedimentary-rock cover, and over dense lavas; gravity lows were measured over

areas of thick, low-density sedimentary rock or sediment cover (McPhee and others, 2007; Langenheim and others, 2010). A surface representing the top of the high-density basement was derived from mathematical inversion of regional gravity measurements (Langenheim and others, 2006 and 2010; McPhee and others, 2007) constrained by outcrop data and well data to assign density values to the formations suspected to underlie a particular area (Nettleton, 1976). This surface was inferred to represent the elevation of pre-Miocene rocks and allowed estimation of the thickness of relatively low-density rocks overlying the basement.

The gravity-mapping technique was used to identify major structural features beneath the SRPW that have no clear surface manifestation. The resulting model of depth to pre-Miocene bedrock for the SRPW defines both the overall basin geometry and the configuration of subbasins that are bounded by internal faults (McPhee and others, 2007; Langenheim and others, 2010). The gravity method allows for good delineation of the basin shapes, but is less precise in the thickness or depth estimations because the density-depth relation at any particular location can be poorly known. Steep-gravity gradients bound many of the basement bedrock highs and lows, which is indicative of some degree of fault control. Such a fault is shown at depth bounding the southwestern side of the Cotati basin on interpretive geologic section B–B' of figure 2.

The analysis of gravity data reveals two deep, steep-sided sedimentary basins: the Windsor basin beneath the northern part of the SRP and the Cotati basin beneath the southern part, which are separated by a buried bedrock ridge (McPhee and others, 2007; Langenheim and others, 2008; fig. 1). The Windsor basin is about 5.5 by 7.5 mi in size and is centered near the town of Windsor. The thickest exposures of the Glen Ellen Formation in the SRPW are observed near this basin in the hills that flank the northeast side of the SRPW (fig. 1). The basin has a roughly triangular form, bounded by the Healdsburg fault segment on the northeast, the Trenton Ridge fault to the south, and a zone of poorly exposed normal faults on the west. Inversion of gravity data indicates the basin is 3,000–6,500 ft deep (Langenheim and others, 2008). The southern and western margins of the Windsor basin appear to have a series of downward steps into the basin (Langenheim and others, 2010), indicating that normal faulting played a role in basin subsidence. Based on outcrop and well data, the deeper parts of the Windsor basin are likely filled with tuff beds and lavas of the Sonoma Volcanics intercalated with sedimentary units of the Petaluma Formation (McLaughlin and others, 2008). Rocks of the Glen Ellen Formation and Quaternary alluvial fan deposits overlie these older rocks.

Analysis of gravity data indicates that the Cotati basin in the southern part of the study area is larger and has a more complex shape than the Windsor basin (Langenheim and others, 2008; fig. 1). The Cotati basin includes two structurally-controlled subbasins with an intervening east-trending basement ridge that is less than 4,000 ft bls (Langenheim and others, 2010). The deepest part of the Cotati basin is beneath Rohnert Park and is modeled as 8,000 to 10,000 ft deep.

Gravity data and lithologic logs from oil exploration boreholes indicate the basin is filled with as much as 6,500 ft of Miocene to Pliocene rocks of the Sonoma Volcanics, Petaluma Formation, and Wilson Grove Formation that are unconformably overlain by about 520 ft of Late Pliocene to Quaternary alluvial and lacustrine deposits (McLaughlin and others, 2008).

The Windsor and Cotati basins are separated by a buried ridge of relatively elevated bedrock that is uplifted by the west-northwest-striking Trenton Ridge fault (fig. 1). The Trenton Ridge fault is exposed only in the northwestern part of the study area, where it strikes northwest to west-northwest, dips to the northeast, and places Franciscan Complex basement rocks over the Pliocene sandstone of the Wilson Grove Formation (Delattre and others, 2008). Seismic reflection surveys and gravity data indicate that basement rocks could be as shallow as 1,000 ft bls in the area between the Windsor and Cotati basins (Williams and others, 2008). Reflections from basement rocks or Pliocene to Quaternary sedimentary rocks show increasing downwarping with depth across the Trenton Ridge fault, indicative of the slow formation of the ridge over time (Williams and others, 2008).

To the east of the southern segment of the Rodgers Creek fault, gravity data and geologic mapping define a shallow, sediment-filled, pull-apart basin underlying most of Rincon Valley and part of Bennett Valley (section B–B' on fig. 2; McLaughlin and others, 2008; Langenheim and others, 2010). This small basin is interpreted to be the result of interaction between the Rodgers Creek fault and the Maacama fault to the northeast (McLaughlin and others, 2008). Gravity data define a large feature in the bedrock of Bennett Valley (Langenheim and others, 2008). The Bennett Valley gravity high extends from the Rodgers Creek fault zone on the west to the Bennett Valley fault zone on the east. Although relatively high-density basement rocks or dense lavas in the Sonoma Volcanics could produce the gravity high, neither of these formations crop out extensively above or near the gravity high. A few small outcrops of serpentine and Franciscan Complex rocks exposed in ravines in the bounding fault zones indicate a Mesozoic basement could lie at shallow depth in this area, however. Langenheim and others (2008 and 2010) also interpreted magnetic data in concert with gravity data to better constrain the formations producing the gravity high and concluded Franciscan Complex greenstone or hydrothermally altered ophiolitic rocks are the most likely source, not Sonoma Volcanics.

Surface-Water Hydrology

In general, the characteristics of streamflow provide an indication of the integrated hydrologic response of the upstream drainage to precipitation. Available streamflow records from gages in the SRPW were analyzed to develop a better understanding of the seasonal distribution of streamflow, along with the response of streamflow to variability in climate. The long-term history of streamflow was evaluated by using an estimated (also referred to as an extended) record of streamflow. The analysis of streamflow is used to help develop the conceptual model of the SRPW hydrologic system presented in *chapter D* of this report.

Surface-Water Drainage Pattern

Throughout much of the Northern California Coast Ranges, the surface-water drainage pattern is strongly influenced by geologic structure. Most of the larger streams, such as the Russian River, flow parallel to the dominant northwest-southeast structural trend of folds and faults for much of their length. In contrast to this regional drainage pattern, many of the streams within the SRPW do not parallel the southeast- to northwest-trending geologic and physiographic structure, but rather cut across the structural trends, flowing east to west from the highlands in the Sonoma and Mayacmas Mountains in the eastern SRPW. The SRPW is mostly within the middle Russian River drainage basin and includes 16 subbasins identified by the California Interagency Watershed Map (CIWM) of 1999 (Cal-Atlas, 2007). The 16 CIWM subbasins define an area of 251 square miles (mi^2) and are grouped into 3 larger drainage-basin areas that are named according to the main stream draining the area: Mark West Creek, Santa Rosa Creek, and Laguna de Santa Rosa drainage basins (figs. 3 and 4).

The USGS National Hydrography Dataset (NHD; Simley and Carswell, 2009) identifies 28 named streams within the SRPW. The Sonoma County Water Agency (SCWA) identifies additional named streams within the SRPW (Sonoma County Water Agency, written commun., 2008). The larger streams identified by NHD and SCWA are labeled in figure 4 as major streams. The three largest stream channels within the SRPW are Mark West Creek, Santa Rosa Creek, and the Laguna de Santa Rosa. According to both the NHD and SCWA, Mark West Creek is the main tributary connecting the SRPW to the Russian River (fig. 3). However, the Laguna de Santa Rosa stream channel is the primary channel draining the SRP (Sloop and others, 2007; Cummings, 2004).

Laguna de Santa Rosa Drainage Basin

The Laguna de Santa Rosa drainage basin is an 88-mi^2 area drained by the Laguna de Santa Rosa upstream of the Santa Rosa Creek tributary (fig. 3). The basin drains the southern and southwestern areas of the SRPW. The Laguna de Santa Rosa is a low-gradient drainage that originates at an altitude of 260 ft, west of the city of Cotati and close to the southern boundary of the SRPW (fig. 4). Much of the Laguna de Santa Rosa upstream of the Mark West Creek confluence is below an altitude of 50 ft. Santa Rosa Creek, which joins the Laguna de Santa Rosa north of the Laguna de Santa Rosa drainage basin, is the largest tributary to the Laguna de Santa Rosa. Other important tributaries to the Laguna de Santa Rosa include Copeland Creek, Crane Creek, Hinebaugh Creek, Five Creek, Colgan Creek, Gossage Creek, Washoe Creek, and Roseland

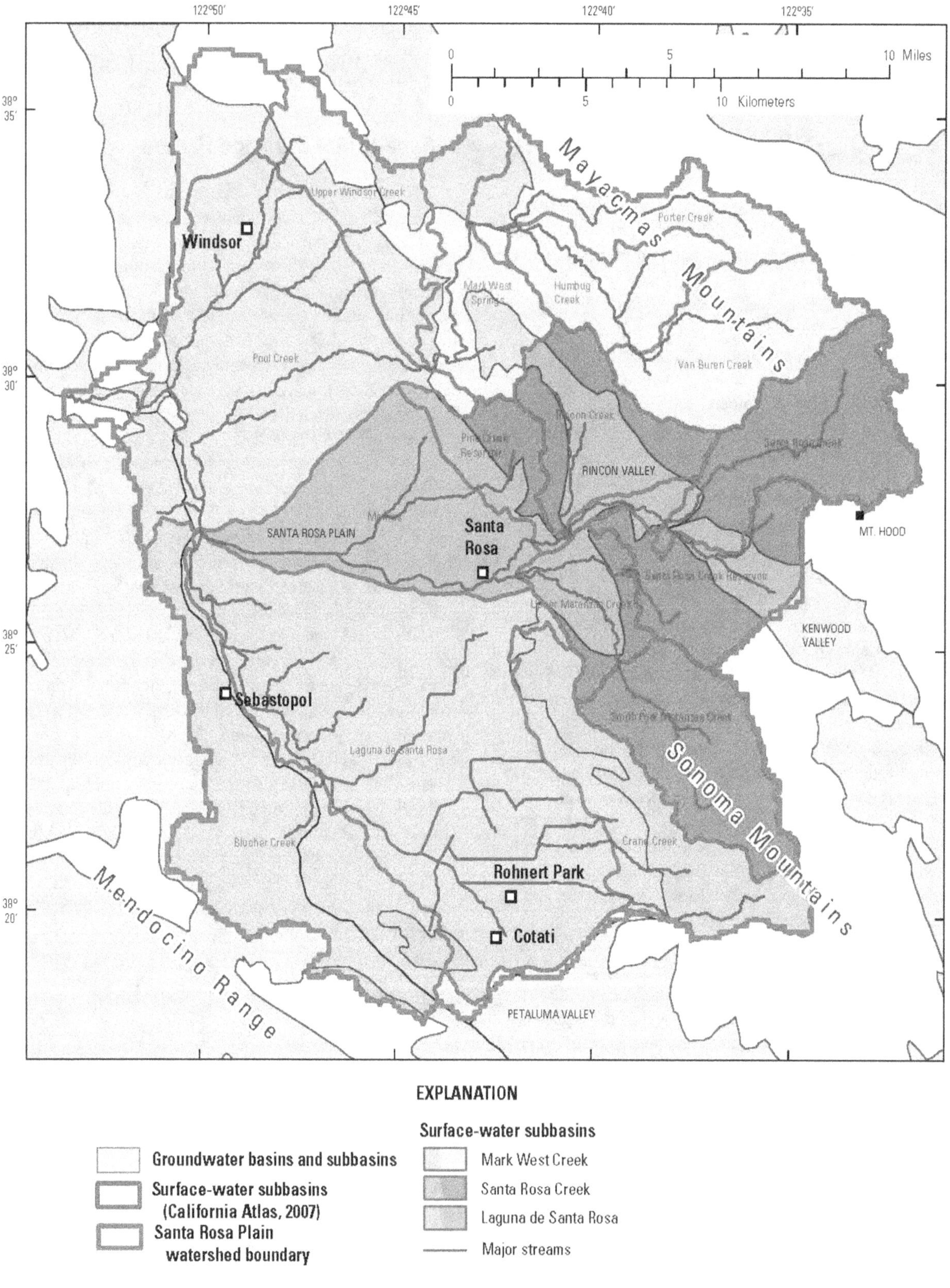

EXPLANATION

	Groundwater basins and subbasins
	Surface-water subbasins (California Atlas, 2007)
	Santa Rosa Plain watershed boundary

Surface-water subbasins

	Mark West Creek
	Santa Rosa Creek
	Laguna de Santa Rosa
—	Major streams

Figure 3.   Santa Rosa Plain watershed boundary and surface-water drainage basins, Sonoma County, California.

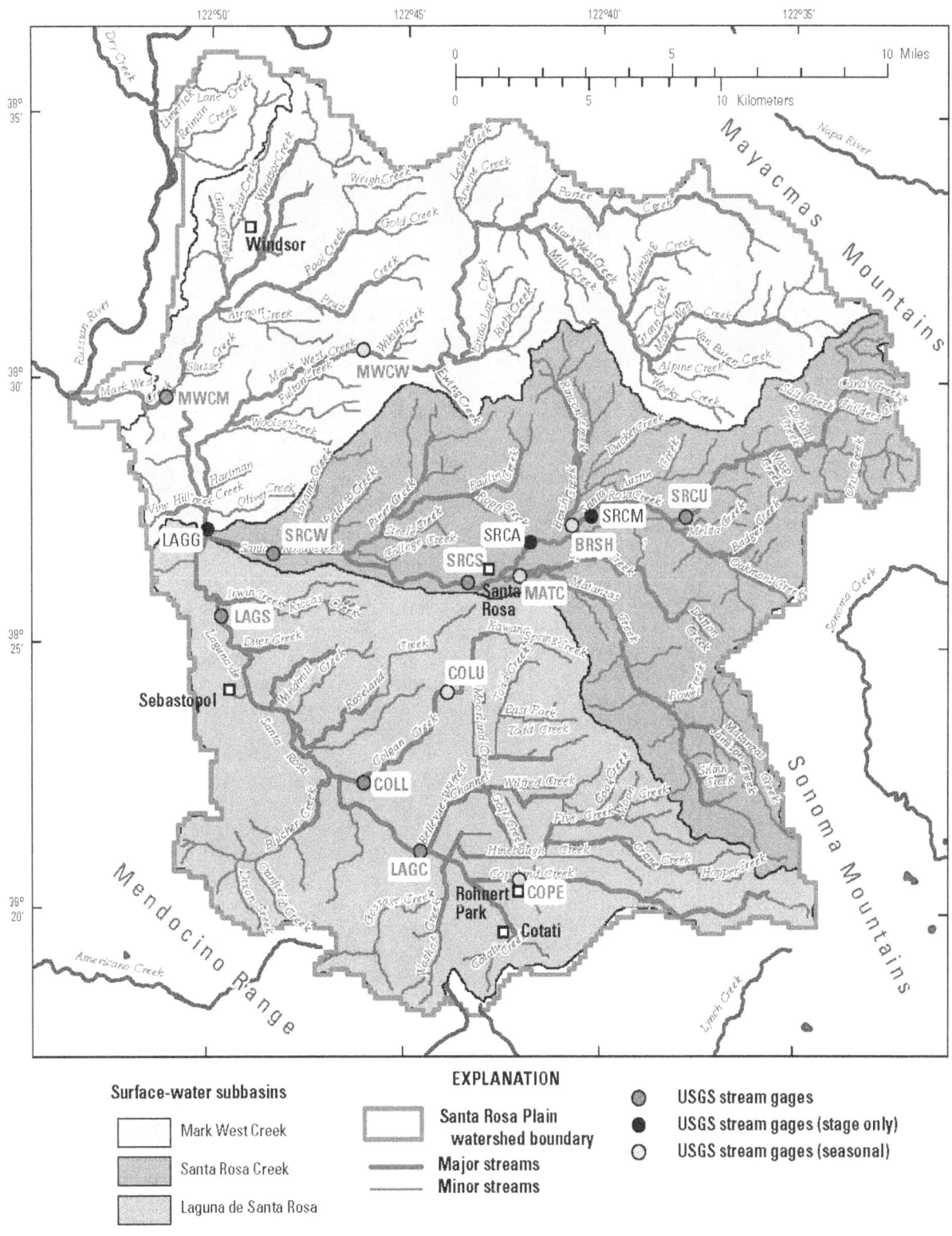

Figure 4. Major and minor stream channels and stream gage locations within the Santa Rosa Plain watershed, Sonoma County, California. (COPE, Copeland Creek at Rohnert Park; LAGC, Laguna de Santa Rosa near Cotati; COLU, Colgan Creek near Santa Rosa; COLL, Colgan Creek near Sebastopol; LAGS, Laguna de Santa Rosa near Sebastopol; LAGG, Laguna de Santa Rosa near Graton; SRCU, Santa Rosa Creek near Santa Rosa; SRCM, Santa Rosa Creek at Mission Blvd; BRSH, Brush Creek at Santa Rosa; SRCA, Santa Rosa Creek at Alderbrook Drive; MATC, Matanzas Creek at Santa Rosa; SRCS, Santa Rosa Creek at Santa Rosa; SRCW, Santa Rosa Creek at Willowside Road; MWCW, Mark West Creek near Windsor; MWCM, Mark West Creek near Mirabel Heights)

Creek (fig. 4). Copeland Creek and Crane Creek have short perennial reaches (Simley and Carswell, 2009) draining the Sonoma Mountains in the southeastern part of the SRPW study area. Copeland Creek is perennial in its upper sections, becomes intermittent as it flows westward across the alluvial fan east of Rohnert Park, and then becomes mostly channelized as it continues flowing westward through Rohnert Park and the city of Cotati before joining the Laguna de Santa Rosa at an altitude of 92 ft.

Crane Creek originates in the northern Sonoma Mountains, at an altitude of 1,535 ft, and flows westward. Crane Creek joins Hinebaugh Creek within the city of Rohnert Park. Hinebaugh Creek is channelized downstream of the confluence with Crane Creek and joins the Laguna de Santa Rosa at an altitude of 85 ft. Washoe Creek is a short, 2 mi-long, perennial channel draining the southern-most part of the SRPW, flowing from south to north. On the western side of the Laguna de Santa Rosa subbasin, Blucher Creek is also characterized as a perennial stream (Simley and Carswell, 2009).

The "Laguna de Santa Rosa" also refers to the general area of wetlands, ponds, and vernal pools within the area of the 1-percent annual exceedance probability (AEP) floodplain surrounding the main Laguna de Santa Rosa channel (Federal Emergency Management Agency, 2002; fig. 5). The Laguna de Santa Rosa channel and floodplain together form a natural overflow basin connecting Santa Rosa Creek, Mark West Creek, and the smaller creeks in the SRPW with the Russian River. The overflow basin, approximately defined by the 1-percent AEP floodplain, is the second largest freshwater wetland area in the coastal northern California region and is valued as an important ecological resource.

During the dry summer season, the Laguna de Santa Rosa overflow basin consists of a winding ribbon of channels and interconnected wetlands. In addition to the swamp and marsh areas connected to the main channel, the area within the floodplain includes numerous vernal pools. During the winter storm season, the channel and wetland areas transform into a series of lakes within the floodplain. When flooding occurs in response to larger storms, the lakes and wetlands coalesce into larger water bodies within the floodplain. During the largest storms, flooding is more extensive and, for short durations (generally no longer than several days), the water bodies in the Laguna de Santa Rosa combine into a nearly continuous lake 10 mi long that can be several hundred feet to over a mile wide (Cardwell, 1958). In addition, during periods of high runoff in the Russian River, directions of flow in the Laguna de Santa Rosa can be either to or from the Russian River. The Laguna de Santa Rosa floodplain acts as a natural regulator of floods on the lower Russian River by temporarily capturing and storing up to 79,000 acre-ft of flood water, thus dampening the peak flows in the Russian River downstream of the Mark West Creek tributary (Sloop and others, 2007). Outflow from the Laguna de Santa Rosa can be partially or completely blocked by backwater effects from the Russian River. As the Russian River flooding recedes, the discharge from the Laguna de Santa Rosa into Mark West Creek can exceed the storm runoff generated within the Mark West Creek drainage basin because of the additional volume of Russian River water included in the discharge.

Santa Rosa Creek Drainage Basin

The Santa Rosa Creek drainage basin is a 77-mi^2 drainage area in the central and eastern parts of the SRPW (fig. 3). Santa Rosa Creek, the main stream in the Santa Rosa Creek drainage basin, is 22 mi long and flows in a westerly direction from drainage divides in the Mayacmas and Sonoma Mountains to the confluence with the Laguna de Santa Rosa (fig. 4). Similar to Mark West Creek, the headwaters of Santa Rosa Creek and its major tributaries begin in steep, relatively undeveloped mountainous terrain that features natural vegetative cover, including both forest and grassland. The middle part of the Santa Rosa Creek drainage includes urbanized areas within the city of Santa Rosa and also includes rolling hills covered in grass, pasture, and crops. Santa Rosa Creek becomes an engineered channel with concrete embankments as it passes through the urbanized areas within the city of Santa Rosa. In the central part of Santa Rosa, a section of the engineered channel is underground. Downstream of Santa Rosa, Santa Rosa Creek is a straightened flood-control channel, designed to drain more rapidly across the alluvial basin and into the Laguna de Santa Rosa floodplain. The lower part of the drainage consists mostly of agricultural land. The lowest point in the drainage has an altitude of approximately 49 ft at the confluence of Santa Rosa Creek with the Laguna de Santa Rosa. Important tributaries to Santa Rosa Creek include Piner Creek, Paulin Creek, Brush Creek, Rincon Creek, and Ducker Creek on the north side of the drainage, and Matanzas Creek, Spring Creek, and Oakmont Creek on the south side (fig. 4).

The upper sections of Santa Rosa Creek and Matanzas Creek are classified as perennial streams (Simley and Carswell, 2009), although, by late summer and fall, flows diminish to less than 2 cubic-feet per second (ft^3/s) throughout much of the drainage. Most of Piner Creek, the lower section of Brush Creek, and sections of Spring Creek are classified as reconstructed channels because of a history of channel modifications as a result of agricultural development, urbanization, and flood control (U.S. Army Corps. of Engineers, 2002). The upper section of Paulin Creek, all of Rincon Creek, Ducker Creek, Spring Creek, Oakmont Creek, and all first-order streams within the Santa Rosa Creek subbasin are classified as intermittent (Simley and Carswell, 2009).

Mark West Creek Drainage Basin

The Mark West Creek drainage basin is an 86-mi^2 drainage area in the northern part of the SRPW (fig. 3). The basin is drained by Mark West Creek, but excludes the Mark West Creek drainage area downstream of the Windsor Creek tributary as defined by Cal-Atlas (2007; fig. 4).

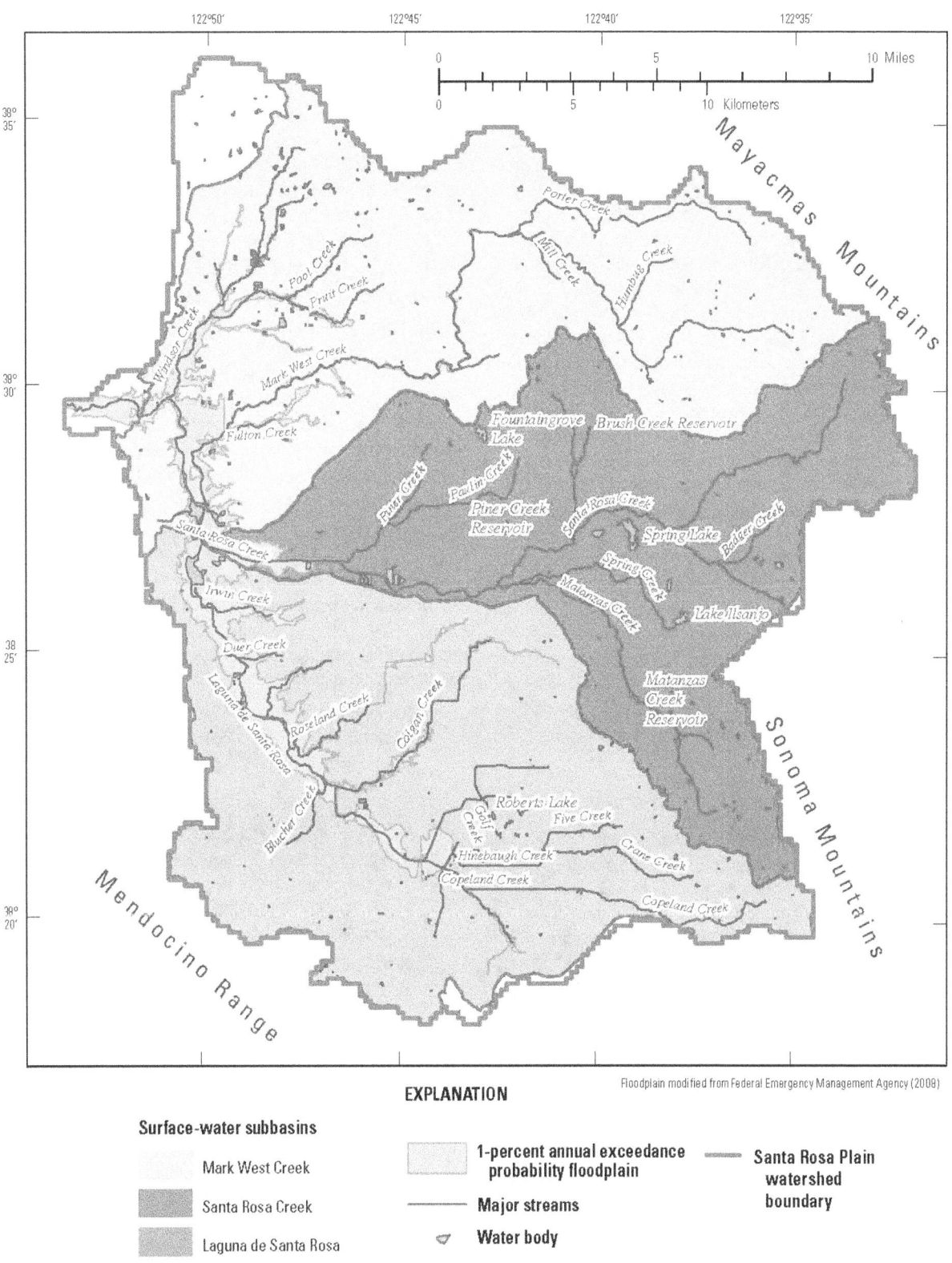

Figure 5. Water bodies within the Santa Rosa Plain watershed, Santa Rosa Plain, Sonoma County, California.

Mark West Creek is the main stream in the Mark West Creek subbasin. The stream channel is 30 mi long and originates at an altitude of 1,922 ft in the Mayacmas Mountains, close to the western-most part of the study area. Mark West Creek has a mountainous, relatively undeveloped and forested upper drainage. In the lower part of the subbasin, Mark West Creek maintains a well-defined channel in mostly agricultural land.

The main tributaries of Mark West Creek are Windsor and Porter Creeks on the north and the Laguna de Santa Rosa on south (fig. 4). Smaller tributaries on the south side of the main branch include Mill Creek, Weeks Creek, and Van Buren Creek. Smaller tributaries on the north side of the main branch include Humbug and Leslie Creeks. Other important named streams within the Mark West Creek drainage basin include Pool and Pruit Creeks, which are tributaries to Windsor Creek.

According to the NHD (Simley and Carswell, 2009), the main channel of Mark West Creek is perennial throughout its length, with summer flows maintained by numerous springs in the Mayacmas Mountains and other locations. Most of the main channel is in its natural state; anthropogenic (engineered) channel modifications have been minor relative to changes to other channels in the SRPW. Much of the riparian vegetation adjacent to the Mark West Creek channel, as well as the channel bed itself, is undeveloped and characteristic of natural channel conditions. The Weeks Creek and Humbug Creek tributaries are also perennial streams, whereas Windsor Creek, Pool Creek, Porter Creek, Mill Creek, and Van Buren Creek are generally intermittent streams that become dry during late spring to late summer (Simley and Carswell, 2009).

Water Bodies

The SRPW includes a total of 403 permanent and semi-permanent water bodies identified on 7.5-minute USGS topographic maps, composing a total area of 982 acres (Simley and Carswell, 2009; fig. 5). The majority of these water bodies have areas less than 10 acres, with a minimum area of 0.02 acres and a mean area of 2.4 acres. The water bodies include 16 intermittent lakes and ponds (with a total area of 12 acres), 291 perennial lakes and ponds (with a total area of 394 acres), 83 reservoirs (with a total area of 315 acres), and 10 swamp and marsh areas, or wetlands (with a total area of 260 acres). The largest water bodies are swamp or marsh areas (wetlands), with an average area of 26 acres, located mostly within the 1-percent AEP floodplain of the Laguna de Santa Rosa. The largest water body within the SRPW is an unnamed swamp/marsh feature of 103 acres located in the main channel of the Laguna de Santa Rosa, east of Sebastopol and directly upstream of the confluence with Santa Rosa Creek.

The SRPW includes eight named water bodies identified by the NHD (Simley and Carswell, 2009; fig. 5). Four of the named water bodies, Brush Creek reservoir, Piner Creek reservoir, Matanzas Creek reservoir, and Spring Lake (also referred to as Santa Rosa Creek reservoir) are flood-control facilities (U.S. Army Corps. of Engineers, 2002). Piner Creek and Brush Creek reservoirs are mostly empty during summer. Spring Lake and Matanzas Creek reservoirs store water throughout the year for recreational purposes and to maintain summer flows in Santa Rosa Creek. Lake Ilsanjo (also referred to as Annadel reservoir), Fountaingrove Lake, Lake Ralphine, and Roberts Lake also store water throughout the year primarily for recreational purposes. Spring Lake is the largest named water body classified as a lake or reservoir, with an area of 72 acres, followed by Fountaingrove Lake (32 acres), Lake Ilsanjo (30 acres), Lake Ralphine (21 acres), Matanzas Creek Reservoir (11 acres), and Roberts Lake (5 acres).

Streamflow

Streamflow records from gages within and adjacent to the SRPW were analyzed to help characterize the surface-water hydrology in the watershed, including correlation of stream discharge with drainage area, monthly and seasonal distribution of streamflow, and climatic variability.

Streamflow Gages

Streamflow or stream-stage records are available at 15 USGS streamflow gaging stations within the SRPW (fig. 4, table 2). To simplify the identification of gages on maps and in the text of this report, the following gage codes were defined for this study (table 2):

Laguna de Santa Rosa drainage basin
- COPE (11465660, Copeland Creek at Rohnert Park)
- LAGC (11465680, Laguna de Santa Rosa near Cotati)
- COLU (11465690, Colgan Creek near Santa Rosa)
- COLL (11465700, Colgan Creek near Sebastopol)
- LAGS (11466200, Laguna de Santa Rosa near Sebastopol)
- LAGG (11466500, Laguna de Santa Rosa near Graton)

Santa Rosa Creek drainage basin
- SRCU (11465800, Santa Rosa Creek near Santa Rosa)
- SRCM (11466050, Santa Rosa Creek at Mission Blvd)
- BRSH (11466065, Brush Creek at Santa Rosa)
- SRCA (11466080, Santa Rosa Creek at Alderbrook Drive)
- MATC (11466170, Matanzas Creek at Santa Rosa)
- SRCS (11466200, Santa Rosa Creek at Santa Rosa)
- SRCW (11466320, Santa Rosa Creek at Willowside Road)

Mark West Creek drainage basin
- MWCW (11465500, Mark West Creek near Windsor)
- MWCM (11466800, Mark West Creek near Mirabel Heights)

Of the 15 gages, 6 (COLL, SRCU, MWCM, LAGS, SRCW, and LAGC) have records that are continuous throughout the entire year, while 5 gages (BRSH, MATC, MWCW, COPE, and COLU) are seasonal-only, with a measurement period from late fall through early spring (generally from October through April) to record winter streamflow. One of the gages (SRCS) has a combination of continuous and seasonal-only data. Three of the gages (SRCM, SRCA, and LAGG) have stage-only records (fig. 4, table 2).

Most streamflow records within the SRPW start in water-year 1998 or later (table 2). Many of the records are short, with an average record length of only 2 to 5 water years (table 3). Only four of the stations within the SRPW (COLL, LAGC, LAGS, and SRCW) have relatively longer continuous streamflow records of 11 to 12 years (fig. 6, table 3). For this study, all streamflow records through water year 2010 (ending September 30, 2010) were analyzed. At the time of this study, nine stream gages within the SRPW were still active (table 2).

Correlation Between Drainage Basin Area and Average Discharge

Drainage basin area was compared to average streamflow for each gage with sufficient record in the SRPW and three gages in the adjacent Sonoma Creek and Napa River drainages (fig. 7). The comparison, using log-log (power function) regression, indicated a strongly positive correlation between drainage basin area and average streamflow for the 12 gages within the SRPW (Pearson correlation coefficient, r, of 0.962 in log-log space) and the 3 gages neighboring the SRPW (r of 0.998 in log-log space; fig. 8). The similarity of the regressions indicated that the factors controlling the rainfall-runoff response are similar for the drainages within and neighboring the SRPW.

Monthly and Seasonal Streamflow

A comparison of monthly mean discharge for water years 1999 to 2010 for the four gages in the SRPW with relatively longer continuous streamflow records of 11 to 12 years (COLL, LAGS, MWCM, and SRCW) showed characteristic seasonal variability with high winter (December–February) and low summer (June–August) flows (fig. 9). For all gages, streamflow varied by at least two orders of magnitude between the high winter and the low summer flows. The highest monthly discharge was measured at the MWCM gage, with monthly mean flows of more than 1,200 ft³/s for the months of January 2006, March 2006, and February 2008. Winter flows for three of the gages (MWCM, SRCW, and LAGS) exceeded 100 ft³/s for at least 1 month for all water years. For all water years, summer flows for all gages were less than 10 ft³/s. The gage with the smallest drainage area (COLL) had zero measured discharge for 1 to 4 month periods during the summer for all water years. Gage LAGS, on the central part of the Laguna de Santa Rosa channel, had zero-flow conditions during 1-month or longer periods in the summer for 2 water years.

Average monthly streamflow records from seven gages in the SRPW with continuous records (COLL, LAGC, and LAGS in the Laguna de Santa Rosa subbasin; SRCU, SRCS, and SRCW in the Santa Rosa Creek subbasin; and MWCM in the Mark West Creek subbasin) indicated a consistent seasonal distribution of streamflow in the SRPW (fig. 10). In general, December, January, and February were the months with the highest streamflow, followed by March, April, and then November. Each of the surface-water drainage basins has a gage (LAGS, SRCW, and MWCM) that had average monthly discharge of at least 120 ft³/s or higher for December, January, February, and March. August and September were the driest months at all gages, with average monthly discharge less than 5 ft³/s at all gages. Gage MWCM in the Mark West Creek drainage basin, with a drainage area that includes almost the entire area of the SRPW, had a maximum average monthly discharge of 900 ft³/s for January and a minimum average monthly discharge of 2 ft³/s for September (fig. 10C). Gage COLL in the Laguna de Santa Rosa drainage basin has one of the smallest drainage areas within the SRPW and had a maximum average monthly discharge of only 30 ft³/s for February and a minimum average monthly discharge was less than 0.01 ft³/s for September (fig. 10A).

The distribution of average monthly discharge for the five seasonal gages was generally consistent with the results obtained for the seven gages having continuous records within the SRPW. In general, the months of December, January, and February had the highest mean discharge at the five gages, with the exception of the COPE gage (fig. 11). The SRCS gage had the highest average monthly discharge of 256 ft³/s in February. October had the lowest mean discharge (about 4 ft³/s or less) for all gages, ranging from a minimum of about 1.2 ft³/s for the BRSH gage to a maximum of about 4 ft³/s for the SRCS gage (fig. 11).

Daily Mean Streamflow Records

Six gages within the SRPW (MWCM, SRCW, SRCU, LAGS, COLL, and LAGC) had continuous or nearly continuous records of daily mean discharge (table 2). Continuous records that span multiple water years are the most valuable for analyzing the water budget, for developing statistical characterizations of streamflow, and for calibrating hydrologic models. Five of the six gages with continuous records (MWCW, SRCW, LAGC, COLL, and LAGS) were still active at the time of this study (September 30, 2010), whereas gage SRCU has been discontinued.

Of the 15 gages within the SRPW, 5 had seasonal-only daily mean discharge records: MWCW, COPE, COLU, BRSH, and MATC (fig. 4, table 2). The seasonal records (October 1 through April 30) provide data for local flood control. A sixth gage, SRCS, had a record that was partly continuous, but mostly seasonal only. Although the seasonal data generally are not as useful as the continuous records for analyzing water budgets and developing statistical characterizations of streamflow, the data are important for characterizing winter-flow

Table 2.   Description of gaging stations and streamflow records within and adjacent to the Santa Rosa Plain watershed, Sonoma County, California.

[Abbreviations: Blvd, boulevard; Dr., drive; ID, identifier; mi², square mile; mm/dd/yy, month/day/year; NAD, North American datum; NGVD, National Geodetic Vertical Datum; Rd., road; SRPW, Santa Rosa Plain watershed; St., street; USGS, U.S. Geological Survey; —, no data; °, degree; ', minute; ", second]

SRPW gage ID	SRPW gage code	USGS gage ID	USGS gage name	North latitude NAD 1927	West longitude NAD 1927	Gage datum NGVD 1929 (feet)	Drainage area (mi²)	Subbasin or drainage basin name	Period of record for this study (mm/dd/yy) Begin	End	Gage status at end of study	Measured parameter	Record type	Record affected by backwater?
								Laguna de Santa Rosa stream gages						
2	COPE	11465660	Copeland Creek at Rohnert Park	38°20'36"	122°42'03"	100.0	5.5	Laguna de Santa Rosa	10/01/06	04/30/10	Active	Discharge	Seasonal	No.
3	LAGC	11465680	Laguna de Santa Rosa near Cotati	38°21'08"	122°44'35"	—	40.8	Laguna de Santa Rosa	11/06/98	09/30/10	Active	Discharge	Continuous	No.
4	COLU	11465690	Colgan Creek near Santa Rosa	38°24'08"	122°43'55"	—	3.4	Laguna de Santa Rosa	10/01/06	04/30/10	Active	Discharge	Seasonal	No.
5	COLL	11465700	Colgan Creek near Sebastopol	38°22'25"	122°46'02"	—	6.8	Laguna de Santa Rosa	11/07/98	09/30/10	Active	Discharge	Continuous	Yes.
6	LAGS	11465750	Laguna de Santa Rosa near Sebastopol	38°25'32"	122°49'41"	—	79.6	Laguna de Santa Rosa	11/18/98	09/30/10	Active	Discharge	Continuous	Yes.
15	LAGG	11466500	Laguna de Santa Rosa near Graton	38°27'10"	122°50'03"	—	166.0	Mark West Creek	11/09/64	09/30/10	Inactive	Stage	Periodic	Yes.
								Santa Rosa Creek stream gages						
7	SRCU	11465800	Santa Rosa Creek near Santa Rosa	38°27'25"	122°37'50"	—	12.5	Santa Rosa Creek	08/01/59	10/13/70	Inactive	Discharge	Continuous	No.
14	SRCM	11466050	Santa Rosa Creek at Mission Blvd	38°27'28"	122°40'16"	220.0	—	Santa Rosa Creek	11/01/97	04/30/10	Active	Stage	Seasonal	No.
8	BRSH	11466065	Brush Creek at Santa Rosa	38°27'18"	122°40'45"	—	10.1	Santa Rosa Creek	10/01/05	04/30/10	Inactive	Discharge	Seasonal	No.
13	SRCA	11466080	Santa Rosa Creek at Alderbrook Dr.	38°26'58"	122°41'50"	170.0	—	Santa Rosa Creek	10/31/97	04/30/10	Inactive	Stage	Seasonal	No.
9	MATC	11466170	Matanzas Creek at Santa Rosa	38°26'20"	122°42'05"	—	21.0	Santa Rosa Creek	10/01/04	04/30/10	Active	Discharge	Seasonal	No.
10	SRCS	11466200	Santa Rosa Creek at Santa Rosa	38°26'12"	122°43'25"	100.0	57.0	Santa Rosa Creek	10/01/39	04/30/10	Active	Discharge	Mixed	No.
11	SRCW	11466320	Santa Rosa Creek at Willowside Rd.	38°26'43"	122°48'22"	—	77.6	Santa Rosa Creek	12/09/98	09/30/10	Active	Discharge	Continuous	Yes.
								Mark West Creek stream gages						
1	MWCW	11465500	Mark West Creek near Windsor	38°30'34"	122°46'07"	140.0	43.0	Mark West Creek	10/01/06	04/30/08	Inactive	Discharge	Seasonal	No.
12	MWCM	11466800	Mark West Creek near Mirabel Heights	38°29'39"	122°51'08"	—	251.0	Mark West Creek	10/01/05	10/05/10	Inactive	Discharge	Continuous	Yes.
								Stream gages adjacent to Santa Rosa Plain watershed						
16	NAPH	11456000	Napa River near St. Helena	38°30'41"	122°27'17"	193.2	78.8	Napa River	10/01/29	09/30/10	Active	Discharge	Continuous	No.
17	NAPN	11458000	Napa River near Napa	38°22'06"	122°18'08"	24.7	218	Napa River	10/01/29	09/30/10	Active	Discharge	Continuous	No.
18	SCAC	11458500	Sonoma Creek at Agua Caliente	38°19'24"	122°29'36"	94.3	58.4	Sonoma Creek	02/01/55	09/30/10	Active	Discharge	Continuous	No.

Table 3. Summary statistics for streamflow records within and neighboring Santa Rosa Plain watershed (SRPW), Sonoma County, California.

[**Abbreviations**: Avg., average; cfs, cubic feet per second; ID, identifier; mm/dd/yy, month/day/year; No., number; USGS, U.S. Geological Survey; —, no data]

SRPW gage ID	SRPW gage code	USGS gage ID	USGS gage name	No. of days with data	No. of water years in record	Peak discharge (cfs)	Date of peak discharge (mm/dd/yy)	Max daily mean discharge (cfs)	Date of max daily mean discharge (mm/dd/yy)	Min daily mean discharge (cfs)	Date of min daily mean discharge (mm/dd/yy)
1	MWCW	11465500	Mark West Creek near Windsor	425	2	3,970	01/04/08	1,750	01/04/08	0.86	10/01/07
2	COPE	11465660	Copeland Creek at Rohnert Park	849	3	512	12/26/06	255	01/20/10	0	Multiple
3	LAGC	11465680	Laguna de Santa Rosa near Cotati	4,347	11	3,980	12/31/05	2,450	12/31/05	0	Multiple
4	COLU	11465690	Colgan Creek near Santa Rosa	846	3	426	01/25/08	135	01/04/08	0	Multiple
5	COLL	11465700	Colgan Creek near Sebastopol	4,346	11	934	12/31/05	647	01/02/02	0	Multiple
6	LAGS	11465750	Laguna de Santa Rosa near Sebastopol	4,333	11	9,690	12/31/05	5,930	01/26/08	−63*	02/22/09
7	SRCU	11465800	Santa Rosa Creek near Santa Rosa	4,092	11	3,200	02/08/60	1,450	02/08/60	0	Multiple
8	BRSH	11466065	Brush Creek at Santa Rosa	1,092	4	2,390	12/31/05	713	12/31/05	0	10/01/08
9	MATC	11466170	Matanzas Creek at Santa Rosa	1,335	5	3,700	12/31/05	2,040	12/31/05	0.01	10/01/09
10	SRCS	11466200	Santa Rosa Creek at Santa Rosa	2,609	8	9,080	02/27/40	4,830	02/27/40	0	Multiple
11	SRCW	11466320	Santa Rosa Creek at Willowside Rd.	4,314	11	6,410	12/31/05	5,200	12/31/05	0.58	10/02/09
12	MWCM	11466800	Mark West Creek near Mirabel Heights	1,708	5	11,300	12/31/05	7,180	01/03/06	0	Multiple
16	NAPH	11456000	Napa River near St. Helena	25,232	68	18,300	12/31/05	13,700	02/17/86	0	Multiple
17	NAPN	11458000	Napa River near Napa	19,724	51	37,100	02/18/86	26,200	02/17/86	0	Multiple
18	SCAC	11458500	Sonoma Creek at Agua Caliente	13,018	35	20,300	12/31/05	8,180	12/31/05	0	Multiple

Table 3. Summary statistics for streamflow records within and neighboring Santa Rosa Plain watershed (SRPW), Sonoma County, California.—Continued

[**Abbreviations**: Avg., average; cfs, cubic feet per second; ID, identifier; mm/dd/yy, month/day/year; No., number; USGS, U.S. Geological Survey; –, no data]

SRPW gage ID	SRPW gage code	USGS gage ID	USGS gage name	No. of days with zero flow	No. of days with reverse flow	Avg. discharge (cfs)	Avg. fall (Oct–Dec) discharge (cfs)	Avg. winter (Jan–Mar) discharge (cfs)	Avg. spring (Apr–Jun) discharge (cfs)	Avg. summer (Jul–Sep) discharge (cfs)
1	MWCW	11465500	Mark West Creek near Windsor	0	0	46.8	13.2	95.1	7.6	—
2	COPE	11465660	Copeland Creek at Rohnert Park	113	0	7.6	2.2	14.3	4.6	—
3	LAGC	11465680	Laguna de Santa Rosa near Cotati	362	0	32.7	36.3	80.8	13.4	0.48
4	COLU	11465690	Colgan Creek near Santa Rosa	206	0	4.7	2.7	7.6	2.1	—
5	COLL	11465700	Colgan Creek near Sebastopol	1,225	0	8.5	10.3	20.7	3.1	0.04
6	LAGS	11465750	Laguna de Santa Rosa near Sebastopol	357	2	80.3	73.7	212.8	32.1	0.61
7	SRCU	11465800	Santa Rosa Creek near Santa Rosa	204	0	18.8	15.0	52.2	9.5	0.35
8	BRSH	11466065	Brush Creek at Santa Rosa	21	0	20.6	14.3	32.5	11.4	—
9	MATC	11466170	Matanzas Creek at Santa Rosa	0	0	31.8	21.8	48.3	20.3	—
10	SRCS	11466200	Santa Rosa Creek at Santa Rosa	68	0	104.5	96.0	195.9	44.0	1.20
11	SRCW	11466320	Santa Rosa Creek at Willowside Rd.	0	0	93.2	102.6	219.9	43.0	4.23
12	MWCM	11466800	Mark West Creek near Mirabel Heights	15	1	288.0	238.3	739.8	175.5	3.60
16	NAPH	11456000	Napa River near St. Helena	838	0	94.9	74.7	262.2	41.5	1.49
17	NAPN	11458000	Napa River near Napa	1,671	0	202.2	130.3	618.6	91.2	3.30
18	SCAC	11458500	Sonoma Creek at Agua Caliente	525	0	71.3	58.2	193.9	30.4	1.15

*Backwater condition.

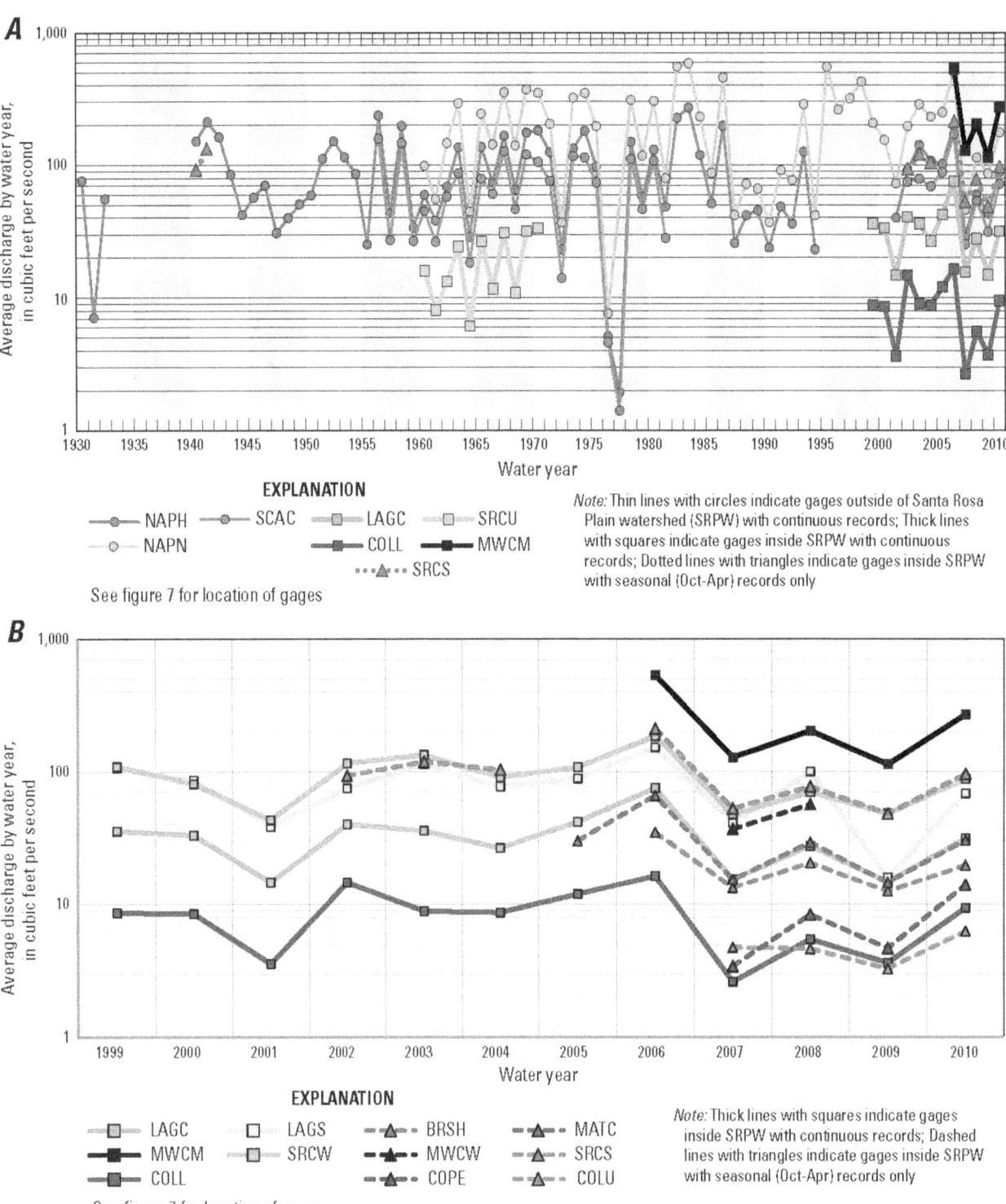

Figure 6. Average water-year discharge for gages within and neighboring the Santa Rosa Plain watershed, Sonoma County, California: *A*, Long-term (water-year 1930–2010) records; *B*, Recent (water-year 2000–10) records. (COPE, Copeland Creek at Rohnert Park; LAGC, Laguna de Santa Rosa near Cotati; COLU, Colgan Creek near Santa Rosa; COLL, Colgan Creek near Sebastopol; LAGS, Laguna de Santa Rosa near Sebastopol; SRCU, Santa Rosa Creek near Santa Rosa; BRSH, Brush Creek at Santa Rosa; MATC, Matanzas Creek at Santa Rosa; SRCS, Santa Rosa Creek at Santa Rosa; SRCW, Santa Rosa Creek at Willowside Road; MWCW, Mark West Creek near Windsor; MWCM, Mark West Creek near Mirabel Heights; NAPH, Napa River near St. Helena; Napa River near Napa; SCAC, Sonoma Creek at Agua Caliente)

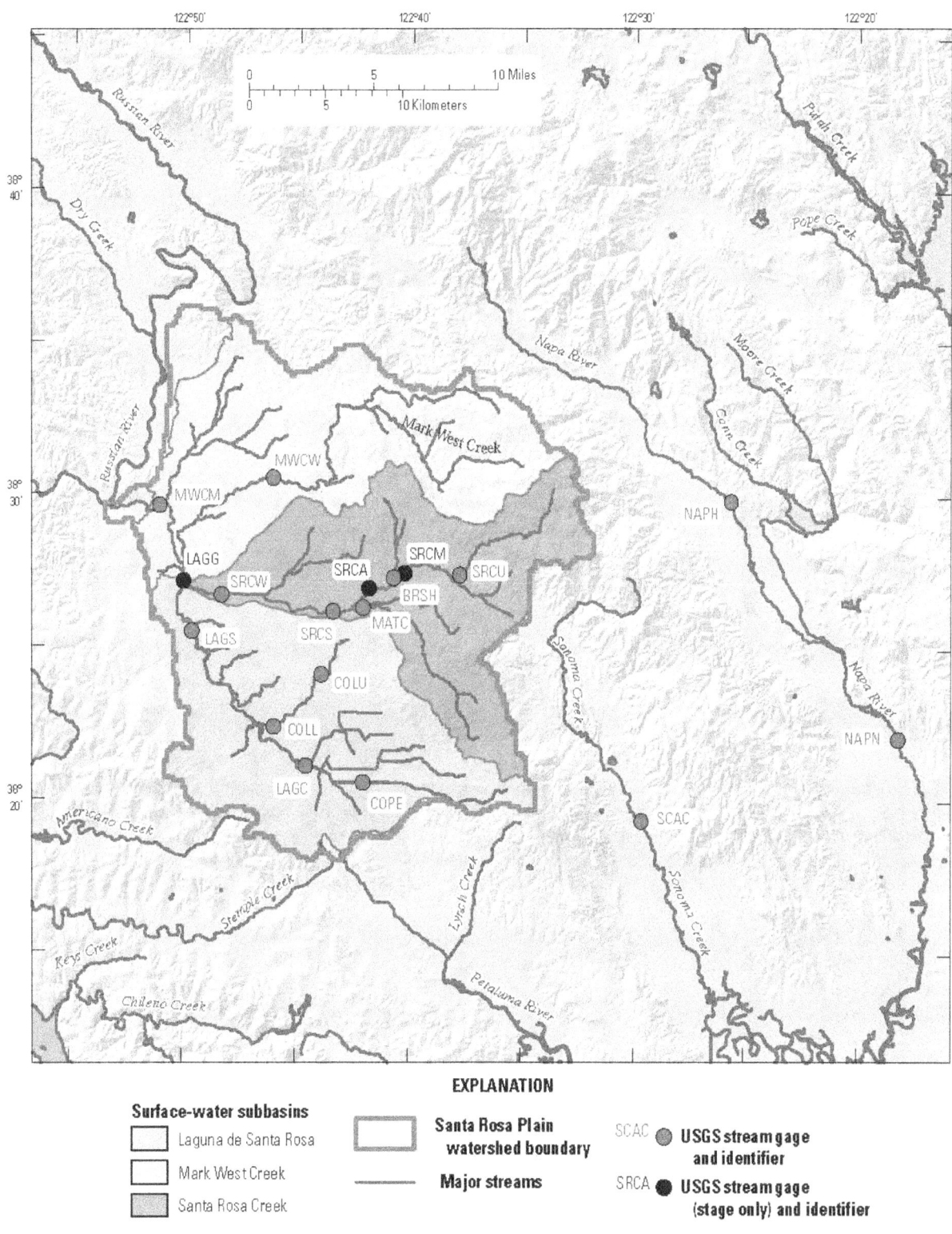

Figure 7.   U.S. Geological Survey stream gages within and neighboring the Santa Rosa Plain watershed, Sonoma County, California. (COPE, Copeland Creek at Rohnert Park; LAGC, Laguna de Santa Rosa near Cotati; COLU, Colgan Creek near Santa Rosa; COLL, Colgan Creek near Sebastopol; LAGS, Laguna de Santa Rosa near Sebastopol; LAGG, Laguna de Santa Rosa near Graton; SRCU, Santa Rosa Creek near Santa Rosa; SRCM, Santa Rosa Creek at Mission Blvd; BRSH, Brush Creek at Santa Rosa; SRCA, Santa Rosa Creek at Alderbrook Drive; MATC, Matanzas Creek at Santa Rosa; SRCS, Santa Rosa Creek at Santa Rosa; SRCW, Santa Rosa Creek at Willowside Road; MWCW, Mark West Creek near Windsor; MWCM, Mark West Creek near Mirabel Heights; NAPH, Napa River near St. Helena; Napa River near Napa; SCAC, Sonoma Creek at Agua Caliente)

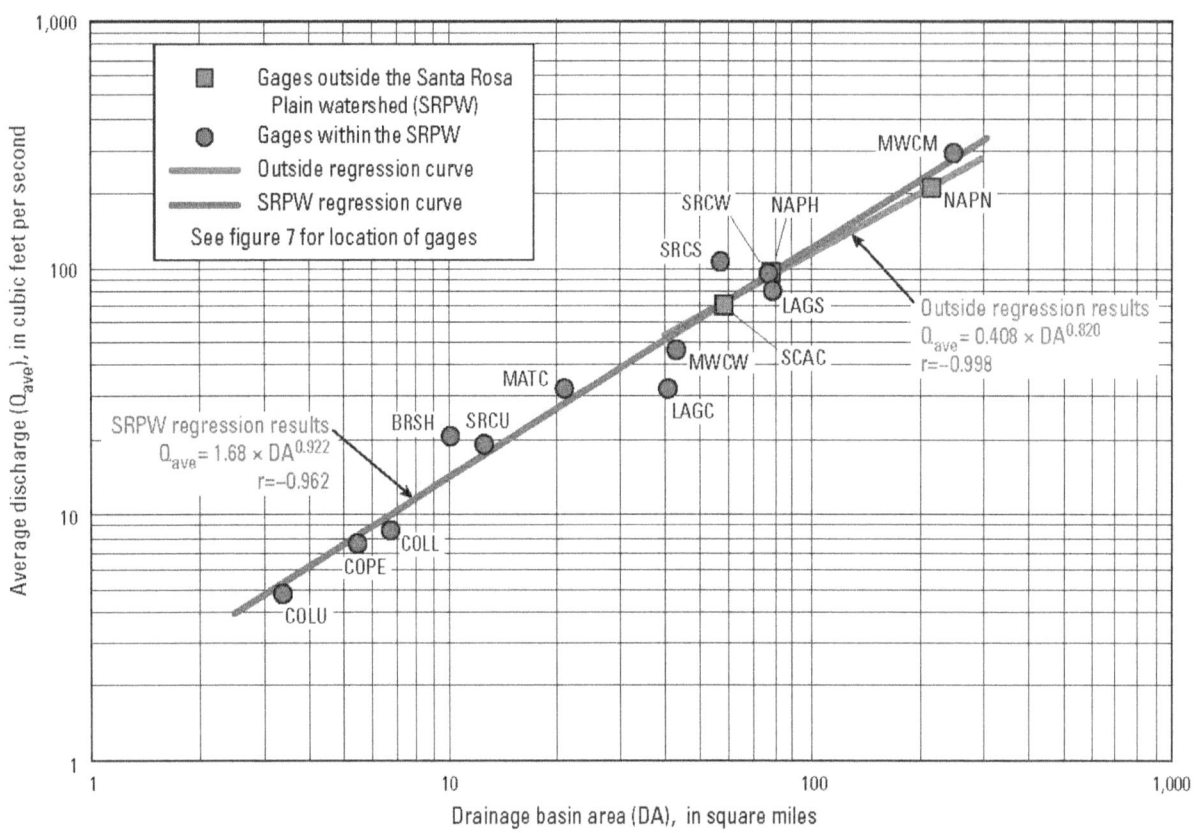

Figure 8.   Correlation between drainage basin area and average discharge for streamflow records within and around the Santa Rosa Plain watershed, Sonoma County, California. (COPE, Copeland Creek at Rohnert Park; LAGC, Laguna de Santa Rosa near Cotati; COLU, Colgan Creek near Santa Rosa; COLL, Colgan Creek near Sebastopol; LAGS, Laguna de Santa Rosa near Sebastopol; SRCU, Santa Rosa Creek near Santa Rosa; BRSH, Brush Creek at Santa Rosa; MATC, Matanzas Creek at Santa Rosa; SRCS, Santa Rosa Creek at Santa Rosa; SRCW, Santa Rosa Creek at Willowside Road; MWCW, Mark West Creek near Windsor; MWCM, Mark West Creek near Mirabel Heights; NAPH, Napa River near St. Helena; Napa River near Napa; SCAC, Sonoma Creek at Agua Caliente)

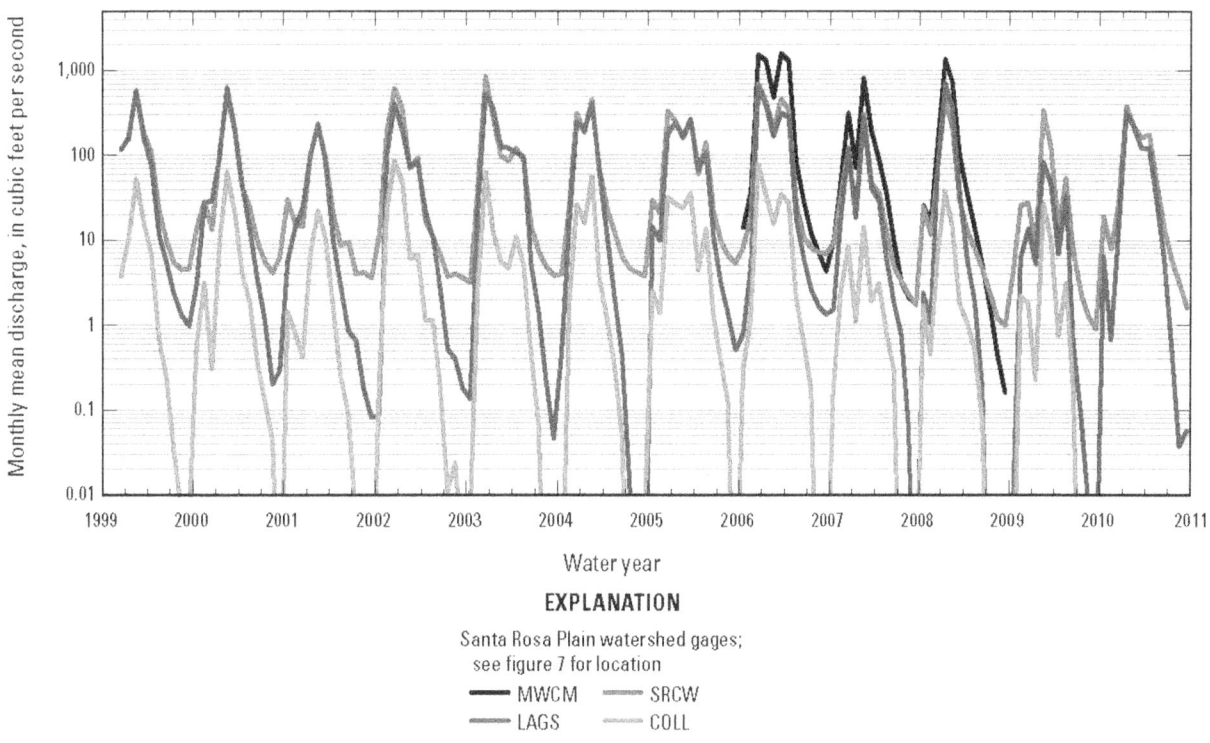

Figure 9.   Monthly mean discharge for four selected stream gages in the Santa Rosa Plain watershed, Sonoma County, California, 1999–2010. Discharge rates of less than 0.01 cubic feet per second represent zero-flow conditions. (COLL, Colgan Creek near Sebastopol; LAGS, Laguna de Santa Rosa near Sebastopol; SRCW, Santa Rosa Creek at Willowside Road; MWCM, Mark West Creek near Mirabel Heights)

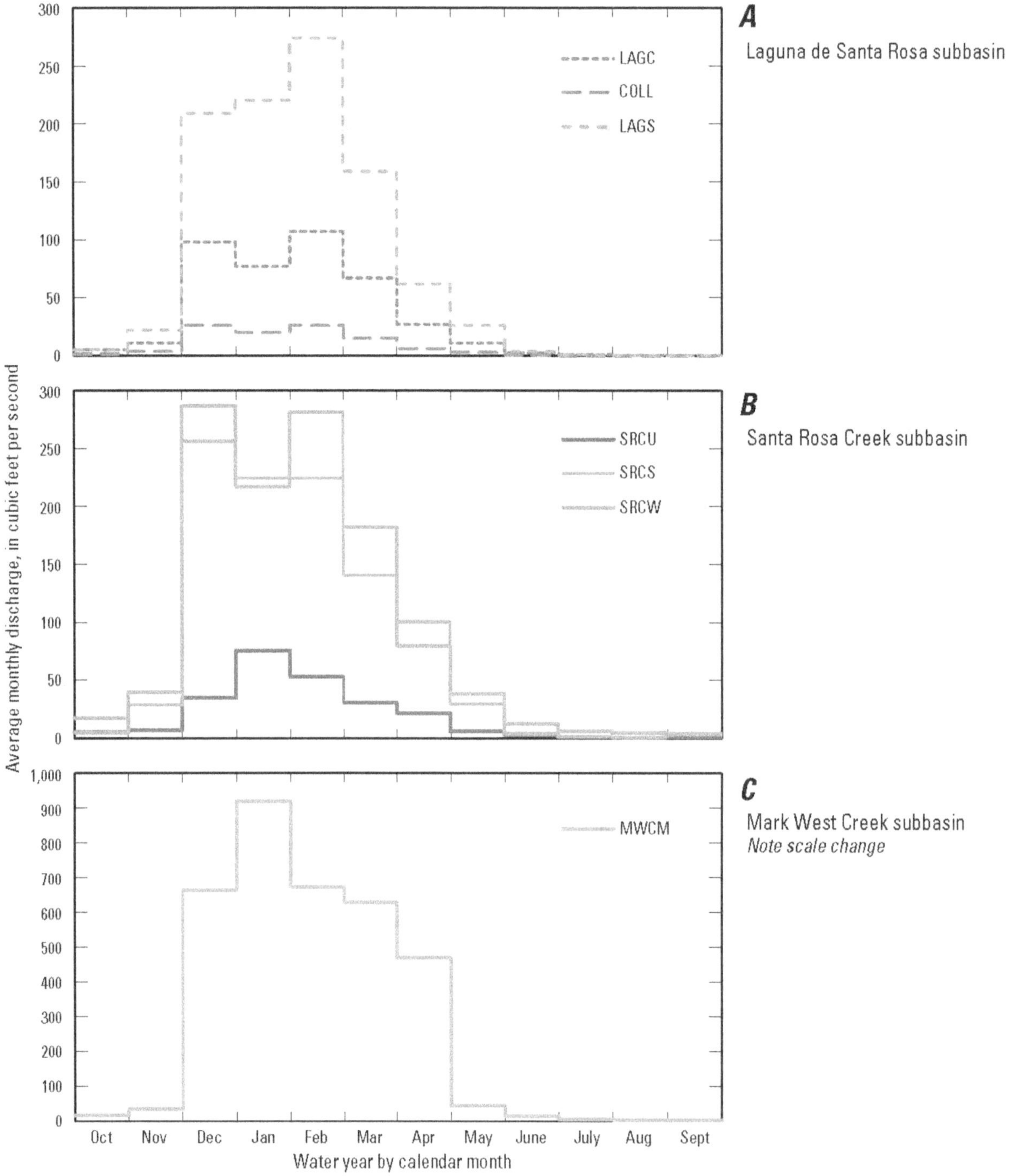

Figure 10. Average monthly discharge for gages having continuous stream discharge records in the Santa Rosa Plain watershed, Sonoma County, California: *A*, Laguna de Santa Rosa subbasin; *B*, Santa Rosa Creek subbasin; and *C*, Mark West Creek subbasin. (LAGC, Laguna de Santa Rosa near Cotati; COLL, Colgan Creek near Sebastopol; LAGS, Laguna de Santa Rosa near Sebastopol; SRCU, Santa Rosa Creek near Santa Rosa; SRCS, Santa Rosa Creek at Santa Rosa; SRCW, Santa Rosa Creek at Willowside Road; MWCM, Mark West Creek near Mirabel Heights)

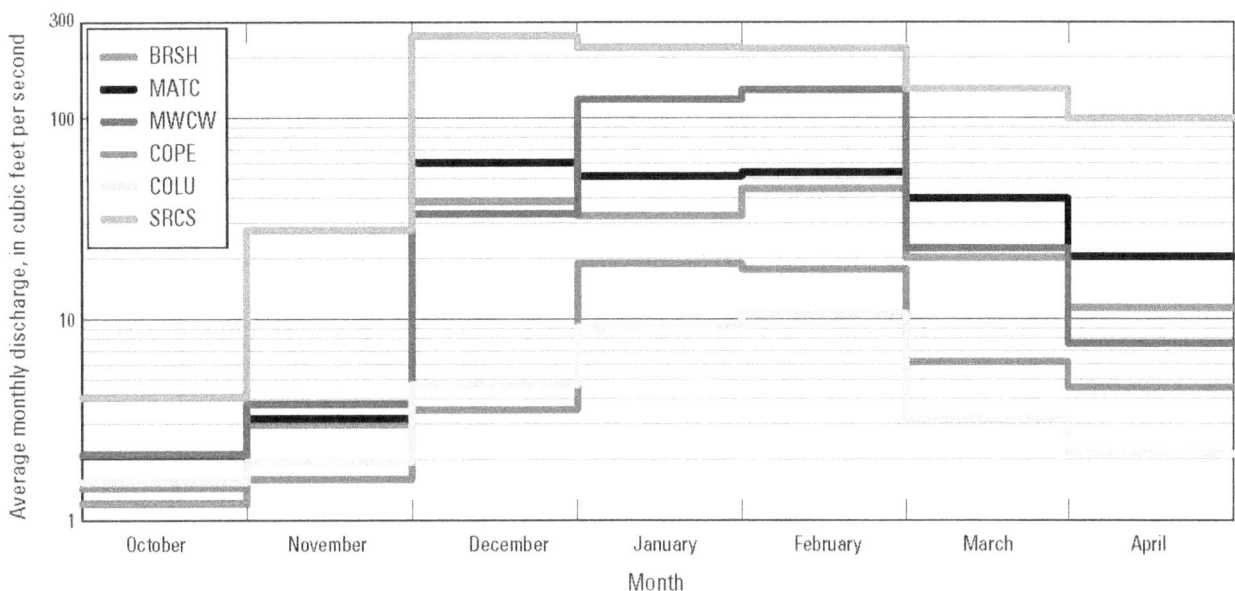

Figure 11. Average monthly discharge for gages having seasonal-only (October–April) records in the Santa Rosa Plain watershed, Sonoma County, California. (COPE, Copeland Creek at Rohnert Park; COLU, Colgan Creek near Santa Rosa; BRSH, Brush Creek at Santa Rosa; MATC, Matanzas Creek at Santa Rosa; SRCS, Santa Rosa Creek at Santa Rosa; MWCW, Mark West Creek near Windsor)

conditions, calibrating hydrologic models to the winter storm-flows, and evaluating the hydrologic response of drainage basins having different basin characteristics within the SRPW.

Laguna de Santa Rosa Drainage Basin

COPE Gage

COPE is a seasonal gage in the urban area of Rohnert Park (fig. 4), but the 5.5 mi² drainage upstream of the gage consists mostly of undeveloped and agricultural land in the eastern headwaters of the Laguna de Santa Rosa drainage basin that drain the western slope of the Sonoma Mountains (fig. 4). According to the NHD, the upper section of Copeland Creek is designated as perennial, whereas the lower section is designated as intermittent (Simley and Carswell, 2009). The gaging station includes a water-stage recorder and a crest-stage gage. The period of record for this gage was October 1, 2006, to April 30, 2010 (fig. 6, table 2). The average October through April discharge measured at this gage was 7.6 ft³/s (table 3). The estimated maximum daily mean discharge of 255 ft³/s was measured on January 20, 2010, and the maximum peak discharge of 512 ft³/s was measured on December 26, 2006. The stream was often dry during October at the location of the gage.

LAGC Gage

LAGC is a continuous gage on the main channel of the Laguna de Santa Rosa upstream of the Colgan Creek tributary in the Laguna de Santa Rosa drainage basin (fig. 4). The gage was active at the time of this study, with a continuous discharge record from November 6, 1998, to present (fig. 6,

table 2). The gage is a water-stage recorder, and the upstream drainage area of 40.8 mi² includes a combination of urban, agricultural, and undeveloped land uses. There is no regulation or diversion upstream of the gage, and discharge is not effected by backwater conditions in the Laguna de Santa Rosa. The mean discharge measured at this gage was 33 ft³/s, with a peak discharge of 3,980 ft³/s on December 31, 2005 (fig. 12, table 3). The maximum daily mean discharge of 2,450 ft³/s was also measured on December 31, 2005. The mean summer discharge was 0.48 ft³/s; however, a minimum daily mean discharge of zero was often measured during the summer months (fig. 12; table 3).

COLU Gage

COLU is a seasonal gage in the upper section of Colgan Creek in the north central part of the Laguna de Santa Rosa drainage basin (fig. 4) and, compared to the other gages, has the smallest drainage area with only 3.4 mi². The gaging station includes a water-stage recorder and a crest-stage gage. Measured streamflow for this small drainage is important for evaluating the hydrologic response of smaller catchments within the developed and urbanized areas of the SRPW. The period of record for the site was from October 2006 through September 30, 2010 (fig. 6), and the site was still active at the time of this study. The mean October through April discharge was 4.7 ft³/s (table 3). The maximum daily mean discharge of 135 ft³/s was measured on January 4, 2008, while the maximum peak flow of 564 ft³/s was measured on January 20, 2010. The mean discharge during October was 1.6 ft³/s, and the record includes many days with zero discharge.

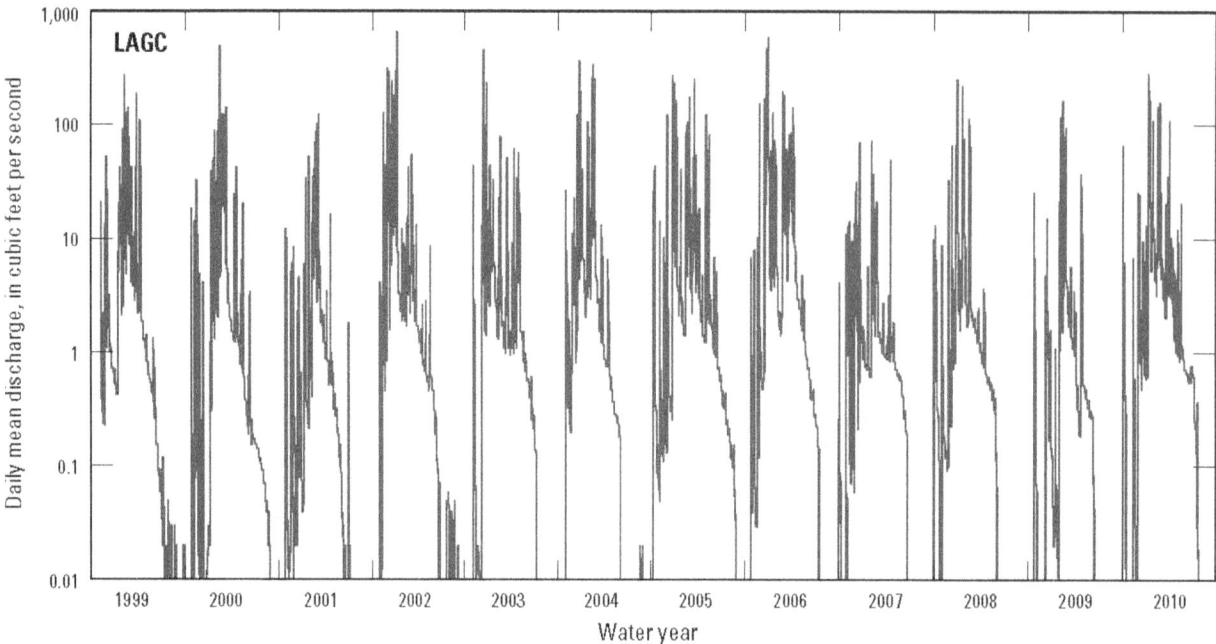

Figure 12.   Daily mean discharge measured at gage 11465680, Laguna de Santa Rosa near Cotati (LAGC), Sonoma County, California, 1999–2010. Discharge rates of less than 0.01 cubic feet per second represent zero-flow conditions.

COLL Gage

COLL is a continuous gage close to the mouth of Colgan Creek in the Laguna de Santa Rosa drainage basin and has a drainage area of 6.8 mi^2 (fig. 4). Colgan Creek is a small tributary on the eastern side of the Laguna de Santa Rosa with a drainage consisting of developed agricultural and urbanized land in the south-central part of the Laguna de Santa Rosa drainage basin (fig. 4). The gaging station uses a water-stage recorder and a Doppler-velocity system. There is no regulation or diversion upstream of the gage; however, high flows are effected by backwater conditions in the Laguna de Santa Rosa. The period of record for the gage was November 1998 to September 30, 2010 (fig. 6, table 2). The mean discharge measured at this gage was 8.5 ft^3/s, with a peak discharge of 934 ft^3/s on December 31, 2005. The maximum daily mean discharge of 647 ft^3/s was measured on January 2, 2002 (fig. 13, table 3). The mean summer discharge was 0.04 ft^3/s; however, a minimum daily mean discharge of zero was often measured during the summer months (fig. 13, table 3).

LAGS Gage

LAGS is a continuous gage in the lower end of the Laguna de Santa Rosa drainage basin, approximately 1.5 mi upstream of the confluence of Santa Rosa Creek and the Laguna de Santa Rosa (fig. 4). With an upstream catchment area of 79.6 mi^2, the streamflow measured at this gage is representative of the surface-water outflow from the Laguna de Santa Rosa drainage basin. The LAGS gage includes a water-stage recorder and a Doppler-velocity system that measures flow direction. There is no diversion or flow regulation upstream of the gage; however, high flows can be strongly

affected by backwater conditions. The mean discharge measured at this gage was 80 ft^3/s, with a peak discharge of 9,690 ft^3/s on December 31, 2005 (fig. 14, table 3). A maximum daily mean discharge of 5,930 ft^3/s was recorded on January 26, 2008. The mean summer discharge at this gage was 0.61 ft^3/s; however, a minimum daily mean discharge of zero was often measured during the summer months (fig. 14, table 3). A minimum daily mean discharge of –63 ft^3/s was measured on February 22, 2009, during a period of significant flow reversal caused by backwater conditions in the Laguna de Santa Rosa.

Santa Rosa Creek Drainage Basin

SRCU Gage

SRCU is a discontinued continuous stream gage in the upper reaches of Santa Rosa Creek, upstream of the city of Santa Rosa, with a drainage area of 12.5 mi^2 (fig. 4). The period of record for the gage is August 1, 1959, through October 13, 1970. The daily mean discharge record consists of 11 complete water years and is representative of the rainfall-runoff characteristics of the more rugged, undeveloped upper drainages in the Mayacmas Mountains along the eastern part of the Santa Rosa Creek drainage basin. The mean discharge measured at this gage was 19 ft^3/s, with a peak discharge of 3,200 ft^3/s measured on February 8, 1960. The maximum daily mean discharge of 1,450 ft^3/s was measured on February 8, 1960 (fig. 15, table 3). The mean summer discharge at this gage was 0.35 ft^3/s; however, a minimum daily mean discharge of zero was often measured during the summer months (fig. 15, table 3).

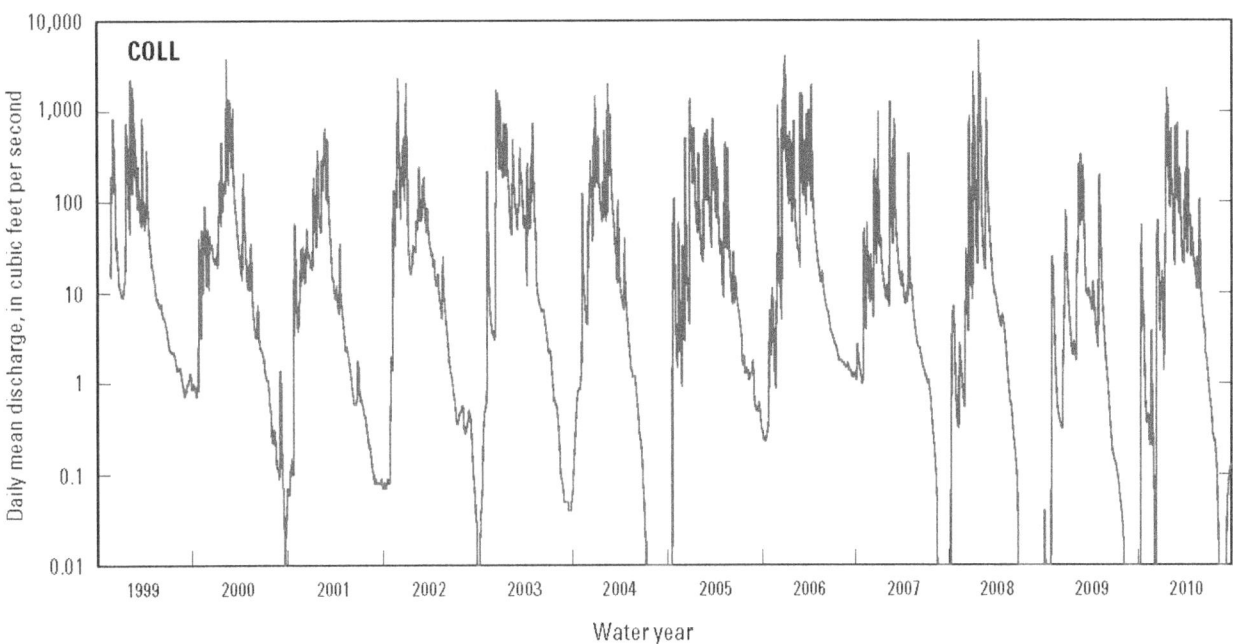

Figure 13. Daily mean discharge measured at gage 11465700, Colgan Creek near Sebastopol (COLL), Sonoma County, California 1999–2010. Discharge rates of less than 0.01 cubic feet per second represent zero-flow conditions.

BRSH Gage

BRSH is a seasonal gage close to the mouth of Brush Creek in the central part of the Santa Rosa Creek drainage basin (fig. 4) that has a drainage area of 10.1 mi^2. The gage records include a combination of stage-only data from November 2002 to May 2005 and both discharge and stage data from October 2005 to April 2010 (fig. 6). Measured streamflow at this gage is useful for characterizing the semi-rural area of Rincon Valley and first-order drainages in the Mayacmas Mountains. The mean discharge for the gage was 21 ft^3/s (table 3). A maximum daily mean discharge of 713 ft^3/s and the maximum peak discharge of 2,390 ft^3/s were both measured on December 31, 2005 (table 3).

MATC Gage

MATC is a seasonal gage on Matanzas Creek, downstream of the Spring Creek tributary in the south central part of the Santa Rosa Creek drainage basin (fig. 4), that has a drainage area of 21.0 mi^2. The drainage area consists mostly of the less developed, more rugged and rural terrain of Bennett Valley and the surrounding Sonoma Mountains. Records for this gage include stage-only data from November 2002 to April 2004 and both stream discharge and stage data from October 2004 to September 30, 2010 (the gage was active at the time of this study; fig. 6B). Measured streamflow at MATC is useful for characterizing recharge and runoff in the upstream drainages and headwater areas of the Sonoma Mountains. The mean discharge for the gage was 32 ft^3/s (table 3). The maximum daily mean discharge of 2,040 ft^3/s and the maximum

peak flow of 3,700 ft^3/s were both recorded on December 31, 2005. Minimum flows of 0.01 ft^3/s were measured during October 2009. No periods of zero discharge were recorded at this gage.

SRCS Gage

SRCS is a seasonal gage on the main channel of Santa Rosa Creek in the south central part of the Santa Rosa Creek drainage basin (fig. 4) that has an upstream drainage area of 57.0 mi^2. Beginning in December 1939, the site has the earliest streamflow record for the SRPW; however, the earlier record includes data only through September 1941 (fig. 6). The gage was inactive from October 1941 to September 2001. The gage was reactivated in October 2001 and operated as a seasonal gage, with some water years including records for September and May. Only stage data is available for water-year 2005. The gage was still active at the time of this study.

Discharge at gage SRCS can be affected by diversions for flood control (diversion of Santa Rosa Creek is diverted into Santa Rosa Creek reservoir) and irrigation (these diversions are not recorded). The maximum daily mean discharge was 4,830 ft^3/s (measured on February 27, 1940), and the minimum daily mean discharge was 0 ft^3/s (measured multiple times; table 3). A maximum peak instantaneous discharge of 9,080 ft^3/s was recorded on February 27, 1940, and a minimum discharge of zero was measured from October 1, 1939, to December 31, 1939 (table 3). The measured average daily mean discharge at this site for water-years 2002 through 2009 (October through April only) was 105 ft^3/s (table 3).

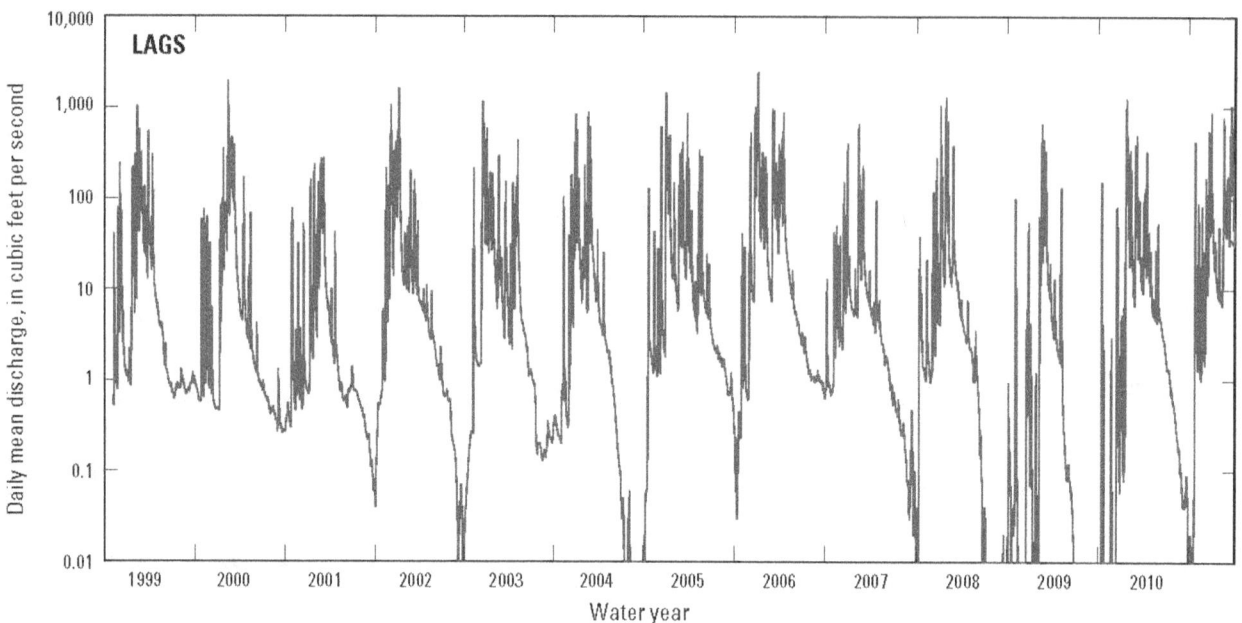

Figure 14. Daily mean discharge measured at gage 11466200, Laguna de Santa Rosa near Sebastopol (LAGS), Sonoma County, California, 1999–2011. Discharge rates of less than 0.01 cubic feet per second represent zero-flow conditions with the exception of February 22, 2009, where the measured discharge equaled –63 cubic feet per second.

SRCW Gage

SRCW is a continuous gage on the main channel of Santa Rosa Creek in the downstream end of the Santa Rosa Creek drainage basin, about 1.6 mi upstream of the confluence with the Laguna de Santa Rosa, that has an upstream drainage of 77.6 mi^2 (fig. 4). At the time of this study, the gage was still active, and approximately 12 years of nearly continuous stream-discharge data were available (from December 9, 1998, to September 30, 2010; fig. 6, table 2). The daily mean discharge records are generally good for this gage; however, some diversion of streamflow for irrigation is possible. Discharge rates of less than 0.01 ft^3/s represent zero-flow conditions. During high flows, backwater effects from flooding in the Laguna de Santa Rosa can affect discharge at the SRCW gage, and there can be diversions for flood control in the upper parts of the drainage during the largest storms. Subjective analysis of the hydrograph indicated that these diversions were insignificant when compared to the natural flows, as indicated by the general shape of the hydrograph (for example, the rising, peaks, and receding limbs).

The mean discharge measured at this gage was 93 ft^3/s, with a peak discharge of 6,410 ft^3/s measured on December 31, 2005. The maximum daily mean discharge of 5,200 ft^3/s was measured on December 31, 2005 (fig. 16). The mean summer (July through September) discharge at this gage was 4.2 ft^3/s; in contrast, the mean winter (January through March) discharge was 220 ft^3/s (table 3).

Mark West Creek Drainage Basin

MWCW Gage

MWCW is a seasonal gage on the main branch of Mark West Creek in south central part of the Mark West Creek drainage basin (fig. 4), approximately 4 mi upstream of gage MWCM, that has a drainage area of 43 mi^2. The MWCW gage has a relatively short record limited to water-years 2007 and 2008 (from October 1, 2006, to January 7, 2009). The drainage basin upstream of the MWCW gage includes 49.9 percent of the Mark West Creek basin (as defined by CIWM) and represents mostly natural, undeveloped conditions. The channel is mostly perennial and includes the largest number of springs of all drainage basins within the SRPW (Simley and Carswell, 2009).

Although the record for this gage is sparse (seasonal data spanning only 2 water years), the measured streamflow at this site is useful for evaluating the rainfall-runoff response and the potential recharge magnitude for the hilly, rugged terrain of the Mayacmas Mountains in the northeastern part of the study area. The average October through April discharge for this site was 47 ft^3/s. A maximum peak discharge of 3,970 ft^3/s was recorded on January 4, 2008. Although summer flows were not measured at this gage, the streamflow hydrographs showed early October flows of 1 to 2 ft^3/s, indicating that flows during much of summer were probably less than 2 ft^3/s.

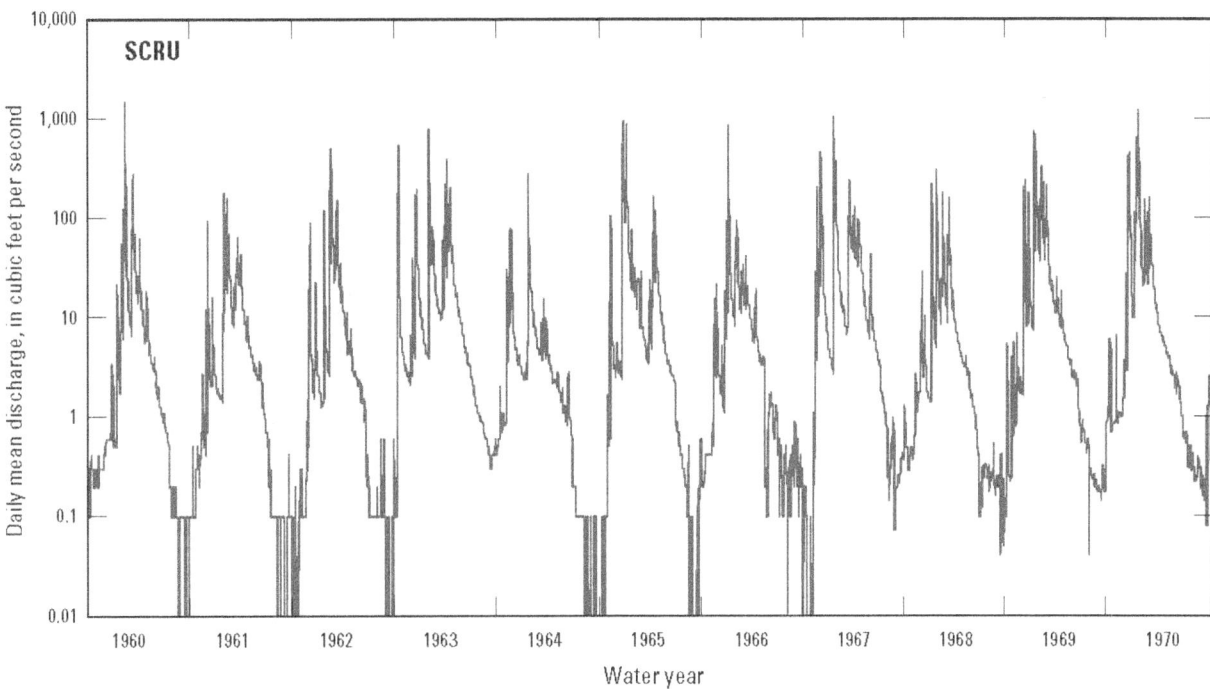

Figure 15.   Daily mean discharge measured at gage 11466200, Santa Rosa Creek near Santa Rosa (SRCU), Sonoma County, California, 1959–1970.

MWCM Gage

MWCM is a continuous gage on the lowermost section of the main branch of Mark West Creek, close to the confluence with the Russian River (fig. 4). Of all available streamflow records for the SRPW, the record at the MWCM gage is the most important in terms of developing the study-area water budget and for calibrating hydrologic models because the 251 mi^2 upstream drainage area includes 96 percent of the SRPW study area. Streamflow measured at the MWCM gage represents the integrated hydrologic response of the study area as a whole to climate, basin characteristics, and anthropogenic factors. The MWCM gage has a nearly continuous record for water-years 2006 through 2010 (fig. 17). With the exception of the flood-control facilities on Santa Rosa Creek, no flow regulation significantly affects streamflow at MWCM. Some flow diversions for irrigation of up to 6,000 acres are possible (these diversions are primarily on the Santa Rosa Creek tributary), but such diversions are not directly detectable in the streamflow records. During very high flow conditions, discharge records for MWCM can be poor as a result of backwater effects from the Russian River.

The mean discharge for the period of continuous data at the MWCM gage (water-years 2006 through 2008) was 288.3 ft³/s, with a peak daily mean discharge of 7,180 ft³/s on January 3, 2006, and a minimum discharge of zero for the period September 4, 2008, through September 18, 2008 (fig. 17). As mentioned previously, backwater effects from the Russian River can impede and occasionally reverse streamflow at the MWCM gage. For example, December 31, 2005, was the date of both the maximum recorded peak discharge

of 7,180 ft³/s and also the extreme minimum discharge of −830 ft³/s, indicating reverse (upstream) flow on Mark West Creek in response to Russian River overflow. A maximum gage height of 69.8 ft was recorded on January 1, 2006, corresponding to peak flood conditions on the Russian River. The maximum daily mean discharge of 7,180 ft³/s was measured for Mark West Creek when flooding in the Russian River subsided, 2 days after the maximum gage height was recorded. Therefore, the maximum daily mean discharge reflected a rapid draining of the large volume of water held in the Laguna de Santa Rosa floodplain, rather than discharge caused by the generation of runoff in response to precipitation within the Mark West Creek drainage basin.

Extension of the MWCM Streamflow Record

The streamflow records from the SRPW gages are too short to show longer-term streamflow characteristics and trends (fig. 6, table 2). A comparison of the time series for annual (water year) average discharge for the three gages neighboring the SRPW with long discharge records (NAPH, NAPN, and SCAC) to gages within the SRPW with the longest discharge records of 11 to 12 years (COLL, LAGC, SRCS, SRCU, and MWCM) indicated the limitations of the shorter-term records in adequately representing the longer-term streamflow characteristics of the SRPW (fig. 6A). For example, the extremely dry period during water-year 1977, and the wettest periods of 1983–1984 and 1995–1998, are missing from the streamflow records within the SRPW.

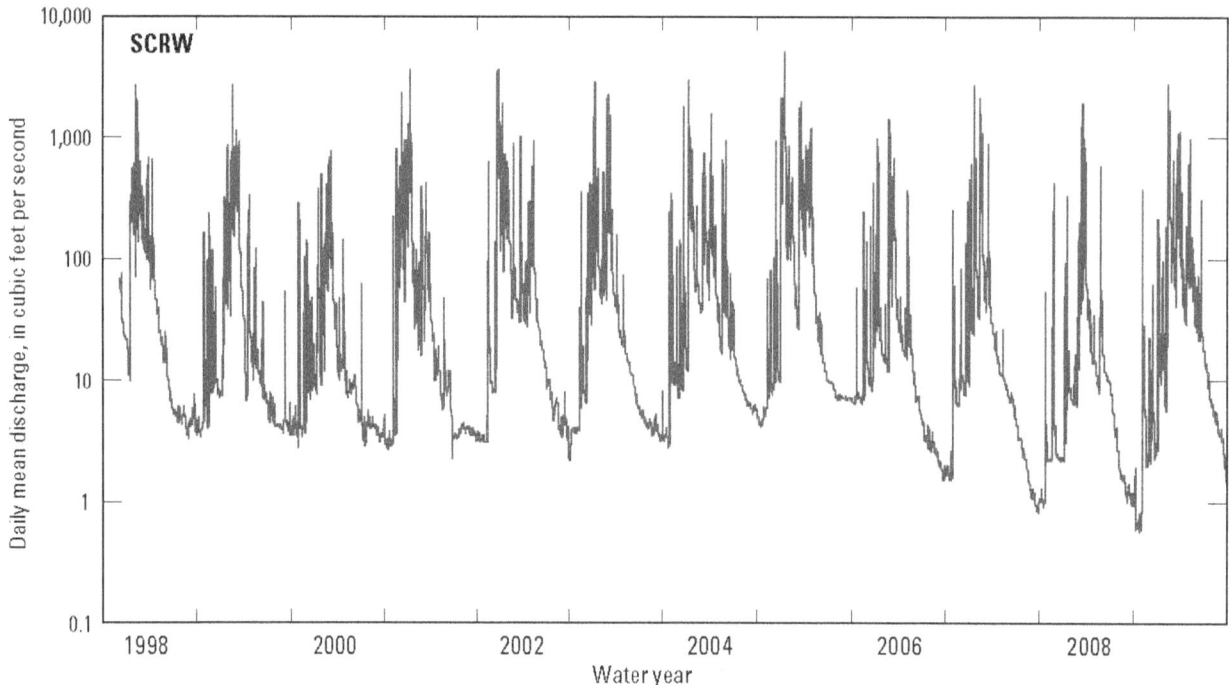

Figure 16.   Daily mean discharge measured at gage 11466320, Santa Rosa Creek at Willowside Road (SRCW), Sonoma County, California, 1998–2010.

To estimate longer-term (greater than 30 year) streamflow variability, the daily mean discharge record for MWCM was extended by using streamflow records for gage NAPN (11458000, Napa River near Napa) on the Napa River (fig. 7). The NAPN gage recorded data from water-years 1930 through 2010, and the MWCM gage recorded data from water years 2006 through 2010 (table 2). Comparison of the streamflow records at NAPN and MWCM indicated a positive correlation for water-years 2006 through 2010 (fig. 18). The daily mean discharge record for MWCM was extended to water years 1960 through 2005 by using the maintenance of variance-extension, type 1 (MOVE.1) method (Hirsch, 1982) and the NAPN gage record (fig. 19). The MOVE.1 technique produces discharge estimates at the short-term gage with a statistical distribution similar to that expected if the discharge had actually been measured (Helsel and Hirsch, 1992, p. 277). It estimates the probability of extreme high or low discharge. To develop the MOVE.1 estimates, the linear correlations between daily mean discharge at the NAPN and MWCM gages were analyzed on a seasonal basis (fig. 18). Linear regression between NAPN and MWCM indicated significant coefficients of determination (r^2) ranging from a maximum of 0.91 for spring to a minimum of 0.59 for fall (fig. 18).

For the purposes of this report, the flow at the MWCM gage at which there is flooding is assumed to range from 6,000 to 8,000 ft³/s, which can result in local flooding of roadways (Michael Webster, U.S. Geological Survey, written commun., 2013). Note that the flowrate at which there is flooding is not only dependent on the flow at the MWCM gage, but also on the flow of the Russian River, as a consequence of backwater effects (see the "Laguna de Santa Rosa Drainage Basin"

section of this chapter). The combined daily mean discharge record (extended and measured) at the MWCM gage included 62 days between October 1, 1959, and September 30, 2010, that had a daily mean discharge of 6,000 ft³/s or greater (fig. 19A). The combined record included 25 days that had a daily mean discharge of 8,000 ft³/s or greater (fig. 19A). In contrast to the high flows, the estimated maximum daily mean discharge during the exceedingly dry water-year of 1977 was only 71.4 ft³/s, which was measured on January 3, 1977. Seven water years, including water-year 1977, had relatively low maximum daily mean discharges of less than 2,000 ft³/s.

The extended daily mean discharge record for MWCM was combined with the measured record to develop a time series of annual discharge for water-years 1960 through 2010 (fig. 19B). The long-term estimated mean discharge for the 51-year time series was 265 ft³/s, compared to the measured mean discharge of 250 ft³/s for water-years 2006 through 2010 and an estimated mean discharge of 266 ft³/s for the 46-year extended record. In the combined record, a maximum estimated annual mean discharge was 663 ft³/s for water-year 1983. In the measured record, a maximum annual mean discharge of 533 ft³/s was measured for water-year 2006. In addition to water-year 1983, two other years were estimated to have a mean discharge greater than 500 ft³/s: 1982 and 1995. Seven water years were estimated to have a mean discharge less than 100 ft³/s: 1961, 1972, 1976, 1977, 1987, 1990, and 1994. Water-years 1976 and 1977 were the two driest years in terms of discharge during the 51-year period, with a mean discharge of 40 ft³/s estimated for 1976 and only about 8 ft³/s estimated for 1977.

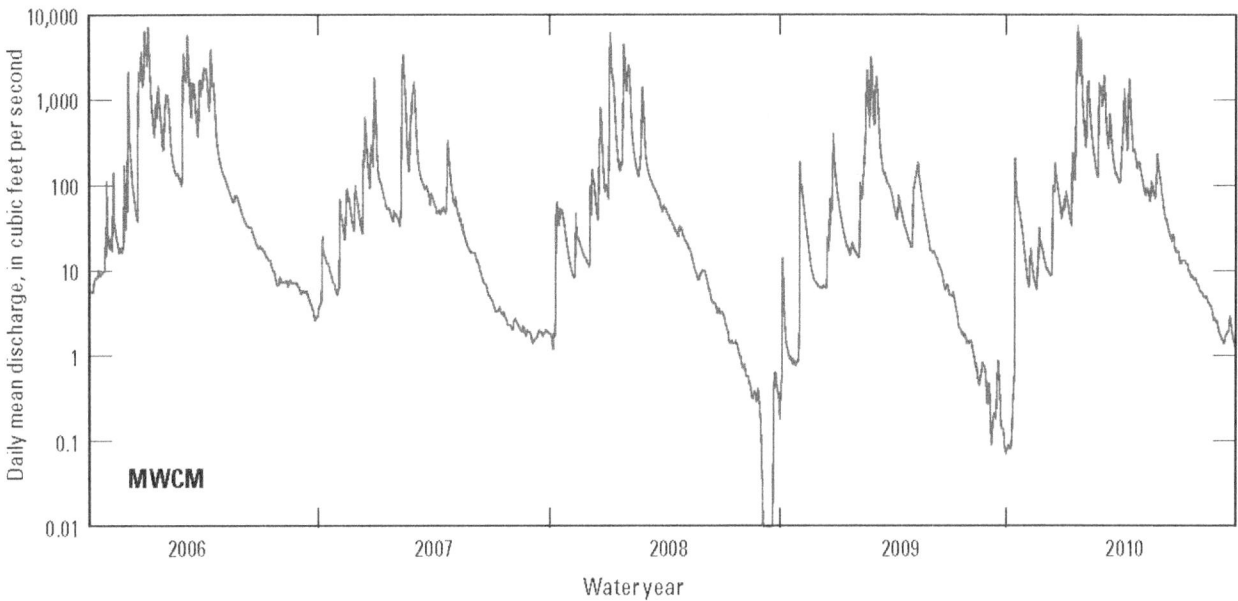

Figure 17.   Daily mean discharge measured at gage 11466800, Mark West Creek near Mirabel Heights (MWCM), Sonoma County, California, 2006–2010. Discharge rates of less than 0.01 cubic feet per second represent zero-flow conditions.

By using the combined record, a cumulative departure from the mean curve was also calculated to help evaluate wet and dry periods in terms of streamflow (fig. 19C). Dry periods are characterized by downward trends (for example, water-years 1960–66 in figure 19C), and wet periods are characterized by upward trends (for example, water-years 1966–75 in figure 19C). Using these definitions of wet and dry periods, the wet periods are water-years 1966–74, 1981–86, 1994–99, and 2002–06. The dry periods are water-years 1960–66, 1976–81, 1986-94, 1999–2002, and 2006–09.

Flow-Duration Analysis

A comparison of flow-duration curves for all continuous daily mean discharge records for the MWCM gage and its estimated flows was used to characterize the distribution of streamflow for the SRPW (fig. 20). Flow duration for the MWCM gage indicated that a daily mean discharge of about 600 ft^3/s was exceeded about 10 percent of the time (fig. 20). The 1-percent daily mean discharge at the MWCM gage (the discharge that is equaled or exceeded 1 percent of the time) was equal to about 3,000 ft^3/s for both the measured and extended record.

The flow-duration curves for the measured and extended record at the MWCM gage were generally well matched for flows that were equaled or exceeded about 92 percent of the time (fig. 20). For the highest flows that occur less than 1 percent of the time, daily mean discharge was higher for the estimated long-term record. The relatively short record for

the MWCM gage probably is inadequate for representing the highest discharges at this site. For the driest 8 percent of flows at MWCM, measured flows were usually greater than the extended record (fig. 20). The measured flows were probably higher than the extended record because baseflow and flows associated with urbanization in recent decades tended to augment streamflow, especially during late summer and fall.

Summary of Streamflow Characteristics

Streamflow within the SRPW is strongly seasonal with high winter flows and mostly intermittent summer flows. Most of the streamflow is runoff generated in response to rainfall, with about 90 percent of the total annual discharge volume from October through May. Streamflow is highly variable, not only on a seasonal basis but also from year to year. Following exceptionally dry winters, streamflow data indicated that streams designated by the NHD as perennial can become dry during the summer.

Winter streamflow is characterized by a relatively rapid response time of overland flow reaching first-order streams in the upper drainages and then continuing on to the main channels. The response time refers to the time of concentration of overland runoff into the main stream channels. The rapid response times are caused by a combination of factors, including both storm characteristics and basin characteristics. An important storm characteristic is that almost all precipitation within the SRPW is rain. Winter streamflow generally is well correlated to rainfall intensity and magnitude. Antecedent

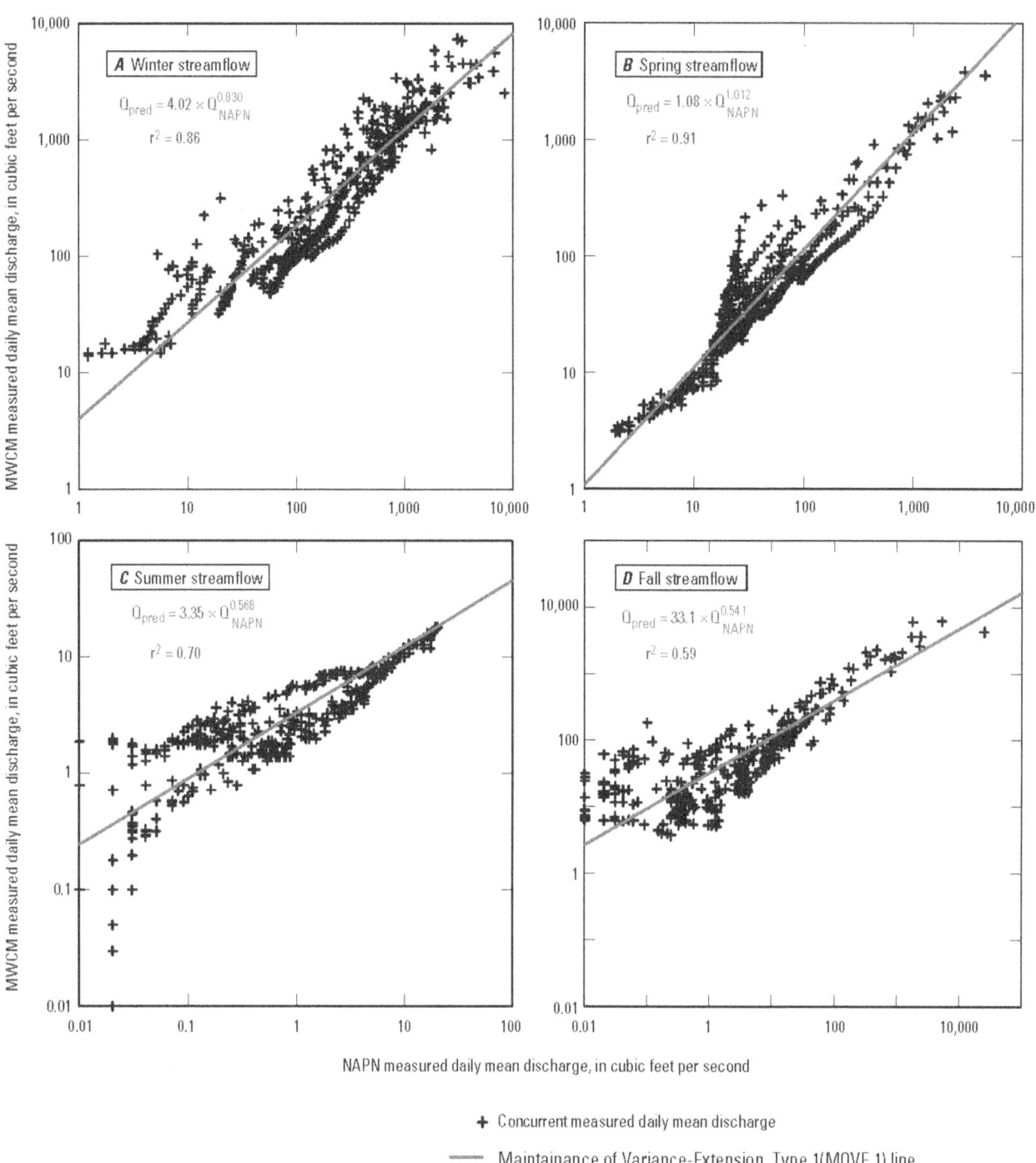

Figure 18.   Correlation between seasonal daily streamflow at NAPN and MWCM gages, along with MOVE.1-estimated streamflow for the Santa Rosa Plain watershed, Sonoma County, California: *A*, winter (January through March); *B*, spring (April through June); *C*, summer (July through September); and *D*, fall (October through December). (MWCM, Mark West Creek near Mirabel Heights; NAPN, Napa River near Napa)

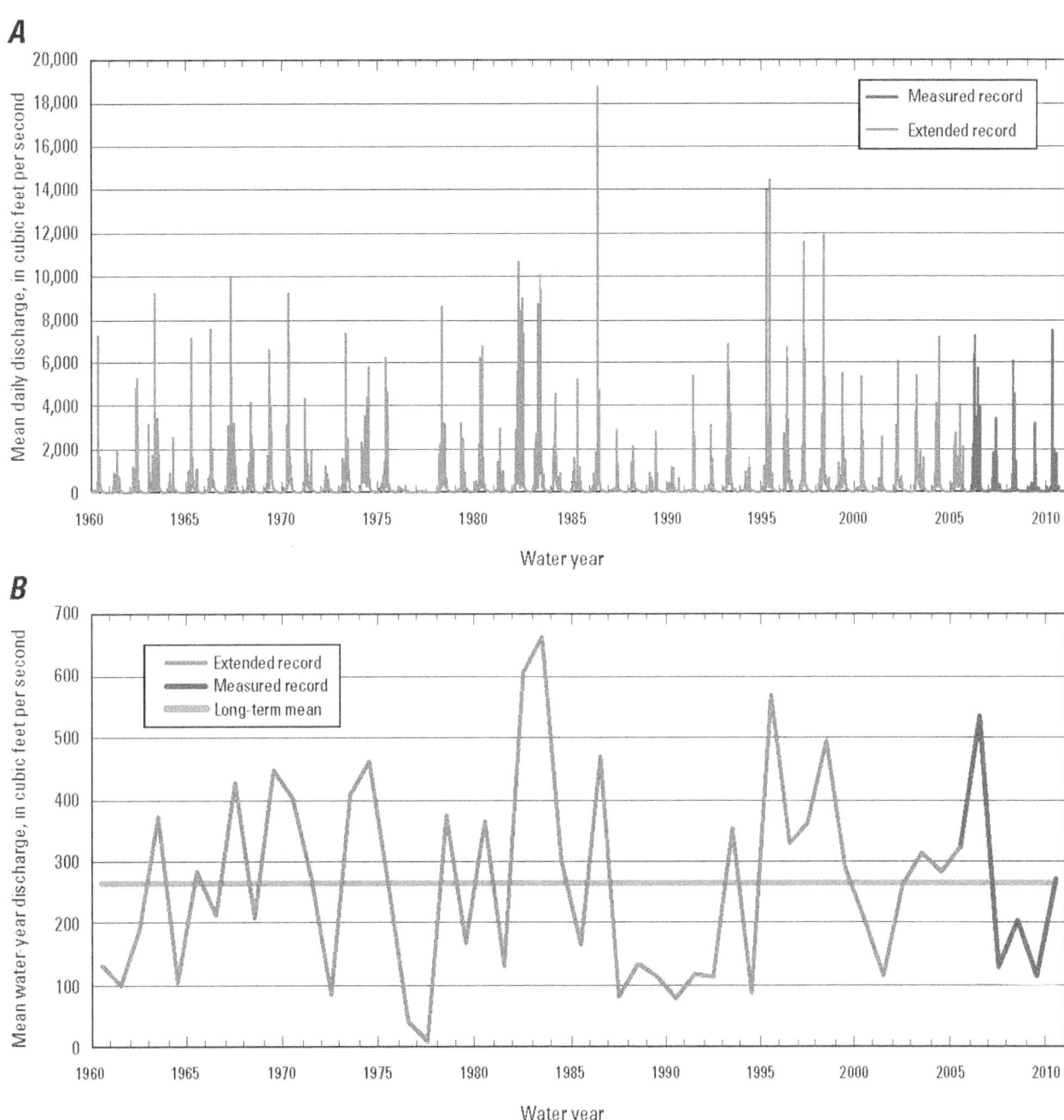

Figure 19. Extended streamflow record estimated at the lower Mark West Creek near Mirabel Heights (MWCM) gage. *A*, Estimated daily mean discharge for water-years 1960–2005 and measured daily mean discharge for water-years 2006–2010. *B*, Estimated annual (water year) discharge for water-years 1960–2005 and measured annual discharge for water-years 2006–2010. *C*, Cumulative departure from the mean curve for water-years 1960–2005, Santa Rosa Plain watershed, Sonoma County, California.

C

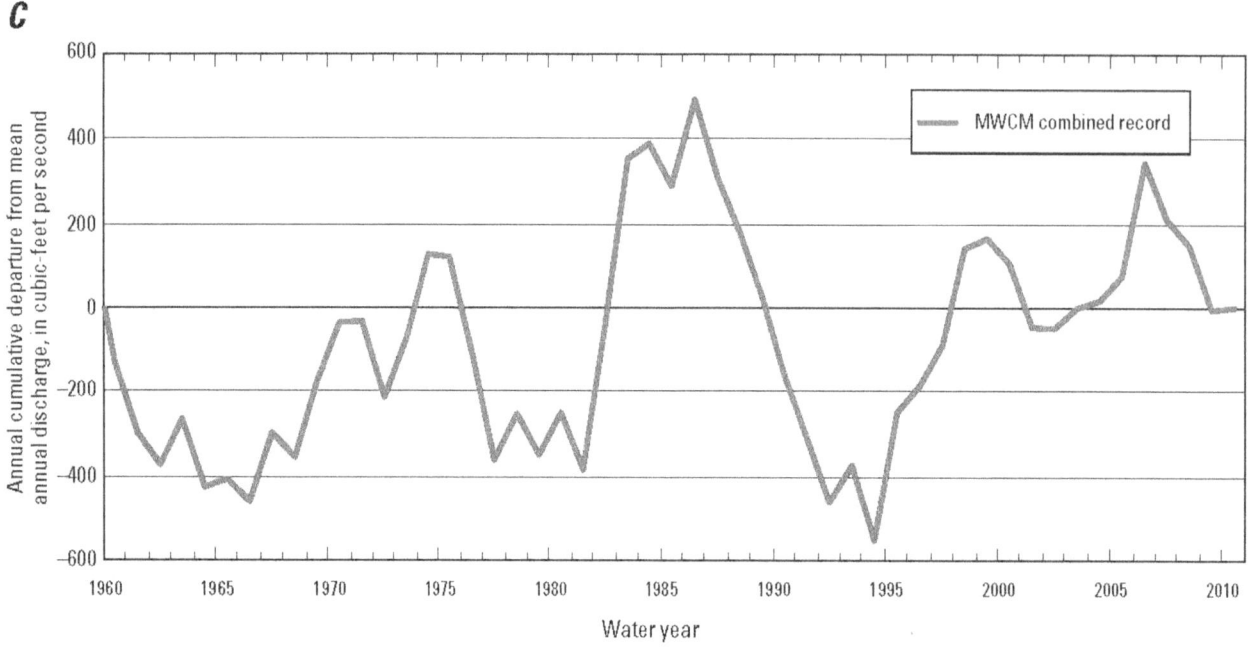

Figure 19. Continued.

conditions also are important in determining runoff; an increase in storm frequency causes wetter soil and ground conditions, resulting in increased runoff in response to rainfall. Steep hillsides, thin soil and vegetation cover, and impervious surfaces in urban areas are basin characteristics that cause more rapid runoff.

There is typically some localized flooding in low-lying areas each winter in response to the largest storms. The rapid response times for most drainages within the SRPW increases the potential for flooding in low lying areas of the basin, especially within the 1-percent AEP floodplain of the Laguna de Santa Rosa (fig. 5). Developed areas, primarily within Windsor, Cotati, and Rohnert Park, have been inundated in response to the larger storms associated with higher rainfall intensities. The potential for urban-area flooding tends to be high in the low-lying, low-gradient sections of drainage basins downstream of areas with a high percentage of impervious land cover (rooftops, parking lots, and roadways).

As stated earlier, streamflow in the Laguna de Santa Rosa, the lower Mark West Creek, and the lower Santa Rosa Creek can be slowed and even reversed because of backwater effects in the Laguna de Santa Rosa floodplain generated by high flows on the Russian River. The Laguna de Santa Rosa floodplain acts as a natural flood retention basin for the Russian River by temporarily capturing and storing up to 79,000 acre-ft of flood water (Sloop and others, 2007).

During summer, low-flow conditions are found throughout the SRPW, with most of the streamflow consisting of baseflow and, in some cases, runoff from irrigation. Streamflow in the upper reaches of Santa Rosa Creek and Mark West

Creek drainages are perennial in many places, especially along sections of Mark West Creek, Matanzas Creek, Spring Creek, and Santa Rosa Creek. By late summer and early autumn, the natural flows in the upper channels diminish to less than 1 ft³/s for most locations. In contrast to winter streamflow, summer streamflow in the lower reaches of the principal drainages is generally less than 5 ft³/s and is mostly limited to the main channels, whereas the smaller tributaries remain dry throughout the summer. Summer streamflow in the lower reaches of Santa Rosa Creek is often a combination of natural baseflow from the upper drainages, urban runoff from the city of Santa Rosa, irrigation return flows, and outflows from reservoirs (Lake Ralphine and Spring Lake). A large percentage of the summer streamflow does not discharge to the Russian River, but rather flows into wetlands within the Laguna de Santa Rosa, where it is lost to the atmosphere through evapotranspiration (ET) or infiltrates through the streambed, which can ultimately become groundwater recharge.

Flood Control, Flow Diversions, and Drainage Modifications

The surface-water hydrology of the SRPW has been affected by a long history of anthropogenic influences on stream channel and watershed characteristics related to land-use change and increasing population (Sloop and others, 2007). From the initial settlements through modern-day urbanization, modifications have been made to stream channels for flood control and to promote drainage across agricultural

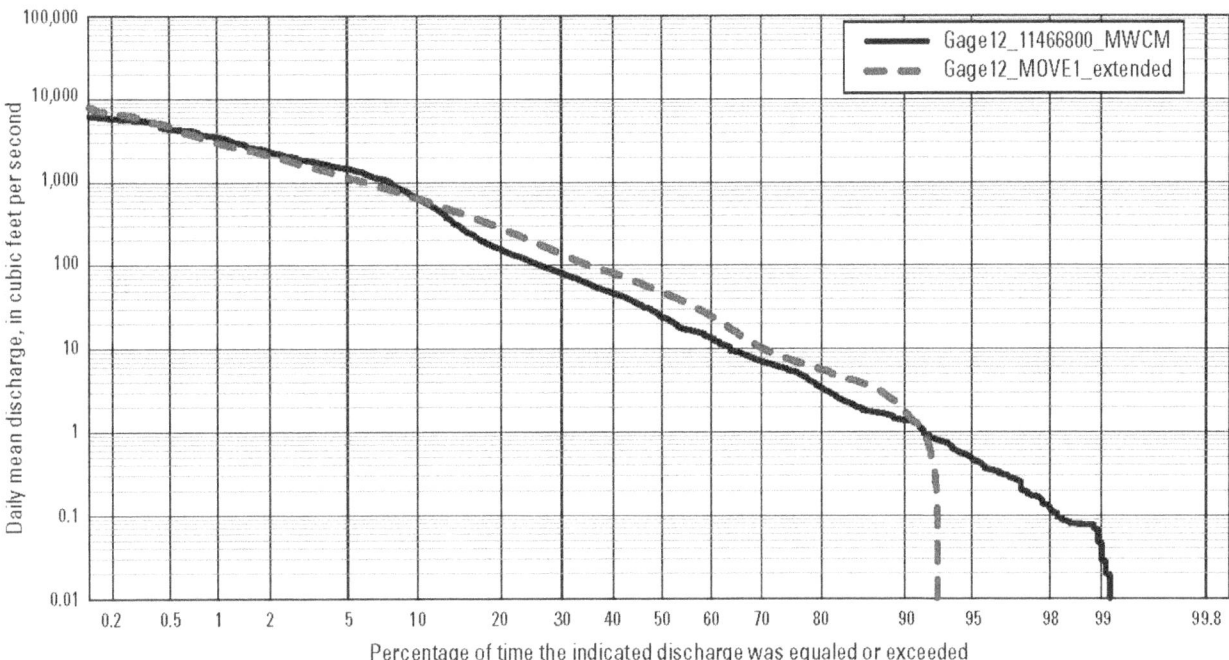

Figure 20.   Flow duration of daily mean discharge and modeled daily mean discharge for the lower Mark West Creek near Mirabel Heights (MWCM) gage, Santa Rosa Plain watershed, Sonoma County, California.

land and developed areas. At the same time, changes in land use have generally resulted in an increase in the potential for flooding (Sloop and others, 2007). An increase in the imperviousness of the land surface caused by an increase in the percentage of area occupied by rooftops, parking lots, and roadways has resulted in more rapid runoff generation and an increase in total runoff volume downstream of urban areas (Sloop and others, 2007). In addition to flood-control modifications, the natural-drainage system has been modified by flow diversions for irrigation and discharge of wastewater effluent. Conversion of land to agriculture, housing developments, and commercial land uses have often been accompanied by stream channel and drainage modifications, such as channel straightening and deepening, to improve the drainage of low-lying areas and to dewater swamp and marsh land. In some cases, canals were constructed to improve the connectivity of the natural drainage system, especially across alluvial fans where natural channels are often not well defined (Sloop and others, 2007; Dawson and Sloop, 2010). Groundwater-resource development can lower water tables enough to decrease summer baseflow in some channel sections, whereas return flows from irrigation and urban-area discharges can increase summer flows in other channel sections.

Flood Control

The SRPW includes five retention basins, all impounded by earthen dams, that are used to mitigate flooding of Santa Rosa Creek as it passes through the city of Santa Rosa. Four of these retention basins were constructed by the Natural

Resources Conservation Service (NRCS) and the Sonoma County Flood Control and Water Conservation District during the early 1960s (U.S. Army Corps of Engineers, 2002) and are owned and operated by SCWA: Spring Lake, Matanzas Creek Reservoir, Piner Creek Reservoir, and Middle Fork Brush Creek Reservoir. A fifth retention basin, Lake Ilsanjo, was constructed in 1956 and is owned and operated by the California Department of Parks and Recreation as part of Annadel State Park.

Spring Lake is located within the city of Santa Rosa (fig. 5) and is the largest flood-control facility in the SRPW with a maximum storage capacity of 3,550 acre-ft and a surface area of 0.24 mi^2 (154 acres). Having a natural drainage area of only 0.55 mi^2 (355 acres), Spring Lake also receives water through two channel diversions that connect the reservoir to the Santa Rosa Creek and Spring Creek drainages. The main flood-control diversion connects to the channel of Santa Rosa Creek about 0.5 mi upstream of the reservoir and uses a concrete control sill along with an 8-foot diameter culvert and control orifice to regulate inflows into the reservoir. The diversion allows the first 840 ft^3/s of natural flow to remain in the Santa Rosa Creek channel. As the flow in Santa Rosa Creek exceeds 840 ft^3/s, a portion of the flow is diverted to the reservoir. There is a maximum diversion of 5,220 ft^3/s for natural flows of 8,250 ft^3/s and greater in the main channel of Santa Rosa Creek. A second diversion connects with the channel of Spring Creek about 1.0 mi south of the reservoir. Flow of about 10 ft^3/s remains in the natural channel of Spring Creek; flows greater than 10 ft^3/s and up to 1,000 ft^3/s are diverted to Spring Lake.

Matanzas Creek Reservoir was built in 1963 and is located on Matanzas Creek in the upper section of the drainage (fig. 5). The reservoir has a maximum surface area of 62 acres, a maximum storage capacity of 1,500 acre-ft, and a catchment area of 11 mi² (7,040 acres). Piner Creek Reservoir was built in 1962 on Paulin Creek (fig. 5) and has a maximum surface area of 19 acres, a maximum storage capacity of 172 acre-ft, and a catchment area of 2.05 mi² (1,312 acres). Middle Fork Brush Creek Reservoir was built in 1961 and is located on the middle fork of Brush Creek (fig. 5). It has a maximum surface area of 20 acres, a maximum storage capacity of 138 acre-ft, and a catchment area of 2.24 mi² (1,434 acres). Lake Ilsanjo is on Spring Creek (fig. 5) and has a maximum surface area of 67 acres, a maximum storage capacity of 395 acre-ft, and a drainage area of 1.71 mi² (1,094 acres).

Several studies have been carried out within the SRPW to evaluate flood risk, specifically for Santa Rosa Creek and the potential effects on the city of Santa Rosa (U.S. Army Corps of Engineers, 2002). By using precipitation-runoff models, a discharge with an AEP of 1 percent being equaled or exceeded was estimated to be 19,600 ft³/s at the mouth of Santa Rosa Creek, and 8,250 ft³/s was estimated at the upper Santa Rosa Creek channel just above the Spring Lake diversion. Additional estimates of discharges with an AEP of 1 percent for Santa Rosa Creek included 8,410 ft³/s above the Matanzas Creek confluence and 14,200 ft³/s above the Piner Creek confluence. Estimated discharges with an AEP of 1 percent for other tributary streams included 1,200 ft³/s at the mouth of Spring Creek, 5,270 ft³/s at the mouth of Brush Creek, 5,500 ft³/s at the mouth of Matanzas Creek, and 4,420 ft³/s at the mouth of Piner Creek (U.S. Army Corps of Engineers, 2002).

Surface-Water Diversions

Surface-water diversions affecting the SRPW include internal diversions within the SRPW and diversions across the SRPW boundary. Internal diversions include flood control (discussed previously) and minor diversions of flow from Mark West Creek and Santa Rosa Creek to irrigate up to 6,000 acres within the SRPW (U.S. Geological Survey, 2011). In the headwater areas, numerous small diversions of runoff from small, unnamed channels are likely on a localized scale to supply water for ponds and irrigation. The total magnitude of these diversions was assumed to be relatively small. The source of irrigation water within the SRPW was assumed to be either groundwater, recycled water, or some combination of the two.

A significant amount of Russian River water is diverted into the SRPW, primarily for use as the municipal water supply for the town of Windsor and the cities of Santa Rosa, Rohnert Park, and Cotati. A portion of the imported municipal water is used for residential landscape irrigation and other uses, which could result in some increases in runoff and recharge in the SRPW. A minor amount of Russian River water (less than about 1,000 acre-ft/year) is used directly

for irrigation within the SRPW (Donald Seymour, Sonoma County Water Agency, written commun., 2010), and this could also cause a small, localized increase in runoff and groundwater recharge. In this study, Russian River water used directly for irrigation was not considered a significant component of the SRPW water balance.

Most of the imported Russian River water is ultimately processed at two wastewater-treatment facilities within the SRPW (fig. 21). Treated wastewater is discharged in a variety of ways: direct discharge to stream channels, land applications (irrigation and wetlands), and deliveries to the Geysers geothermal are outside of the SRPW (for power generation). As of January 2008, more than 14,000 acre-ft/yr of treated wastewater was delivered to the Geysers (City of Santa Rosa, 2012); this is the largest use of treated wastewater. Most of the remaining water is applied on designated land parcels (fig. 21), as described later, to irrigate hay fields, grapevines, golf courses, and urban parks during the spring and summer (Sloop and others, 2007).

Monthly records on the application of treated wastewater from the town of Windsor and the city of Santa Rosa used for irrigation, also referred to as reclaimed water, were available for water-years 1990 through 2009 (fig. 22). For the most part, land irrigated with treated wastewater was within the Laguna de Santa Rosa 1-percent AEP floodplain (fig. 21). Total monthly treated wastewater used for irrigation varied from zero during winter months to a maximum of about 3,000 acre-ft during the summer months of water-years 1994 and 1995 (fig. 22A). The annual volume of treated wastewater used for irrigation averaged about 10,200 acre-ft, with a maximum of 14,100 acre-ft used during water-year 2001, and a minimum of 7,400 acre-ft used during water-year 2009 (fig. 22B). At the time of this study, the volume of treated wastewater used for irrigation in water year 2010 was not available.

Drainage Modifications

With the onset of more intensive agriculture from the early 1800s, many stream channels were modified to promote more rapid drainage of wetlands and vernal pools that developed on the alluvial fans during the wet winter season (Dawson and Sloop, 2010). Channels that were formerly disconnected on the alluvial fans became straightened and more connected by a network of roadside ditches and canals. In their natural state, stream channels shifted periodically across the alluvial fans during the wet season, with Copeland Creek occasionally switching watersheds between the Russian River and the Petaluma River drainage systems (Dawson and Sloop, 2010). With the conversion of land to ranching and agricultural uses, the stream network changed. Streams draining the mountains on the eastern side of the valley that fed seasonal wetlands and did not join with the Laguna de Santa Rosa, such as Copeland and Crane Creeks, were instead redirected by straight canals and drainage ditches into the main channel of the Laguna de Santa Rosa as early

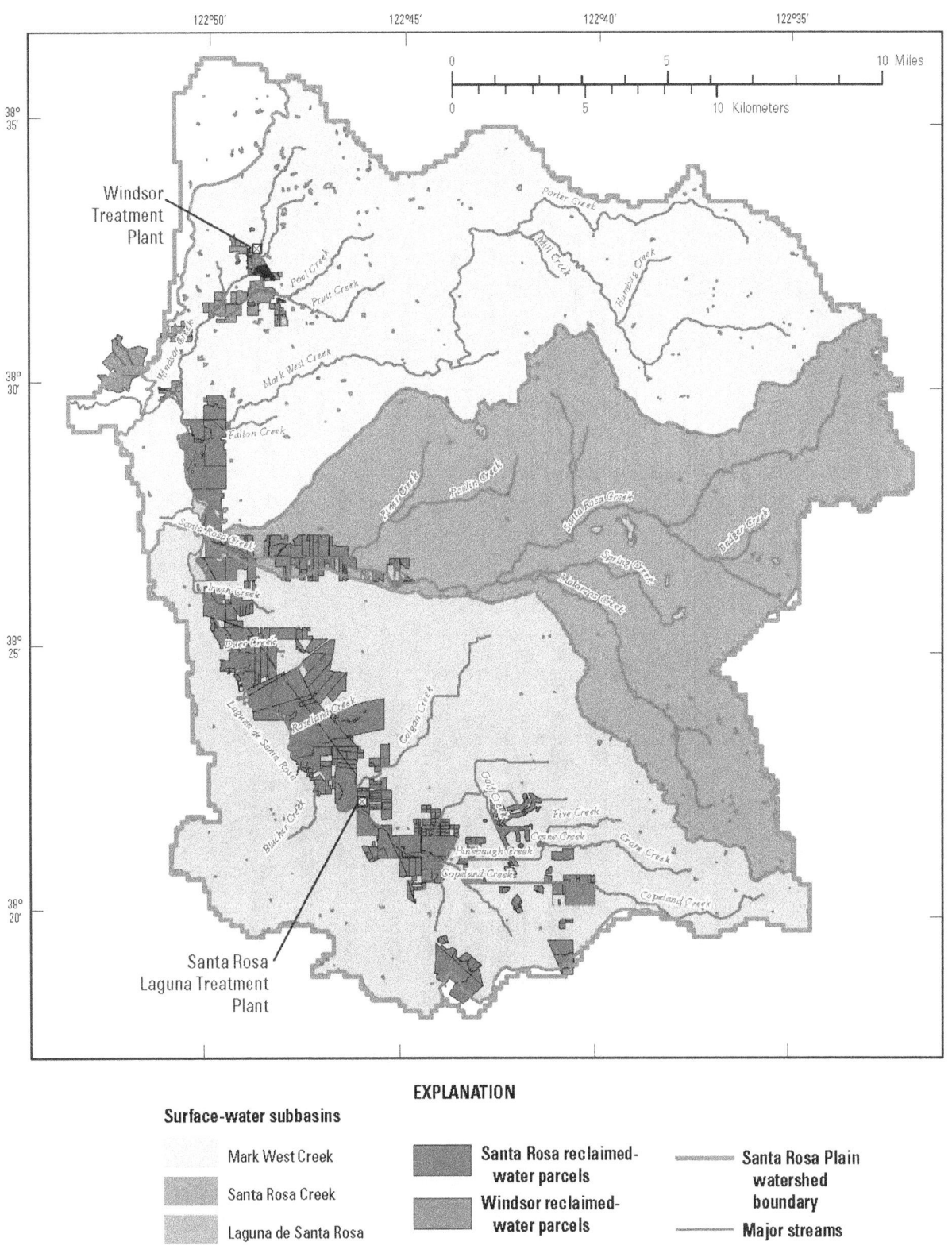

Figure 21. Location of land parcels where reclaimed water is applied for irrigation of crops and urban areas, Santa Rosa Plain watershed, Sonoma County, California.

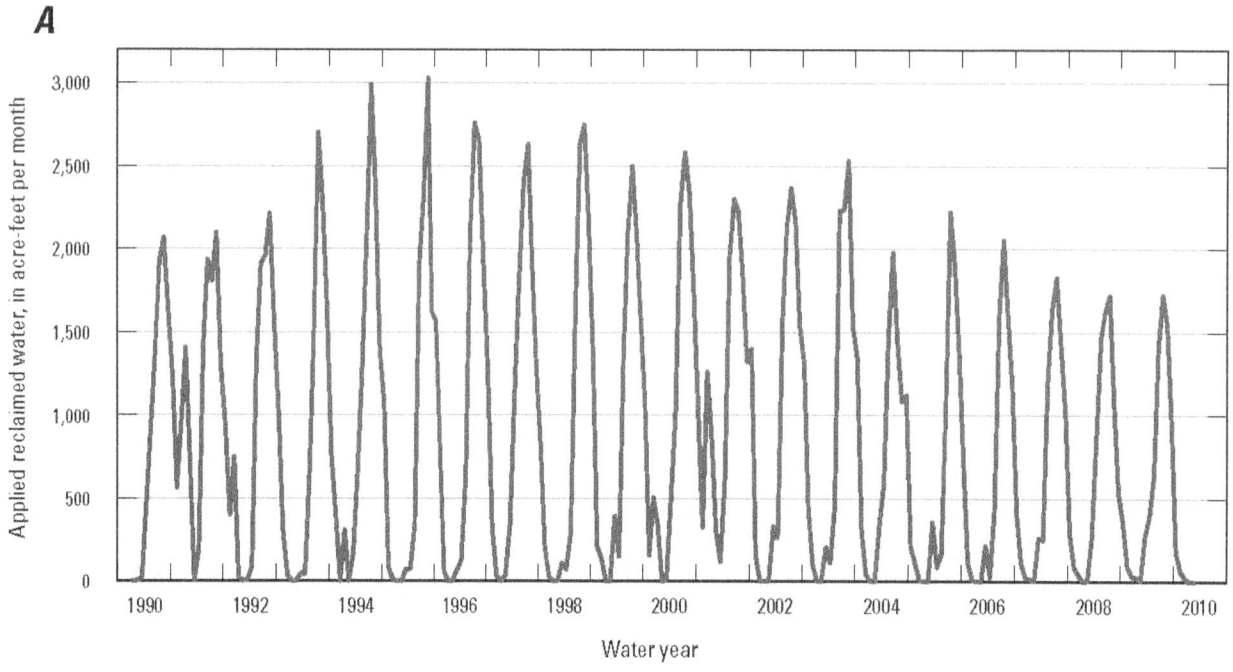

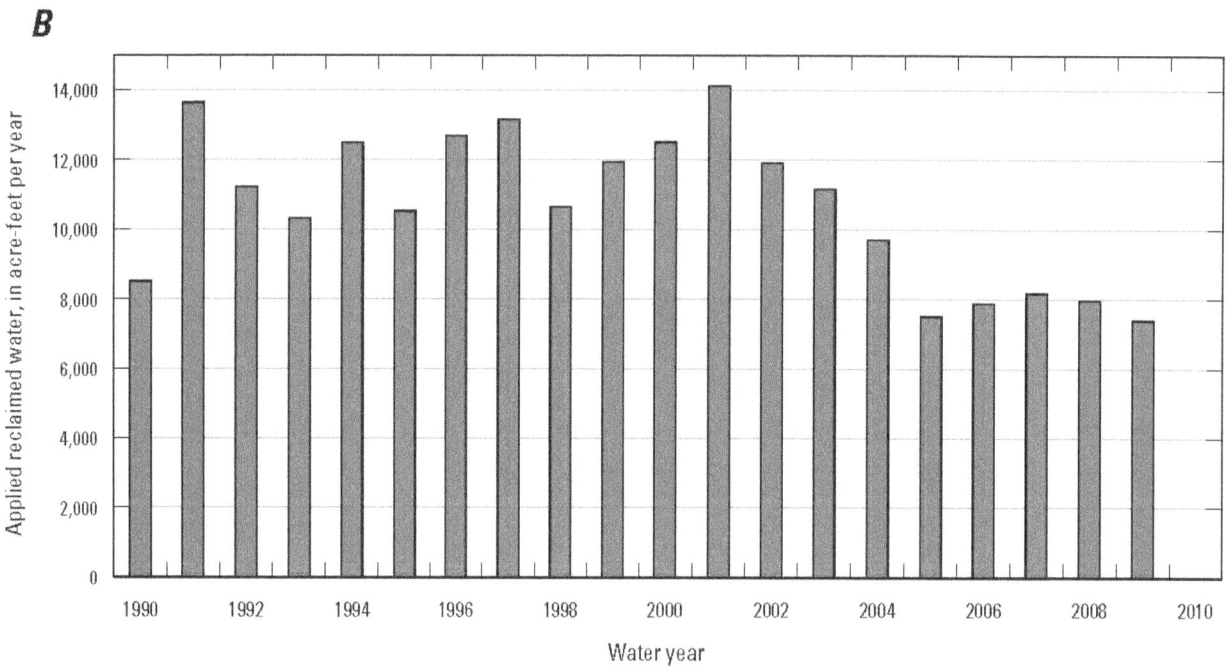

Figure 22.   Reclaimed water applied as irrigation in Santa Rosa Plain watershed, Sonoma County, California, for water-years 1990–2009: *A*, monthly totals; *B*, annual (water-year) totals.

as the 1870s (Dawson and Sloop, 2010). The trend toward increasing connectivity of the drainage network is ongoing, with storm drains installed in housing developments and drainage tile placed under vineyards. The consequences of these historical changes probably include habitat loss, decreased groundwater recharge, decreased water quality, and increased erosion and flooding downstream (Dawson and Sloop, 2010).

More recently, channel modifications have included channel restoration efforts to mitigate invasive vegetation, channel erosion, and sedimentation, and to improve riparian habitat and water quality. The ongoing channel restoration has included the stabilization of eroding channel banks by using riprap and native vegetation cover and the conversion of riparian areas to recreational uses, which includes the removal of underbrush.

Groundwater Hydrology

Groundwater systems in the study area include the saturated sedimentary rocks and sediments underlying the floor of SRP and surrounding lowlands, as well as the volcanic rocks underlying the mountains in the eastern SRPW, where such rocks are sufficiently permeable to yield water. Beneath the floor of SRP, the principal groundwater system is lithologically heterogeneous, but consists of one continuous body of saturated material. Groundwater in the principal aquifer is contained in the pore spaces of the Quaternary alluvial materials and Tertiary sedimentary rocks, including the Glen Ellen Formation, the Wilson Grove Formation, and the Petaluma Formation. Groundwater is also contained in locally permeable intervals within the Sonoma Volcanics and, to a much lesser extent, in fractured bedrock (fig. 23). Data compiled largely from previous investigations (Cardwell, 1958; Ford, 1975; Herbst and others, 1982; Sweetkind and others, 2010) are used in this report to describe the aquifer system, aquifer properties, recharge and discharge, and groundwater flow in the SRPW.

Groundwater Subbasins and Storage Units

The SRPW has been divided into groundwater subbasins by the California Department of Water Resources (CDWR; Swartz and Hauge, 2003). Groundwater storage units were defined for this study to help describe the groundwater hydrology of the SRPW.

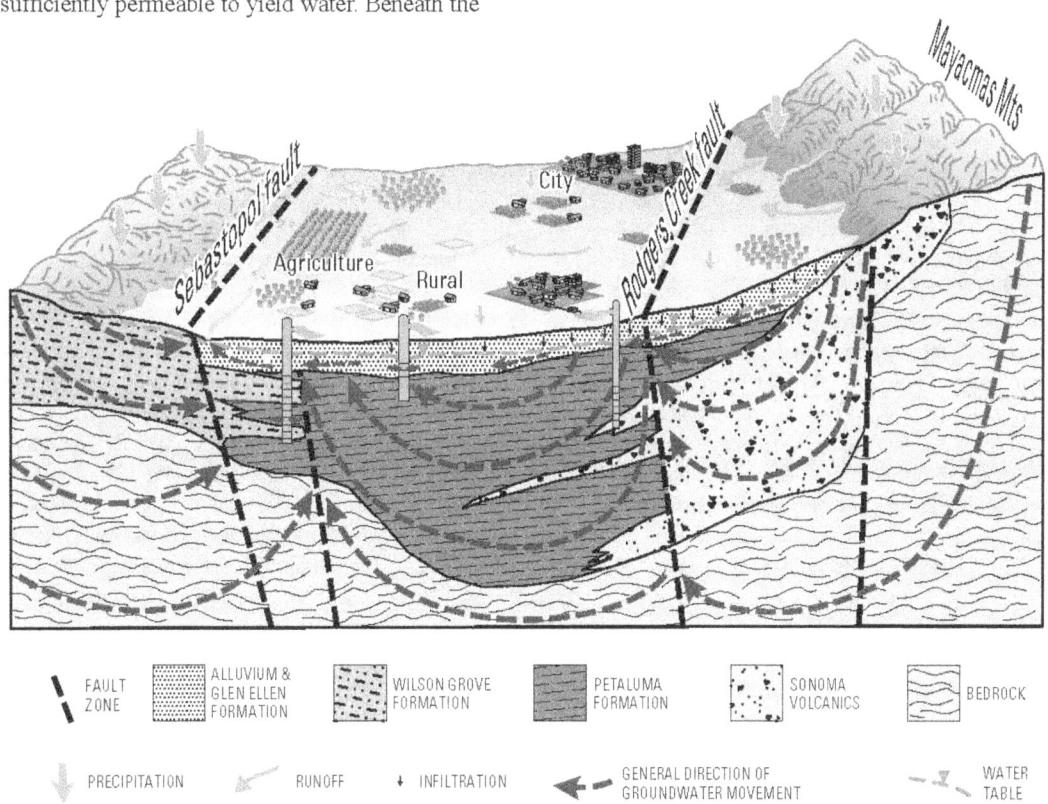

Figure 23. Conceptual diagram of the groundwater system, Santa Rosa Plain watershed, Sonoma County, California.

Groundwater Subbasins

A groundwater basin is defined as an aquifer or aquifer system bounded laterally, and at some depth, by rocks or geologic structures, which, by virtue of a distinct permeability contrast, hydraulically separate groundwater within the basin from that beneath neighboring land (California Department of Water Resources, 2003). Groundwater subbasins are subdivisions of groundwater basins delineated on the basis of geologic, hydrologic, or institutional boundaries. The study area includes the SRP and Rincon Valley groundwater subbasins, the eastern portion of the Wilson Grove Formation Highlands groundwater basin, the western half of the Kenwood Valley groundwater basin, the southern part of the Healdsburg area groundwater subbasin, and the southern part of the Alexander Valley groundwater subbasin (Swartz and Hauge, 2003; fig. 24).

The SRP groundwater subbasin is the largest groundwater subbasin in the study area and includes about 125 mi^2 that covers most of SRP (fig. 24). As mapped by the California Department of Water Resources (2003), the SRP groundwater subbasin includes water-bearing strata only within the Cenozoic sedimentary formations, and not within the Tertiary volcanic rocks, which are primarily the Sonoma Volcanics; however, the volcanic rocks provide substantial quantities of water to wells on the valley floor and in the mountains along the eastern side of the valley. The groundwater-storage capacity of the SRP groundwater subbasin within the depth range of 10 to 200 ft was estimated to be about 950,000 acre-ft (Cardwell, 1958). The storage capacity over an average thickness of 400 ft was estimated to be about 4,300,000 acre-ft (Herbst and others, 1982). Neither estimate includes the amount of groundwater contained in the Sonoma Volcanics. The Sonoma Volcanics, however, contain discontinuous aquifers that in some of the more porous and permeable lithologic packages can yield several hundred to several thousand gallons of water per minute to wells (Cardwell, 1958). Note that the sustainable yield of the basin is less than the storage capacity because of potential negative effects of withdrawing all groundwater in storage, such as, land subsidence, depletion of environmental flows, and pumping of poor quality groundwater.

The Rincon Valley groundwater subbasin includes about 5,760 acres of relatively flat land in a structural depression that lies about 1 to 2 mi east of the SRP (fig. 24). The small subbasin is bounded mostly by outcrops of the Sonoma Volcanics, but is hydrologically connected with the SRP groundwater subbasin through a topographically narrow gap underlain by alluvial deposits. Santa Rosa Creek flows westward through the gap after receiving tributary water from Brush Creek, which flows intermittently and drains most of Rincon Valley. The groundwater-storage capacity of the Rincon Valley subbasin within a depth range of 10 to 200 ft has been estimated to be about 21,000 acre-ft (Cardwell, 1958). Cardwell's (1958) estimate of the average specific yield was 5.5 percent.

The Wilson Grove Formation Highlands groundwater basin covers about 81,280 acres of hilly terrain that extends westward from the western edge of SRP to the Pacific Ocean (fig. 24). Only a small part of this basin along its eastern side is within the study area. The basin is underlain by sandstone of the Wilson Grove Formation, which rests unconformably on Mesozoic basement rocks. No published estimates of storage are available for the portion of the Wilson Grove Highlands groundwater basin within the SRPW.

Groundwater Storage Units

For the purposes of this study, five groundwater storage units were defined on the basis of the work by Cardwell (1958), hydrogeology, and fault locations (fig. 24). The Windsor Basin (WB) storage unit is located north of the Trenton Ridge fault, west of the Mayacmas Mountain foothills, and east of the Sebastopol fault (fig. 24). The Cotati Basin (CB) storage unit is located south of the Trenton Ridge fault, west of the Sonoma Mountain foothills, and east of the Sebastopol fault (fig. 24). The Wilson Grove (WG) storage unit is located between the Mendocino Range and the Sebastopol fault (fig. 24). The Valley (VAL) storage unit includes the alluvial fill of the Rincon Valley, Bennett Valley, and the northern half of Kenwood Valley (fig. 24). The Uplands (UPL) storage unit includes the Sonoma Volcanics in the Mayacmas and Sonoma mountains east of the Rodgers Creek fault zone, but excludes the VAL storage unit (fig. 24).

Definition of Aquifer System

The Glen Ellen, Wilson Grove, and Petaluma Formations, and the Sonoma Volcanics have distinct aquifer properties and constitute the four principal water-bearing aquifer units in the study area (fig. 1). In the following discussion, the Glen Ellen is defined as a combination of the Quaternary alluvial deposits and Glen Ellen Formation (Sweetkind and others, 2010). The distribution, subsurface extent, and interfingering relations among the four principal formations reflect the history of uplift and basin development in the SRPW, tectonic activity including offset along major basin-bounding faults, and the interaction between continental and marine sedimentation. Previous groundwater-resource investigations of the SRPW (Cardwell, 1958; Ford, 1975; Herbst and others, 1982) defined the geology of the groundwater-flow system. Subsequent geologic mapping and geophysical studies, and interpretation of borehole data, have refined the understanding of the basin geometry and the identity and location of major basin-filling units (Blake and others, 2002; Wagner and others, 2010; Graymer and others, 2007; McLaughlin and others, 2008; Langenheim and others, 2010; Sweetkind and others, 2010). Three-dimensional lithologic and stratigraphic models of the SRPW constructed from borehole data, geologic map, and geophysical data delineated the thickness, extent, and the distribution of subsurface geologic units (Sweetkind and others, 2010). These 3D lithologic and statigraphic models were used in this study to define the SRPW aquifer system.

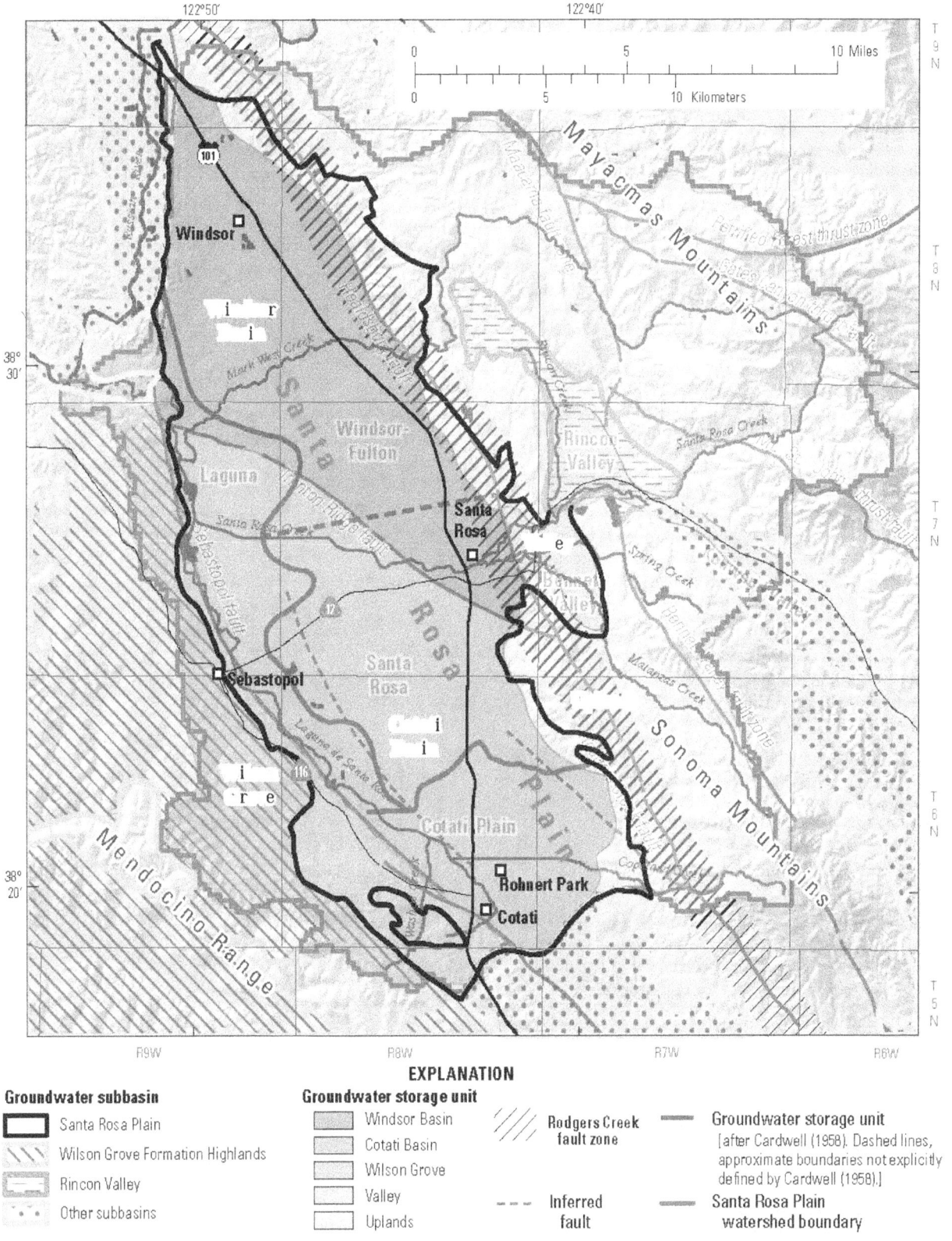

Figure 24.   Groundwater basins, subbasins, and storage units in the Santa Rosa Plain watershed, Sonoma County, California.

Aquifer Extent and Distribution of Material Properties

Data from 2,683 selected boreholes gathered from multiple sources were used to define the subsurface stratigraphy and lithologic heterogeneity of the four principal aquifer units of the SRPW (Sweetkind and others, 2010). The boreholes were selected from a much more numerous set of water-well data by selecting about 10 representative boreholes from each of the 36 sections within a township and range, or about 10 boreholes within each square mile of the study area. In parts of the study area where the population density was low, drill-hole data were correspondingly sparse and fewer than 10 boreholes per square mile were available for analysis. The boreholes that were used represented those that contained the greatest amount of detail in the description of each interval, had a large number of downhole intervals described (as opposed to a single, long interval of "sand and gravel" or "alluvium"), were representative of downhole lithology of nearby holes, and represented a distribution of holes that were not clustered but were approximately equally distributed over the study area.

Lithologic descriptions from the 2,683 boreholes were simplified to 19 internally consistent lithologic classes (Sweetkind and others, 2010). The source of the lithologic data is mostly from drillers' descriptions of material recovered at land surface as well drilling proceeded (drill cuttings). The lack of any uniform protocol in sampling and describing the drill cuttings, such as color charts, grain-size analysis, mineralogical examination, as well as the use of local terms for some materials, required interpretation to translate the original descriptions into geologically consistent terminology.

The resulting lithologic data from the selected boreholes were used to construct a 3D model of lithologic variations within the basin by interpolating and extrapolating data from boreholes by using a nearest-neighbor 3D-gridding process (Sweetkind and others, 2010). The 3D-gridding process is a cell-based solid-modeling approach where cell nodes with dimensions of about 820 ft in the horizontal dimensions and about 33 ft in the vertical dimension are sequentially assigned properties by looking outward horizontally from each node in search circles of ever-increasing diameter. The aspect ratio specified for the gridding process emphasized the horizontal dimension over the vertical, thereby rendering interpolated drill-hole data as horizontal stratigraphic units (fig. 25). The approach is a simple spatial interpolation method that does not consider spatial structure of the data. Within the shallow basin fill, where data are abundant, local lithologic variability is incorporated, and the 3D lithologic variability of the basin fill can be reasonably estimated. Where borehole data are less abundant, such as at greater depths or in uplands, the lithologic model is more dependent on a smaller number of boreholes, and there is greater uncertainty in the understanding of lithologic variability. The resultant 3D lithologic model delineated the variability of lithologic character of each of the four principal water-bearing aquifer units (Sweetkind and others, 2010).

Vertical sections cut through the 3D lithologic model display a west-to-east transition from dominantly fine-grained marine sands of the Wilson Grove Formation, on the west, to heterogeneous continental sediments of the Glen Ellen and Petaluma Formations, beneath the SRP, to Sonoma Volcanics interbedded with the Petaluma Formation in the uplands east of the Rodgers Creek fault zone (figs. 2 and 25). Thin (100 to 150 ft thick) and irregularly distributed coarse-grained deposits (sand and gravel), interpreted to be mainly Glen Ellen Formation (Sweetkind and other, 2010), overlie clay-rich deposits, interpreted to be mainly Petaluma Formation (Sweetkind and others, 2010), throughout the SRP (figs. 2 and 25). The modeled widespread subsurface extent of the Petaluma Formation contrasts with the relatively limited outcrop expression of the Petaluma Formation around the margins of the SRPW (fig. 2).

In order to tie the basin-fill lithology to a stratigraphic context and to mapped surface exposures, a second 3D model was constructed that represents the configuration and relative elevation of the top of each hydrogeologic unit (Sweetkind and others, 2010). This digital 3D hydrogeologic framework model (3D HFM) of the SRPW was constructed by using multiple geologic data sets including geologic maps, surface traces of faults, interpreted subsurface stratigraphic contacts from drill-hole data, and the results of geophysical models (Sweetkind and others, 2010). Subsurface stratigraphy was defined from the borehole lithologic data through the identification of distinctive lithologic packages tied, where possible, to high-quality well control and to surface exposures (Sweetkind and others, 2010). Assignment of stratigraphic tops was fundamentally lithology-based, rather than fossil or age-based, and, as such, was "rock-stratigraphic," rather than time stratigraphic. Mappable lithologic sequences were identified in well data by analyzing numerous serial cross sections across the SRPW and making stratigraphic correlations on the basis of rock type, bedding and sorting characteristics, stratigraphic succession, and an understanding of the relationship between the mapped geologic units and their lithologic characteristics.

The 3D HFM of the SRPW was constructed by standard subsurface mapping methods of creating isopach and structure-contour maps for each of the four principal aquifer units. The elevation of the tops of stratigraphic units and thickness for each of the four major aquifer units were contoured from map and well data by using simplified fault traces to bound contoured regions (Sweetkind and others, 2010). The 3D HFM was constructed from gridded surfaces that defined the top and base of each stratigraphic horizon, which were then stacked in stratigraphic sequence to form a 3D digital solid. The 3D stacking was guided by rules that controlled stratigraphic onlap, truncation of units, and minimum thickness. For computational simplicity, the 3D HFM generalizes complex stratigraphic interfingering and repeated units. As a result, vertical sections derived from the 3D HFM differ in detail from published geologic cross sections (McLaughlin and others, 2008).

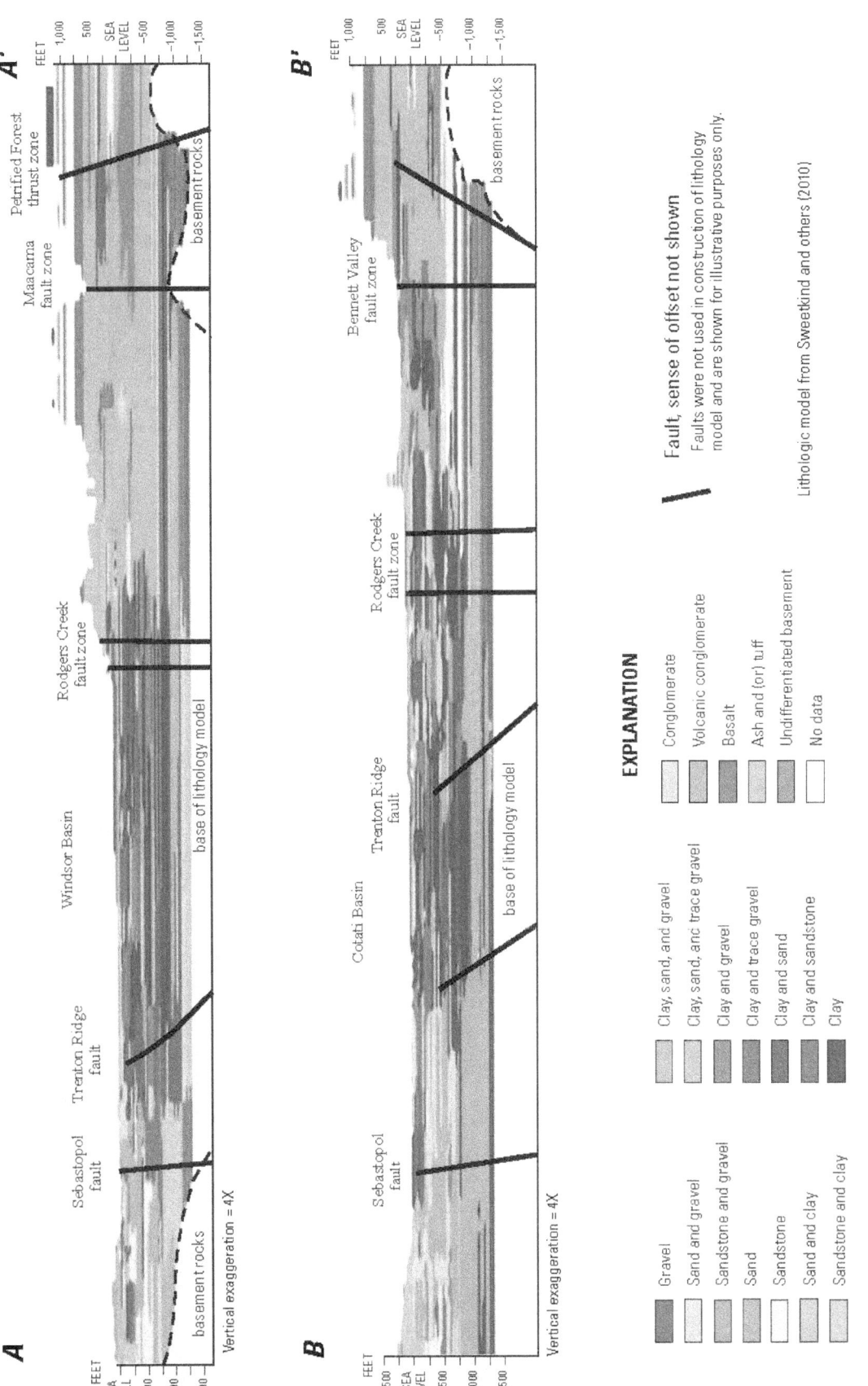

Figure 25. Vertical sections from three-dimensional lithologic model of the Santa Rosa Plain watershed, Sonoma County, California.

Interpretive geologic sections based on the 3D HFM indicated that the Glen Ellen Formation is relatively thin (less than 200 ft) throughout the SRP (section A–A' on fig. 2), in contrast to previous estimates by Cardwell (1958) of 3,000 ft. Previous studies have generally interpreted Wilson Grove Formation and Neogene volcanic rocks to underlie the SRP (Cardwell, 1958; Ford, 1975; Herbst and others, 1982). In contrast, the 3D lithologic model and the 3D HFM of the SRP emphasize the lateral extent of clay-rich lithologies, interpreted to be mainly Petaluma Formation (Sweetkind and others, 2010), throughout the deeper parts of the basins that underlie the SRP (fig. 2).

The groundwater-flow model developed during the next phase of this project required spatially distributed estimates of aquifer properties. Stratigraphic and textural data from the 3D models were used to assess geologic factors that could affect hydraulic conductivity and storage properties. Lateral and vertical variations of sediment texture, including grain size, sorting, and bedding, can affect the direction and magnitude of groundwater flow. The results of the 3D lithology model were classed by sediment texture to help characterize grain-size variations of the aquifer system (Sweetkind and others, 2010; fig. 26). Textural classes were based on the percentage of coarse-grained lithologic classes and on degree of sorting in each cell; relative proportion of clay matrix was considered an important variable. Resultant texture classes included coarse-grained, intermediate, and fine-grained (fig. 26); volcanic rocks did not fit into this scheme and were retained as two additional classes: tuff and basalt.

Hydrogeologic Characteristics of Aquifers

The water-bearing properties of the geologic units in the SRPW vary considerably because of geologic heterogeneity; estimation of these properties allows quantitative prediction of the hydraulic response of the aquifer to changes in recharge, pumping, and other stresses. Previous studies of the SRPW (Cardwell, 1958; Herbst and others, 1982) reported estimates of specific yield of various lithology types and geologic formations by using a variety of methods (table 4). Specific yield values in the SRPW have been estimated to range from 0 to 25 percent (table 4). Coarse-grained, well-sorted sedimentary materials have high values of specific yield because of the large amount of connected pore volume in the material; cemented deposits and clay-rich deposits have smaller total pore volumes and lower specific yields (table 4).

Although the thickness of valley-fill deposits in the northern and southern parts of the SRPW are known to be several thousands of feet thick on the basis of gravity measurements and sparse deep petroleum exploration holes, little is known about the physical characteristics or hydraulic properties of the sedimentary rocks below depths commonly drilled for water wells. Most of the water wells drilled on the valley floor through 2010 are 1,500 ft or less in depth. Of the wells compiled for the geologic framework and textural models of the SRP, only 14 were drilled to depths greater than 1,500 ft.

Table 4. Reported specific-yield values by lithology and geologic formation, Santa Rosa Plain watershed, Sonoma County, California.

[**Abbreviations**: Sy, specific yield; <, less than; —, no data]

Lithology	Sy (percentage)[1]	Geologic formation	Sy (percentage)[2]
Adobe, clay shale	3	Alluvial fan deposits	8 to 17
Cemented gravel, cemented sand, clay and gravel	5	Wilson Grove Formation	10 to 20
Cemented gravel, cemented sand, clay and gravel	5	Sonoma Volcanics	0 to 3
Silt, clay, sand and gravel, fine sand, quicksand, sand and clay	10	Petaluma Formation	3 to 7
Silt, clay, sand and gravel, fine sand, quicksand, sand and clay	10	Franciscan Complex	<3
Coarse sand, loose sand, medium sand	20	Glen Ellen	3 to 7
Gravel, sand and gravel	25	—	—

[1]Specific yield values from Caldwell, 1958.

[2]Specific yield values from Herbst and others, 1982.

As such, well data provide sufficient lithologic information to enable a useful semi-quantitative description of the aquifer system in the study area to a depth of about 1,500 ft.

5,127 drillers' reports for water wells drilled in the study area provided data on well tests, including discharge rate, water-level drawdown, and the length of test. These data allow the calculation of specific capacity (fig. 27A), which is a measure of well productivity and is given in terms of gallons per minute per foot of drawdown (gpm/ft). These wells were widely distributed across the valley floor and included a more sparse distribution of wells in the mountainous areas. The range of specific capacities is from 0.01 to 1,990 gpm/ft; however, only 8 values exceeded 100 gpm/ft, and only 161 exceeded 10 gpm/ft (fig. 27B). The median value of specific capacity for the group of 5,127 wells was 0.67 gpm/ft (fig. 27A), which equates to a transmissivity of about 130 square feet per day (ft^2/d; Driscoll, 1986). The predominantly low specific capacity for wells in the study area can be, in part, explained by the prevalence of fine-grained materials present as thick beds and as matrix within coarser-grained materials (fig. 25). In addition, values for discharge rate and water-level drawdown were obtained from poorly-controlled and short-duration tests of wells shortly after drilling; such tests typically result in lower specific-capacity values than those obtained from more rigorous aquifers tests. For most of the wells, the material penetrated in the subsurface

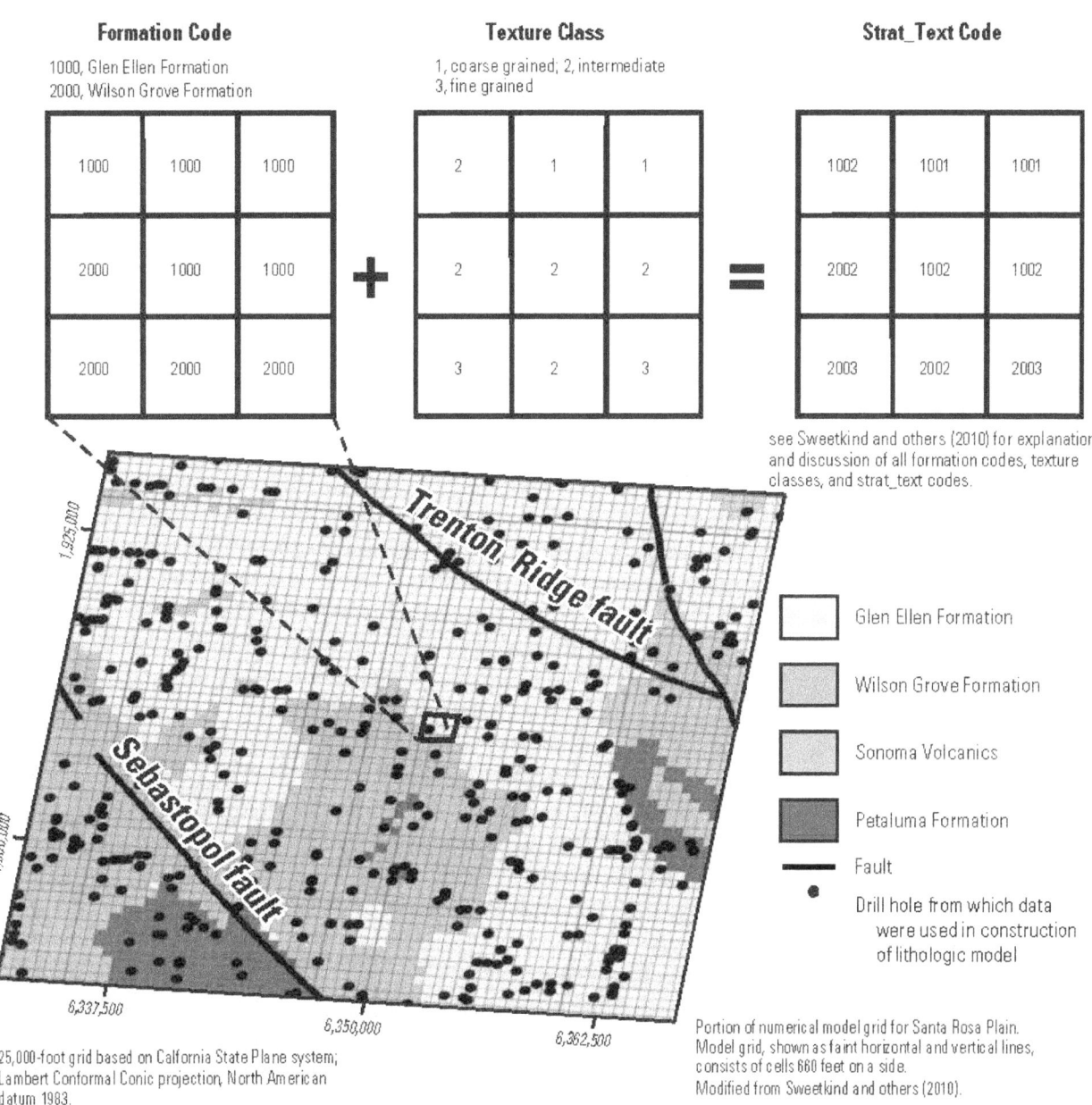

Figure 26.   Assignment of sediment texture attributes, Santa Rosa Plain watershed, Sonoma County, California.

could not be confidently assigned to particular geologic formations. For a large percentage of the wells, it is likely, based
on the well depth, that the wells were completed in more than
one formation, resulting in a total transmissivity value that is
controlled by coarse-grained deposits.

The most reliable well-based estimation of hydraulic
properties in the study area comes from the limited number of aquifer tests that followed established procedures,
most of which are large-capacity public-supply wells used
for municipal water systems. The results of the pumping
tests for 14 wells drilled in the Quaternary alluvial deposits, Wilson Grove Formation, and Glen Ellen formation are
given in table 5, and the locations are shown in figure 28. The
pumping-test results are included with the discussion of the
hydraulic properties of the main water-bearing aquifer units of
the SRPW.

Quaternary Alluvial Deposits

The generally coarse Quaternary alluvial deposits, and
their close proximity to modern streams, allow for rapid
recharge of precipitation and runoff to the groundwater system and exchanges between groundwater and surface water.
Groundwater is unconfined in most places within the alluvial
deposits, but semi-confined conditions exist beneath thick layers of clay or silt in some areas.

The alluvial deposits compose minor aquifers of limited
areal extent along major streams and beneath the alluvial fans
on the eastern side of the SRP. The alluvial deposits that blan-
ket the valley floor are generally poorly sorted and have large
fractions of clay, and are not considered to be a major aquifer.
Within the study area, yields from wells that are completed
only in alluvial deposits ranged from 1 to 650 gpm (Cardwell,
1958). The highest yields were from wells on alluvial fans in
the northern part of the study area near Mark West Creek.

Reported specific yield values of the alluvial deposits ranged from 8 to 17 percent (Herbst and others, 1982).
This range is higher than the range of specific-yield values that
Cardwell (1958) estimated—between 6.8 and 11 percent—for
the upper 200 ft of material on the basis of drillers' logs for
over 900 wells in the SRPW. The two ranges of specific-yield
values are not directly comparable because the 200-ft depth
range used by Cardwell could, in places, include unspecified
thicknesses of underlying formations that have lower specific
yield values than the Quaternary alluvial deposits.

The hydraulic conductivity values from the two available
aquifer tests range from 2 to about 51 feet per day (ft/d), and
storativity values range from about 0.0013 to 0.19 (table 5).
The large range of hydraulic properties is consistent with the
lithologic heterogeneity and varying degree of confinement of
alluvial fan deposits.

Glen Ellen Formation

Previous reports describing the Glen Ellen Formation
(for example, Cardwell, 1958, and Ford, 1975) reported that it
is as thick as 3,000 ft; however, Sweetkind and others (2010)

reported that formation is a few hundred feet thick. Therefore,
well yields, specific-capacity data, and hydraulic parameters
reported by previous authors are to be regarded with caution.
The thickness of the Glen Ellen Formation reported by Sweetkind and others (2010) is used in this report.

Ford (1975) reported that wells perforated in the Glen
Ellen Formation typically yield 15 to 30 gpm. Most wells
in which the Glen Ellen Formation is the principal water-
bearing unit have specific capacities of 10 gpm/ft or less;
the highest specific capacity is 30 gpm/ft in a well drilled in
the northern part of the study area on the fan between Windsor Creek and Mark West Creek. On the basis of the specific
capacity of selected wells, and the saturated thickness of Glen
Ellen Formation they penetrated, Cardwell (1958) estimated
the hydraulic conductivity of the upper part of the Glen
Ellen Formation to be in the range of 13–23 ft/d. Kadir and
McGuire (1987) reported the results of a pumping tests on a
well (8N/9W-12Q2) perforated in the Glen Ellen Formation in
which the hydraulic conductivity was estimated as 5 ft/d. The
upper part of the Glen Ellen Formation, although finer grained
than the deeper part, is less compacted or cemented and, therefore, has higher hydraulic-conductivity values. The upper part
of the Glen Ellen Formation is within the depth range common for wells located on the alluvial fans along the east side
of the SRP. In the eastern part of the SRP, in Rincon Valley
and the lower end of Bennett Valley, the Glen Ellen Formation
has been deformed by folding and faulting (Cardwell, 1958;
McLaughlin and others, 2008). In these areas, most wells
that pump water from the Glen Ellen Formation do so from
stratigraphically lower units of coarser, but more compacted
and cemented, materials; thus, well yields are correspondingly
lower.

The large amount of clay in the Glen Ellen Formation is
in thin to thick beds of nearly pure clay-sized material; clay is
also common in the matrix of coarser-grained units. The ubiq-
uitous clay-sized material, degree of compaction, and cementation all limit the permeability of the Glen Ellen Formation.
The specific-yield range for the Glen Ellen Formation given
by Herbst and others (1982) is between 3 and 7 percent. The
large amount of clay-sized material in some areas is sufficient
to cause locally confined conditions. In the lowland area east
of Santa Rosa, the water in some wells drilled into the Glen
Ellen Formation stood above the water table or even flowed to
land surface in the 1950s (Cardwell, 1958).

Wilson Grove Formation

The thick, sand-dominated Wilson Grove Formation is
exposed in the low hills west of SRP and also is continuous to
the east for an uncertain distance beneath SRP, where it is concealed by variable thicknesses of alluvial materials and Glen
Ellen Formation and interfingers with the Petaluma Formation
(section B-B' on fig. 2). Numerous wells obtain water from the
Wilson Grove Formation in the western part of the study area.
Cardwell (1958) cites a range of specific capacities between

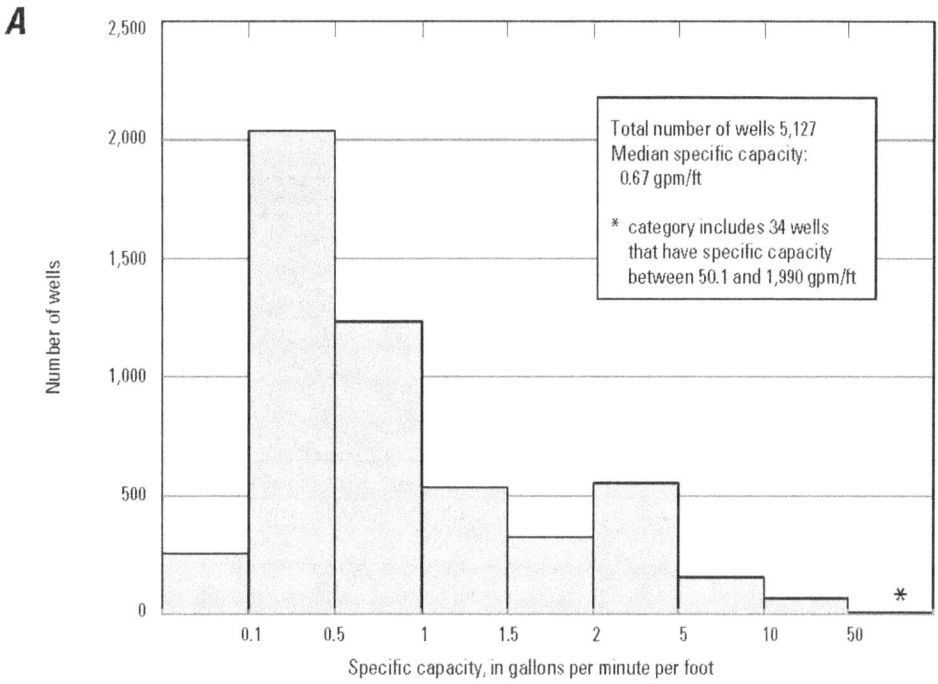

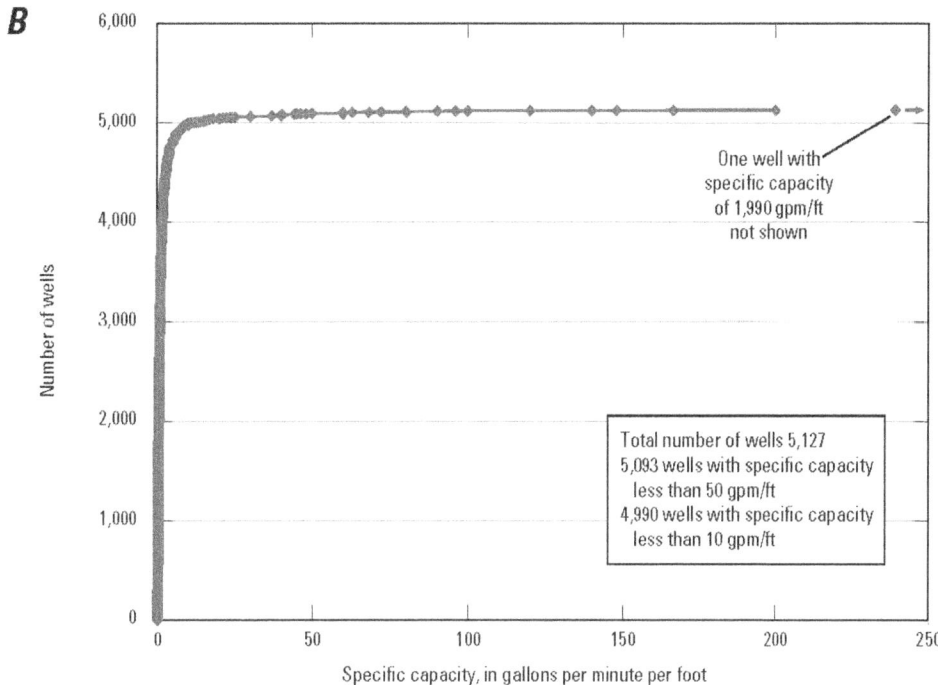

Figure 27. Specific-capacity data for wells in the Santa Rosa Plain watershed, Sonoma County, California: *A*, histogram; *B*, cumulative frequency.

Table 5. Hydraulic properties from aquifer tests, Santa Rosa Plain watershed, Sonoma County, California.

[**Abbreviations**: ft, feet; gpd; gallon per day; gpm, gallon per minute; N/A, not applicable; —, no data]

State well number	Hole depth (ft)	Cased depth (ft)	Top perf (ft)	Bot perf (ft)	Pumping rate (gpm)	Aquifer thickness (ft)	Specific capacity (gpm/ft)	Transmissivity (gpd/ft)	Hydraulic conductivity (ft/d)	Storativity	Notes
Well pumping tests in Quarternary alluvial material											
7N/8W-24A4[1]	1,200	912	295	912	1,000	230	—	72,580–88,000	42.2–51.2	0.0013–0.0017	Analyzed using Jacob method and (or) Hantush and Jacob method.
8N/8W-33K1[2]	400	400	100	400	106	300	7.9	4,825	2	0.19	Analyzed using Cooper-Jacob method.
Well pumping tests in Wilson Grove Formation											
7N/9W-15Q2[1]	642	405	122	394	1,200	[1]500	14.8	26,800	7	0.0011	Analyzed using Jacob method.
7N/9W-25M1[3]	790	770	310	750	2,070	440	17.4	26,100	—	0.00095	N/A.
7N/9W-25E1[1]	1,521	830	410	805	1,200	[1]560	40	24,800	6	N/A	Analyzed using Jacob method.
6N/8W-7A2[1]	815	—	650	800	1,500	[1]160	N/A	10,100	8	0.0015	Analyzed using Hantush and Jacob method.
6N/8W-7A2[1]	815	—	650	800	1,500	[1]160	N/A	15,200	13	0.001	Analyzed using Jacob method.
7N/9W-36K1[3]	1,506	1,065	425	1,045	1,248	—	6.3	9,450	—	—	N/A.
7N/9W-36K1[3]	1,506	1,065	425	1,045	1,650	—	27.5	41,250	—	—	N/A.
7N/9W-36K1[3]	1,506	1,065	425	1,045	1,709	—	13.9	20,850	—	—	N/A.
7N/9W-36K2[3]	1,074	1,040	410	1,020	2,746	[3]900	18.4	27,600	4	0.001–0.08	N/A.
6N/9W-2C1[4]	600	600	332	600	444	268	8.9–25	13,350–37,500	7–19	—	N/A.
6N/9W-2H1[4]	776	530	237	468	925–1,050	231	23.6–75	35,400–112,500	21–65	—	N/A.
6N/9W-2B1[4]	647	528	138	528	850	390	18.5–37.2	27,750–55,800	10–19	—	N/A.
6N/9W-1M3[4]	1,015	572	172	552	1,210–2,017	380	23.2–69.9	34,800–104,850	12–37	—	N/A.
6N/9W-1N_[4]	760	690	270	670	842	400	7–7.8	10,500–11,700	4	—	N/A.
6N/9W-1N_[4]	760	690	270	670	1,220	400	6.3	9,450	3	—	N/A.
Well pumping tests in Glen Ellen Formation											
8N/9W-12Q2[1]	500	500	270	495	200	[1]150	N/A	5,870	5	N/A	Analyzed using Jacob method.

[1]Kadir and McGuire (1987).

[2]Written communication, ENGEO, Inc. (2006).

[3]Written communication, Sonoma County Water Agency (2009).

[4]Written communication, city of Sebastapol (2009).

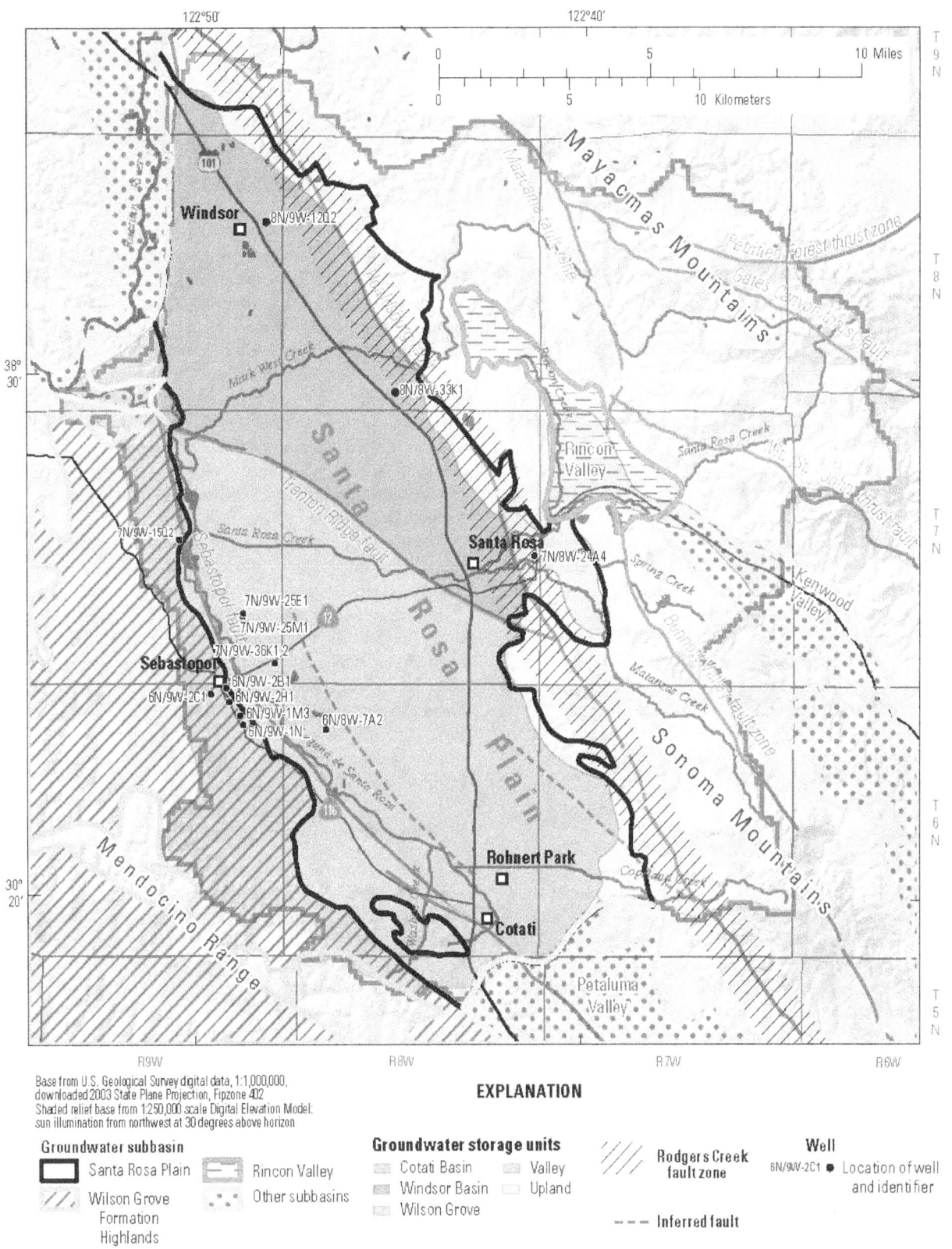

Figure 28. Locations of wells with pumping tests, Santa Rosa Plain watershed, Sonoma County, California.

2 and 29 gpm/ft for the Wilson Grove that were based mostly on short-duration pumping tests. Cardwell's range of values, however, included wells to the west of the study area screened within relatively deep stratigraphic horizons of the Wilson Grove that are finer-grained because they were deposited in deeper water. Within the study area, most wells screened partially or totally in the Wilson Grove Formation are within the upper stratigraphic horizons, which are coarser grained and more permeable than deep deposits to the west. Cardwell (1958) notes that the specific capacity of wells perforated in the lower part of the Wilson Grove is generally 1 gpm/ft or less, whereas wells perforated in the upper part have specific capacities of 5 to 6 gpm/ft. On the basis of short-term pumping tests and the saturated thickness open to wells, Cardwell (1958) estimated the hydraulic conductivity of the Wilson Grove Formation to range from 2 to 33 ft/d, with a mean of about 13 ft/d.

The predominance of relatively clean sand and the low degree of cementation in the Wilson Grove Formation result in moderate to high storativity. Herbst and others (1982) reported the specific yield to be in the range of 10 to 20 percent, higher than any of the other rocks or sediments in the study area.

The hydraulic properties of Wilson Grove Formation have been determined from pumping tests in 11 wells located in the western part of the SRPW, in or near Sebastopol, since the late 1980s (table 5). The range of hydraulic-conductivity values from the 11 wells was from about 3 to 65 ft/d and storativity values ranged from 0.00095 to 0.08. The reason for the difference between the storage values reported by Herbst and others (1982) and the pumping tests in table 5 could be that the values reported Herbst and others (1982) reflected unconfined conditions, while the pumping tests reflected confined conditions.

Petaluma Formation

The Petaluma Formation is dominated by fine-grained materials, either in thick beds or as interstitial material in poorly sorted silty and clayey sands or gravels. Also, the Petaluma Formation is at least 3,000 ft thick in places within the study area, including a large part of the southern SRPW. Even though the formation is dominated by clay, thin, moderately to poorly sorted beds of sands and gravels can be encountered in sufficient quantity by deeper wells that yields greater than 100 gpm are possible. Detailed stratigraphic analyses by Allen (2003) and Holland and others (2009) have identified three distinct members (lower, middle, and upper) in the Petaluma Formation based, in part, on the dominant grain-size and sorting. This stratigraphic subdivision has utility for understanding the variations in hydraulic properties over the full depth range of the aquifer. The lower member is up to 750 ft thick and is predominantly dense beds of mudstone that have the lowest hydraulic conductivity within the formation. The formation coarsens in the 3,500-ft thick middle and upper parts, in which beds of poorly sorted sands and gravels result in increased hydraulic conductivity. In general, the beds

of coarser materials are thin and not of great lateral extent. The productivity of wells drilled in the Petaluma Formation depends mostly on the total thickness of the thin, poorly sorted beds of sand and gravel perforated by the well. In general, the upper member of the Petaluma Formation is the most productive. Domestic wells drilled into the Petaluma Formation yield an average of about 20 gpm and range from 10 to 50 gpm (California Department of Water Resources, 1979).

In the study area, the Petaluma Formation can be thought of as an aquifer of last resort—widely distributed, but with a relatively low productivity that is tapped for water when no better option is available at a particular location. Because of the complicated interfingering stratigraphic relations of the Petaluma Formation with the Wilson Grove Formation and Sonoma Volcanics, some wells can pass from one formation into another more than once. The interfingering of the three formations can also place relatively impermeable lavas or clay beds above more permeable sand or gravel beds, producing confined groundwater conditions. Wells spanning unconfined and confined layers, however, can provide pathways for groundwater to flow between layers that could affect both the hydraulics and water quality of these areas.

Because of the prevalence of silt- and clay-sized particles, specific yields are low, ranging from 3 to 7 percent, and well yields are generally low across the study area (Cardwell, 1958; Swartz and Hague, 2003). At the time of Cardwell's study the Petaluma Formation was thought to be an important aquifer only in the northern part of Petaluma Valley. Estimates of transmissivity based on specific capacities of Rohnert Park municipal wells range from 130 to 1,600 ft²/d (City of Rohnert Park, 2007); however, some of the wells also tap interbeds of Wilson Grove Formation, Sonoma Volcanics, or both (California Department of Water Resources, 1979).

Sonoma Volcanics

The Sonoma Volcanics represent an important aquifer in parts of the SRPW, Sonoma Valley, and Napa Valley. The Sonoma Volcanics comprise a heterogeneous assemblage of lithologic types that have a broad range of hydraulic properties. Lithologies within the Sonoma Volcanics with the lowest specific yields and hydraulic conductivities include unfractured zones in welded tuffs and lavas, thick diatomaceous deposits, and some clay-rich lahar deposits. Hydrothermally altered volcanic rocks are generally rich in clay; although these alteration zones have limited areal extent, they also have low capacity to yield water. Lithologies with the greatest permeability include rubble zones between lava flows, beds of scoria and coarse tephra, air-fall tuffs, and some coarse-grained facies of volcaniclastic units. The rubble zones between lava flows and air-fall tuffs could have wide areal distributions and, therefore, could constitute significant aquifers in some areas. Fractured welded tuffs and lavas have low porosity and, therefore, store little water, but, in some cases, these units have relatively high transmissivity values where the fracture network is extensive.

Water production from wells drilled into thick air-fall pumice units can exceed a few hundred gpm, but wells drilled into unfractured lavas or welded tuffs can produce less than 10 gpm, and dry holes are sometimes encountered. For wells penetrating the Sonoma Volcanics, Ford (1975) gives a range of well yields between 10 and 50 gpm; however, some of the wells penetrate more than one formation, and the relative contributions are unknown.

Ford (1975) also reported specific-capacity values for the Sonoma Volcanics ranging from 0.004 to 26.2 gpm/ft, which equates to a transmissivity range of 0.8 to 5,300 ft²/d. Herbst and others (1982) reported the specific yield of the Sonoma Volcanics to be in the range of 0 to 15 percent (table 4). Results from a 72-hour aquifer test performed in 1975 on a 739-ft deep well located in Bennett Valley provided an estimated transmissivity of 500 ft²/d (Ford, 1975), or an average hydraulic conductivity of 0.68 ft/d.

Basement Rocks

Wells drilled into the basement rocks, which include the Great Valley Group, Franciscan Complex, and Coast Range ophiolite, generally produce small amounts of water (Cardwell, 1958; Kunkel and Upson, 1960; Page, 1986). Previous studies that characterized the basement rocks as non-water-bearing (Cardwell, 1958; Ford, 1975; Herbst and others, 1982; Kadir and McGuire, 1987) did so on the basis of a comparison with the overlying Tertiary and Quaternary formations that generally store, transmit, and yield much greater quantities of water to wells. Most of the permeability in the basement rocks is afforded by fracture networks that have developed in response to folding and faulting. In areas underlain by no aquifers more favorable than the basement, the most productive targets for drilling are highly fractured zones in well-indurated sedimentary rocks within the Great Valley Group (Kl on fig. 1) or the Franciscan Complex (Kjf on fig. 1). Many successful domestic wells of low capacity (5 gpm or less) have been completed in fractured basement rocks in the hills and mountains within the study area.

No water wells have been completed in basement rocks beneath the valley floor that could provide knowledge of the hydraulic properties of these rocks where deeply buried. However, the deeply buried basement rocks are probably more compacted and cemented than those exposed at the surface or residing at shallow depth and, therefore, are less likely to produce significant quantities of water to wells.

Faults and Groundwater Flow

The most important hydrologic aspect of faults in the SRPW is the role they played in the development of the inland valleys and the great depth of some of the now sediment-filled basins within them, especially the SRP and Sonoma Valley. Several faults cross the SRPW and serve as the main boundaries for the sedimentary basins beneath the SRPW. On a local scale, the faults are hydrologically important because they are planar features or zones across or within which the movement

of groundwater can be inhibited or preferentially increased (Heyenkamp and others, 1999). Faulting breaks indurated rocks, producing zones of fractures that increase permeability and can provide preferential paths for groundwater flow. After some length of time, however, the movement of groundwater through fractures can cause chemical weathering and cementation that reduce the permeability and convert the fault plane or zone into a groundwater barrier (Kharaka and others, 1999; Nelson and others, 1999). Faulting in unconsolidated sediments or indurated rocks can produce zones of fine-grained fault gouge with low permeability that act as a groundwater barrier or zone of restricted groundwater flow (Wong and Zhu, 1999). Faults also can displace rocks or sediments in such ways that formations with very different hydraulic properties are made adjacent.

Some faults can allow fluids to move vertically and allow deep waters to move to the surface or into shallow formations (Tamanyu, 1999). For example, hydrothermal systems that are active in Napa, Sonoma, and the SRP tend to have alignments of thermal springs and wells along and near valley-bounding faults (Youngs and others, 1983).

In general, faults are interpreted to form barriers to groundwater movement; however, to date, this has not been shown conclusively for faults in the SRPW. For example, Kadir and McGuire (1987) reported the results of a pumping test using well 6N/8W-7A2 and neighboring piezometers 6N/8W-7A4-6. The authors stated that it could not be determined if the Sebastopol fault was a barrier to flow; however, it was treated as a barrier in a groundwater-flow model developed by the authors. Herbst and others (1982) reported that the Sebastopol fault does act as a barrier to flow. Along any particular fault, the hydraulic characteristics of the materials in the fault zone and the width of the zone can vary considerably, so that a fault can be a barrier along part of its length, but elsewhere allow or even enhance groundwater flow across it.

Groundwater Recharge and Discharge

Sources of groundwater recharge in the SRPW are infiltration of precipitation, streamflow, septic-tank effluent, and irrigation. Groundwater discharges as baseflow in streams, discharge from springs, ET from phreatophytes, and groundwater pumpage. Groundwater recharge to and discharge from the SRPW also can be underflow in the saturated zone across the SRPW boundary, with flows either into or out of neighboring groundwater basins.

Groundwater Recharge

The principal sources of recharge to the groundwater system are direct infiltration of precipitation and infiltration from streams within the SRPW. The 1951 groundwater-level contour map indicates that immediately west of the Rodgers Creek fault zone, in the WB and CB storage units, Santa Rosa Creek could be losing water to (recharging) the groundwater system (fig. 29A).

Herbst and others (1982) reported the average annual recharge flux for the SRPW between 1960 and 1975 was about 29,300 acre-ft. Assuming this estimate is correct, the average annual recharge for the SRPW would be greater than this value because the SRPW includes areas not included in the 1982 estimate: Kenwood Valley, the area west of the Sebastopol Fault, and the mountains that border the Santa Rosa Plain.

Other potential sources of groundwater recharge include underflow from the neighboring Petaluma area, Healdsburg area, and Wilson Grove Formation Highlands groundwater basins and subbasin (fig. 28). Woolfenden and others (2011) estimated groundwater underflow into the SRPW by using a preliminary integrated hydrologic model of the study area. Total estimated average annual groundwater underflow into the SRPW ranged from about 1,100 to 1,300 acre-ft (Woolfenden and others, 2011).

Infiltration from septic tanks, leaking water-supply pipes, irrigation water in excess of crop requirements, and crop frost-protection applications are assumed to be minor sources of recharge. Although recharge from excess irrigation sometimes can be a significant part of total recharge within some basins, within this study area, it is assumed to be minor.

Groundwater Discharge

Groundwater discharges as baseflow in streams in parts of the SRPW. The 1951 groundwater-level contour map indicates that east of the Rodgers Creek fault zone in the UPL and VAL storage units, the Santa Rosa, Spring, and Matanzas Creeks are receiving water from the groundwater system (gaining streams). About 5 mi west of the Rodgers Creek fault zone, water-level contour lines again indicate that Santa Rosa Creek is a gaining stream in the western part of the SRP (fig. 29A).

According to USGS 7.5 minute topographic maps and CDWR records, there are 28 mapped springs in the study area, only 3 of which are named (fig. 30). The mapped springs include both current and historical springs; some of the historical springs possibly are not currently (2012) flowing or no longer exist as a result of anthropogenic land-surface modifications. In addition to the mapped springs, Cardwell (1958) noted that groundwater discharged from the Wilson Grove Formation on the west side of the SRPW through springs and seeps. Most of the springs in the study area are gravity springs found on the steeper slopes or in gullies, where the water table intersects the land surface; however, some are contact springs found along the outcrop of the contact between a permeable and a low-permeability bed. Contact springs are relatively common in the Sonoma Volcanics and Glen Ellen Formation (Cardwell, 1958). Groundwater discharged from springs is a source of baseflow for streams or is lost to ET. Springs are sensitive to changes in groundwater levels caused by natural variations in climate or by the development of groundwater resources, and their flows often diminish or stop during dry periods or when nearby wells are pumped. Detailed inventories of springs in the study area are not available to substantiate any long-term discharge decreases in response to pumping from the large number of wells that have been installed during the past 60 years.

Cardwell (1958) estimated that the ET discharge from the Laguna de Santa Rosa ranged from 4,000 to 6,000 acre-ft/yr. Cardwell (1958) did not estimate the baseflow contribution to the Laguna de Santa Rosa. Groundwater discharge to the Laguna de Santa Rosa in excess of that used by plants or lost to the atmosphere by evaporation flows to the lower reach of Mark West Creek, which flows out of the study area.

There could be groundwater underflow out of the SRPW to the Healdsburg area groundwater subbasin, the Wilson Grove Formation Highlands groundwater basin, and Petaluma Valley groundwater basin. Woolfenden and others (2011) estimated groundwater underflow out of the SRPW by using a preliminary integrated hydrologic model of the study area. Total estimated average annual groundwater underflow out of the SRPW ranged from about 75 to 200 acre-ft (Woolfenden and others, 2011).

The first drilled wells were completed in the study area around 1875 (Cardwell, 1958). Cardwell estimated that, by 1951, there were about 8,500 wells in the SRP. In the 1950s, the water for municipal supply in the cities of Sebastopol and Cotati was obtained solely from groundwater. During this period, the city of Santa Rosa derived its supply from both groundwater and surface water. Cardwell (1958) estimated that in 1949 about 2,800 acre-ft of groundwater were used for public supply in the Santa Rosa Valley. Except for relatively minor amounts of surface water used on fields near Mark West Creek, Santa Rosa Creek, and the Laguna de Santa Rosa, almost all irrigation utilized groundwater (Cardwell, 1958). Cardwell (1958) estimated that in 1949 about 3,400 acre-ft of groundwater were used for irrigation in the Santa Rosa Valley, Bennett Valley, Rincon Valley, and Kenwood area.

As of 2010, groundwater was being used primarily for domestic, public-supply, and agricultural water supply in the study area. Pumping from private domestic and agricultural wells was not reported. Public-supply wells are concentrated near Rohnert Park and Sebastopol (fig. 31). The total annual pumping from public-supply wells for water years 1975–2010 ranged from about 3,900 acre-ft in water year 1975 to about 10,100 acre-ft in water year 2001 (fig. 32). Hevesi and others (2011) estimated agricultural pumpage for water years 1974–2009 by using a calibrated watershed model of the study area with land-use data and monthly crop coefficients. Daily-irrigation demand was estimated and used to estimate the spatial and temporal distribution of average monthly agricultural pumping for 1,072 agricultural wells. Total estimated agricultural water demand ranged from 9,000 acre-ft in water year 1974 to 46,600 acre-ft in water year 2008 (Hevesi and others, 2011). Domestic pumpage was estimated for 1974–2010 by using population density and census tracts in rural areas, and an assumed per capita consumptive use factor of 0.19 acre-ft per person (California Department of Water Resources, 1994). Annual domestic pumpage estimates for water years 1975–2010 ranged from 12,100 in water year 1975

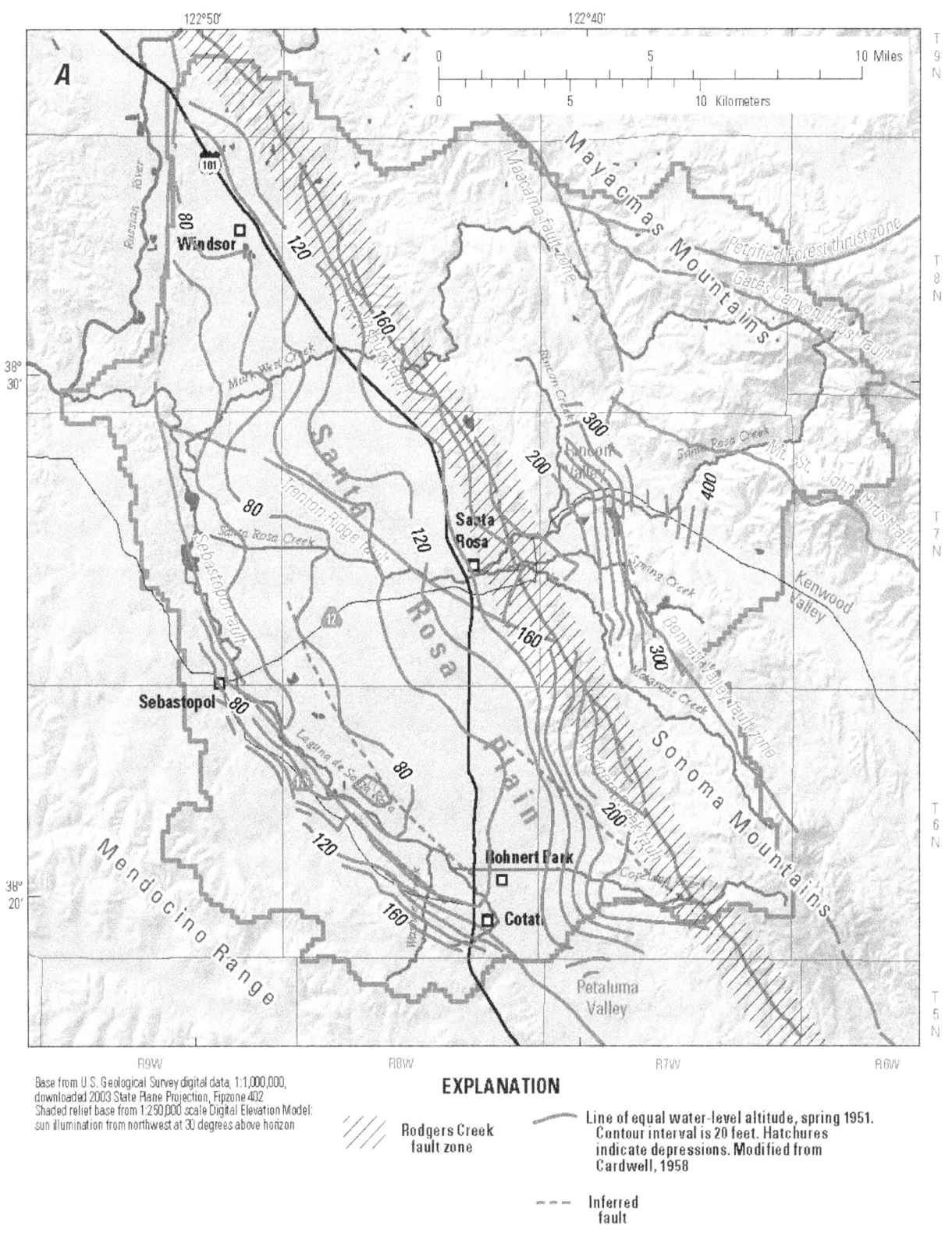

Figure 29.   Groundwater-level contours for spring or fall, Santa Rosa Plain watershed, Sonoma County, California: *A,* spring 1951; *B,* spring 1974; *C,* spring 1980; *D,* fall 1980; *E,* spring 1990;. *F,* fall 1990; *G,* fall 2001;. *H,* spring 2007; and *I,* fall 2007.

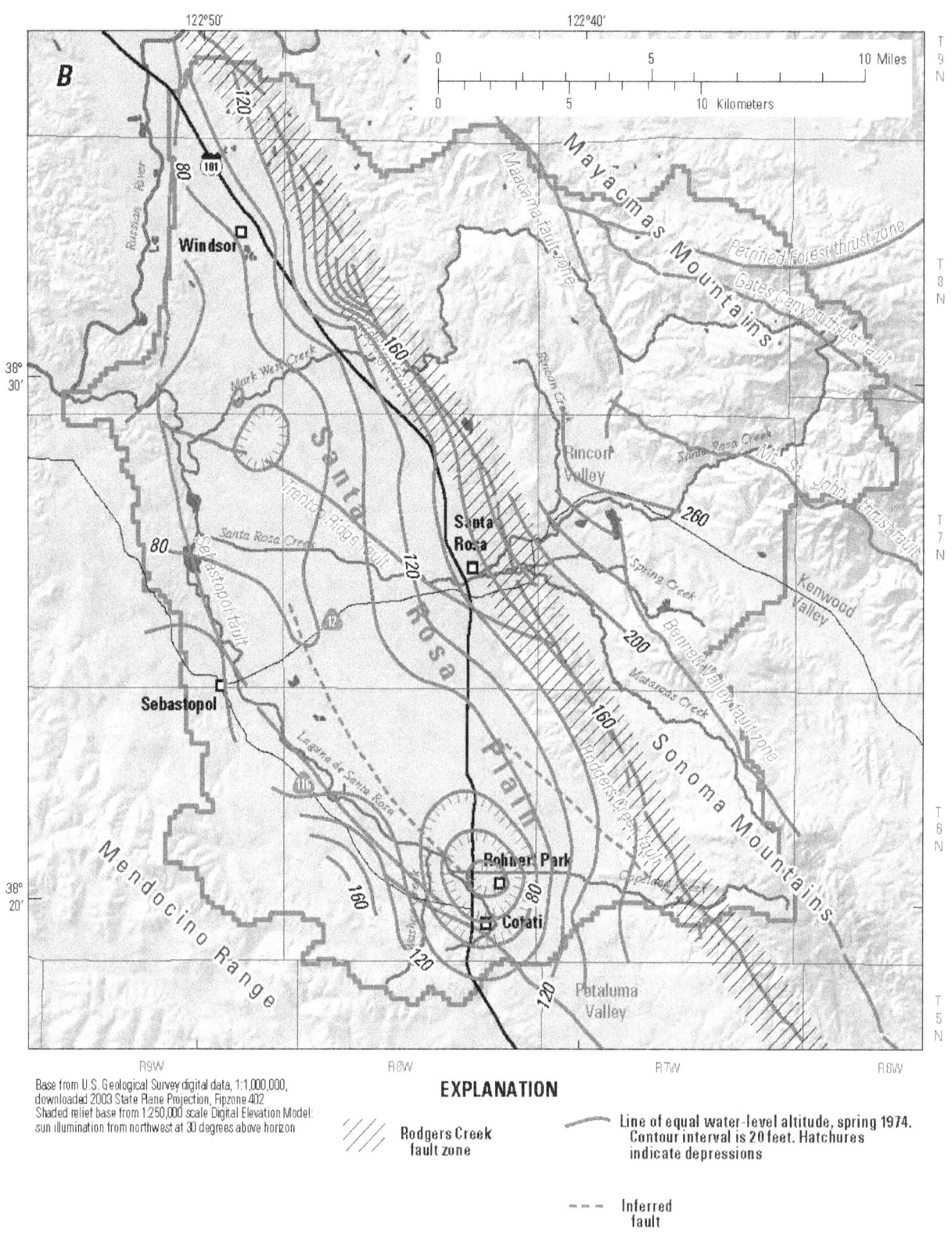

Base from U.S. Geological Survey digital data, 1:1,000,000,
downloaded 2003 State Plane Projection, Fipzone 402
Shaded relief base from 1:250,000 scale Digital Elevation Model;
sun illumination from northwest at 30 degrees above horizon

EXPLANATION

///// Rodgers Creek
fault zone

Line of equal water-level altitude, spring 1974.
Contour interval is 20 feet. Hatchures
indicate depressions

– – – Inferred
fault

Figure 29. Continued.

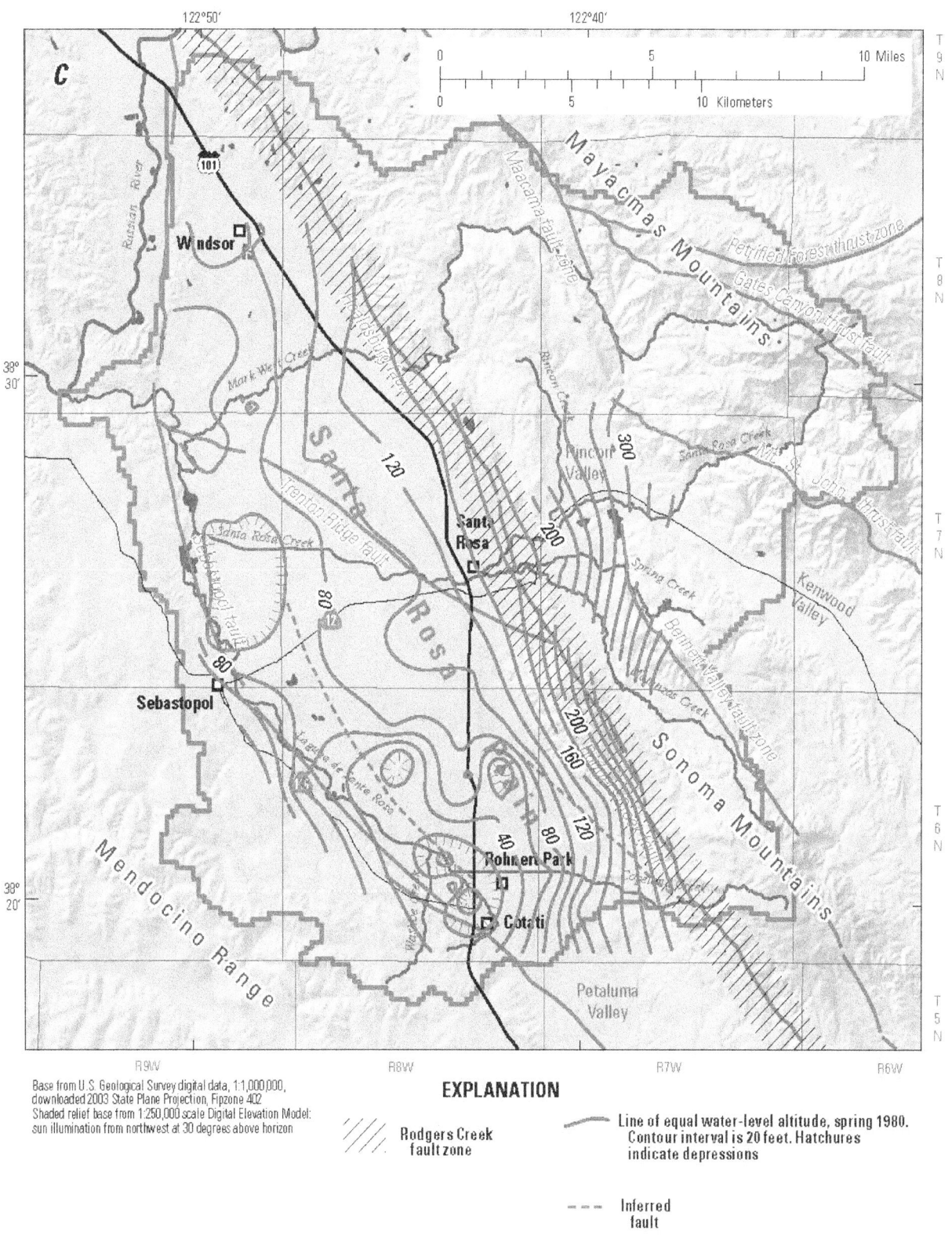

Base from U.S. Geological Survey digital data, 1:1,000,000,
downloaded 2003 State Plane Projection, Fipzone 402
Shaded relief base from 1:250,000 scale Digital Elevation Model:
sun illumination from northwest at 30 degrees above horizon

EXPLANATION

///// Rodgers Creek
       fault zone

——— Line of equal water-level altitude, spring 1980.
       Contour interval is 20 feet. Hatchures
       indicate depressions

- - - Inferred
       fault

Figure 29.   Continued.

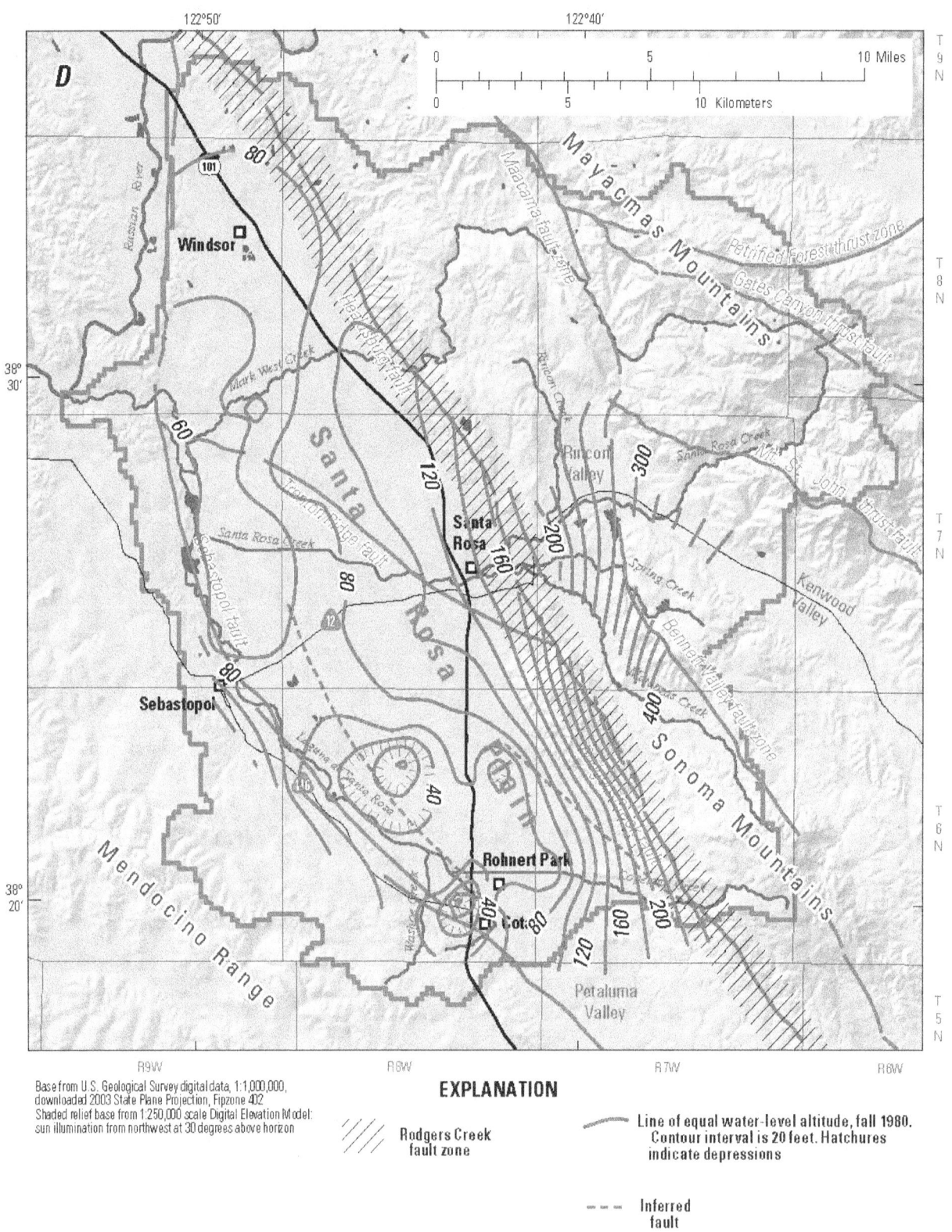

Base from U.S. Geological Survey digital data, 1:1,000,000,
downloaded 2003 State Plane Projection, Fipzone 402
Shaded relief base from 1:250,000 scale Digital Elevation Model:
sun illumination from northwest at 30 degrees above horizon

EXPLANATION

///// Rodgers Creek
     fault zone

Line of equal water-level altitude, fall 1980.
Contour interval is 20 feet. Hatchures
indicate depressions

--- Inferred
    fault

Figure 29.   Continued.

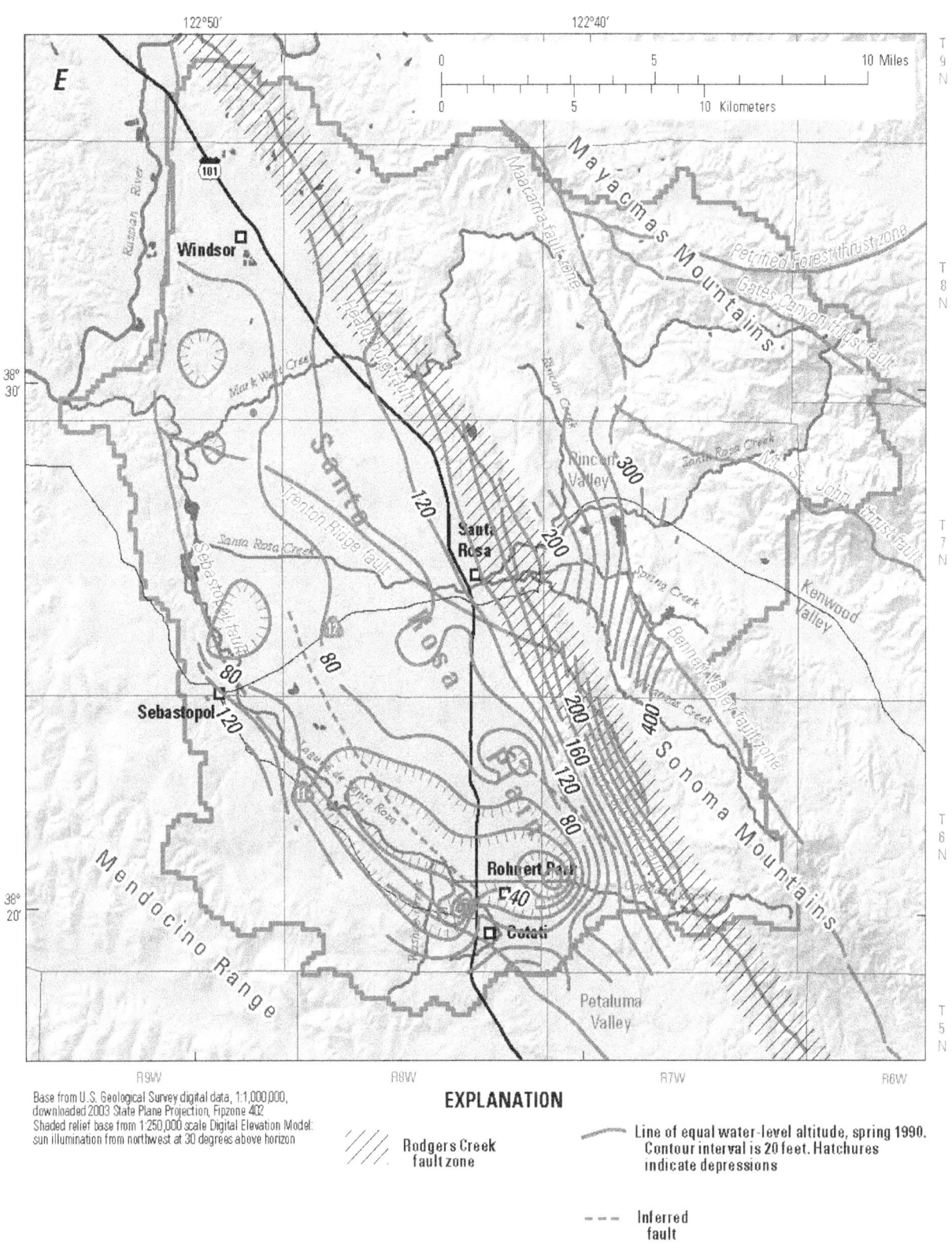

Base from U.S. Geological Survey digital data, 1:1,000,000,
downloaded 2003 State Plane Projection, Fipzone 402
Shaded relief base from 1:250,000 scale Digital Elevation Model:
sun illumination from northwest at 30 degrees above horizon

EXPLANATION

///// Rodgers Creek
       fault zone

Line of equal water-level altitude, spring 1990.
Contour interval is 20 feet. Hatchures
indicate depressions

- - -   Inferred
        fault

Figure 29. Continued.

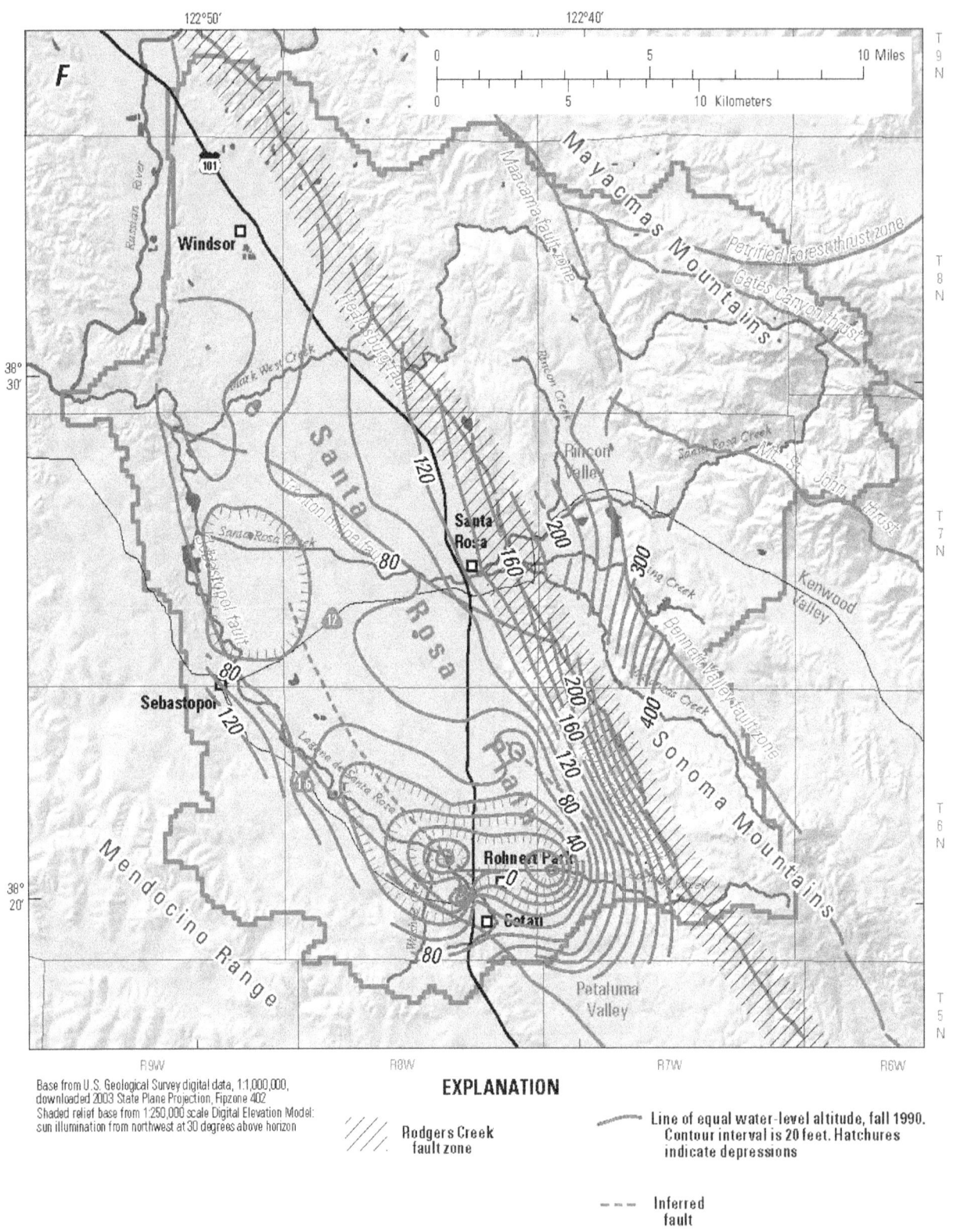

Base from U.S. Geological Survey digital data, 1:1,000,000,
downloaded 2003 State Plane Projection, Fipzone 402
Shaded relief base from 1:250,000 scale Digital Elevation Model:
sun illumination from northwest at 30 degrees above horizon

EXPLANATION

///// Rodgers Creek
///// fault zone

⌒⌒⌒ Line of equal water-level altitude, fall 1990.
Contour interval is 20 feet. Hatchures
indicate depressions

– – – Inferred
fault

Figure 29. Continued.

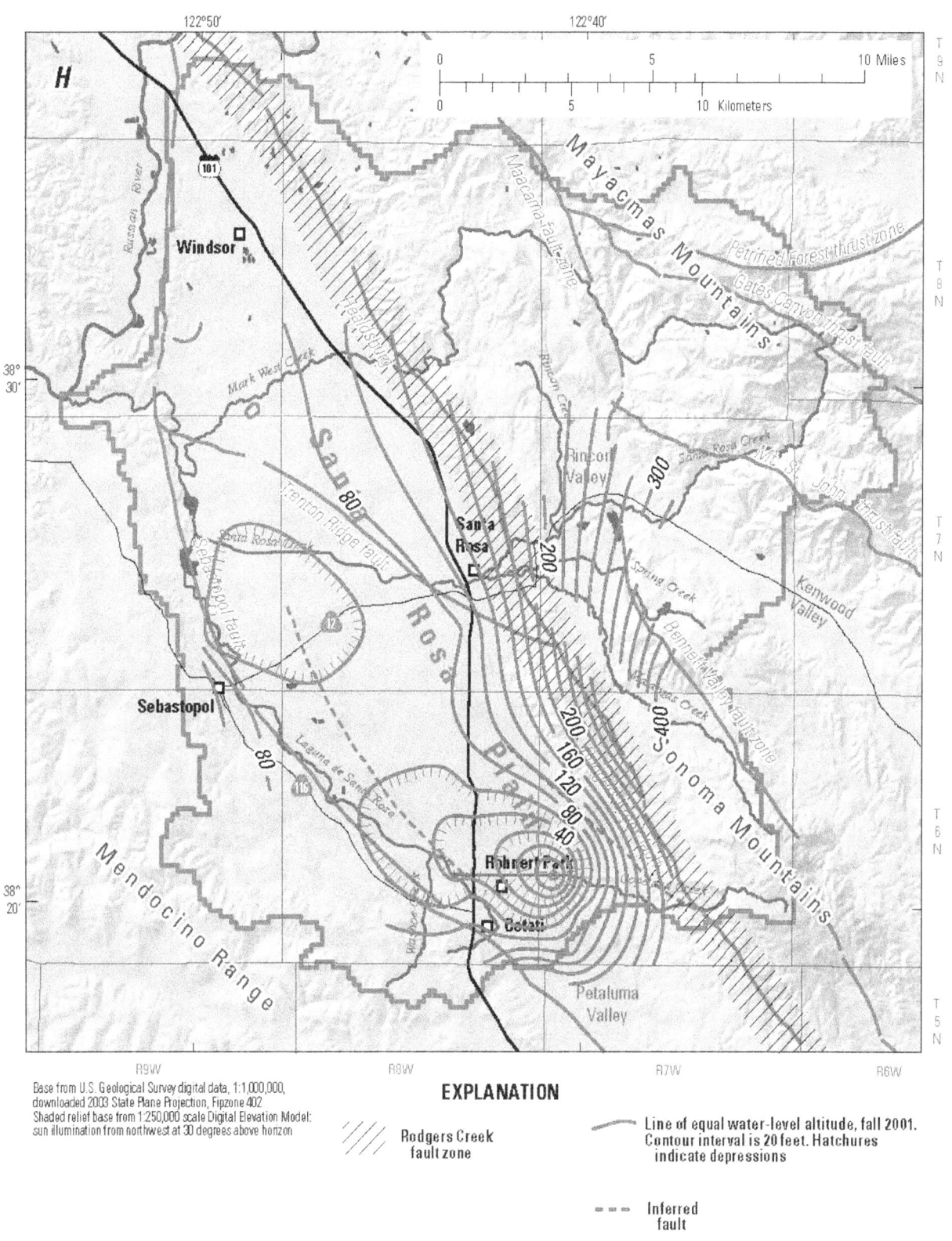

EXPLANATION

///   Rodgers Creek
    fault zone

——   Line of equal water-level altitude, fall 2001.
     Contour interval is 20 feet. Hatchures
     indicate depressions

- - -   Inferred
    fault

Figure 29.   Continued.

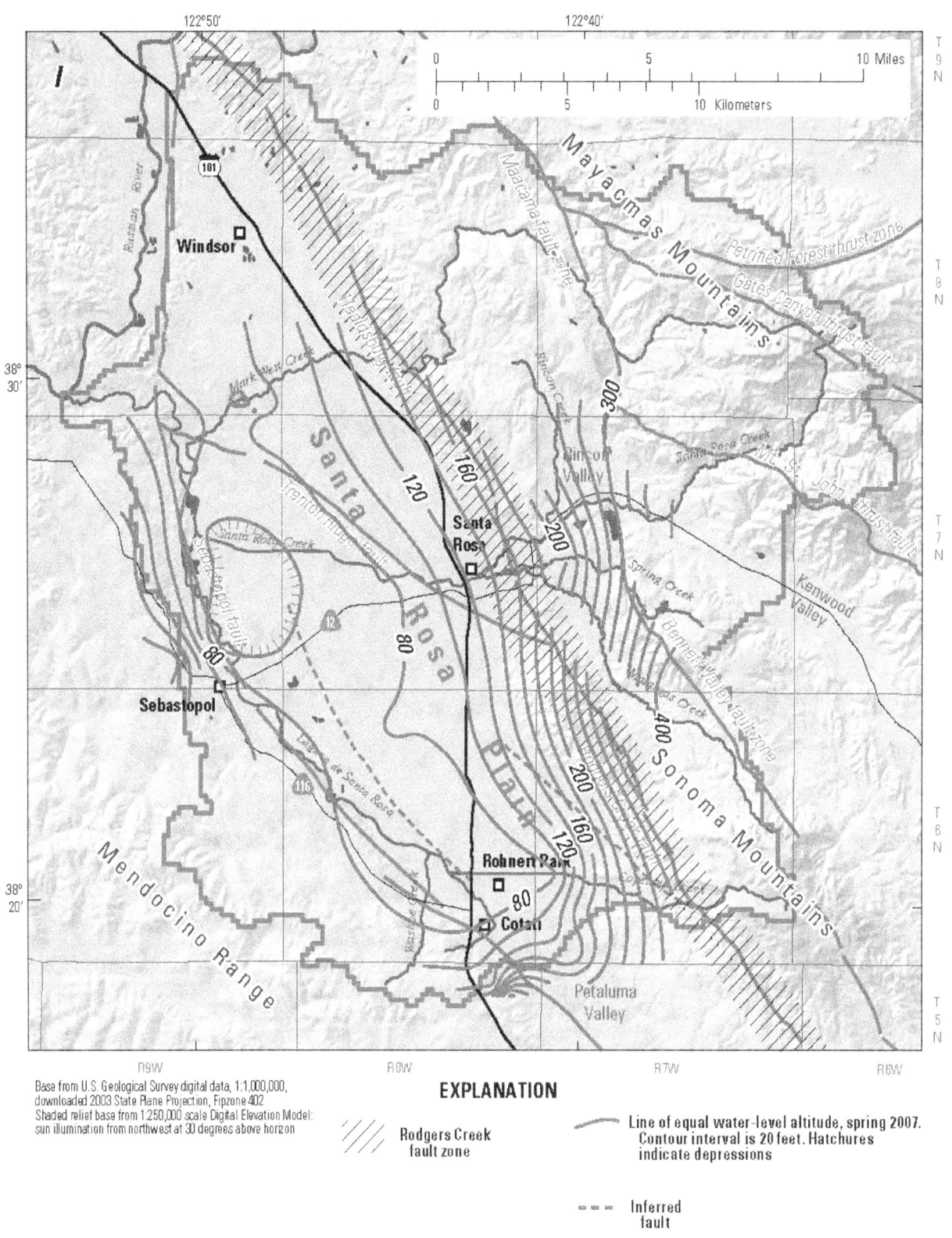

Base from U.S. Geological Survey digital data, 1:1,000,000,
downloaded 2003 State Plane Projection, Fipzone 402
Shaded relief base from 1:250,000 scale Digital Elevation Model:
sun illumination from northwest at 30 degrees above horizon

EXPLANATION

///// Rodgers Creek
       fault zone

~~~~~ Line of equal water-level altitude, spring 2007.
       Contour interval is 20 feet. Hatchures
       indicate depressions

- - - Inferred
       fault

Figure 29.   Continued.

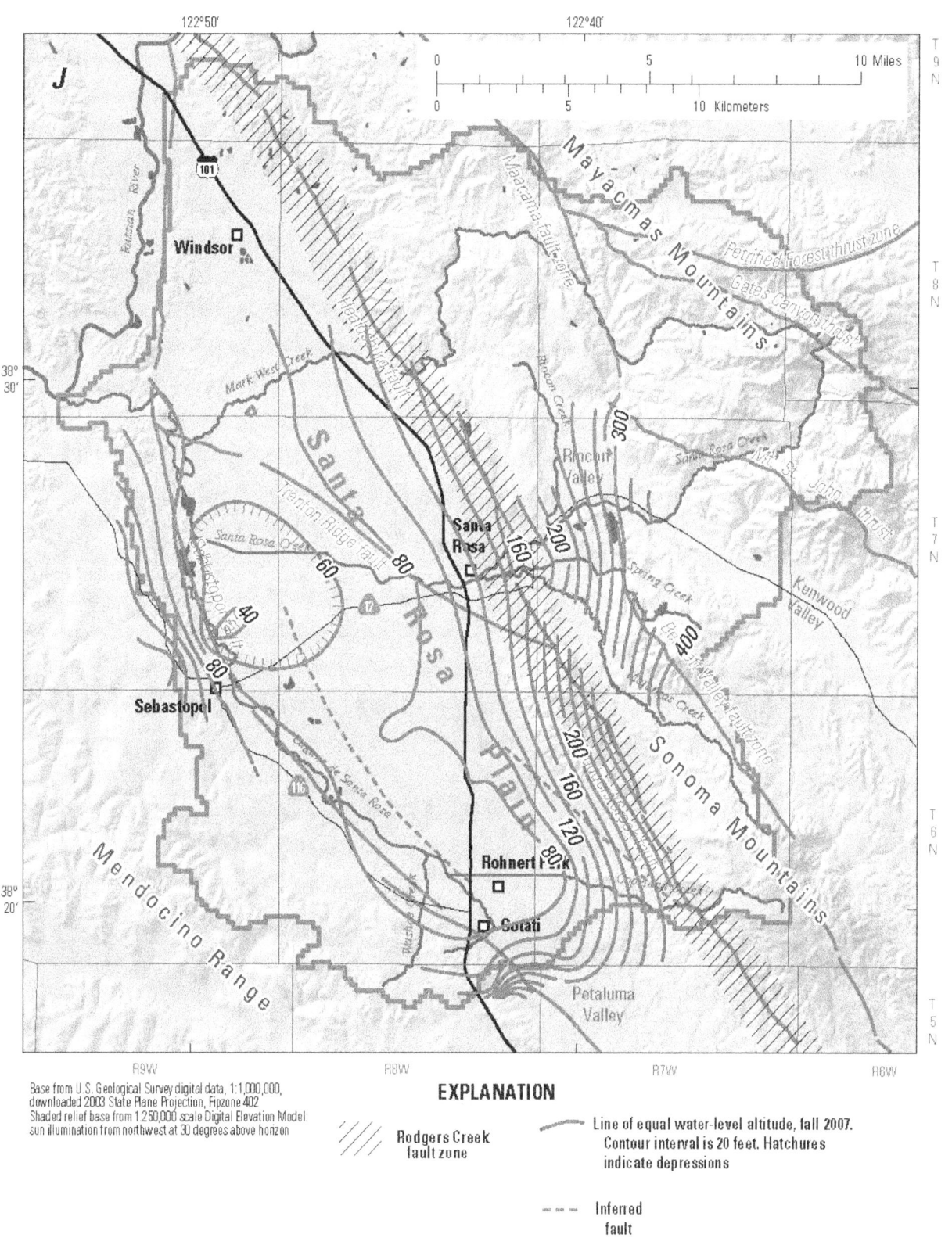

Figure 29. Continued.

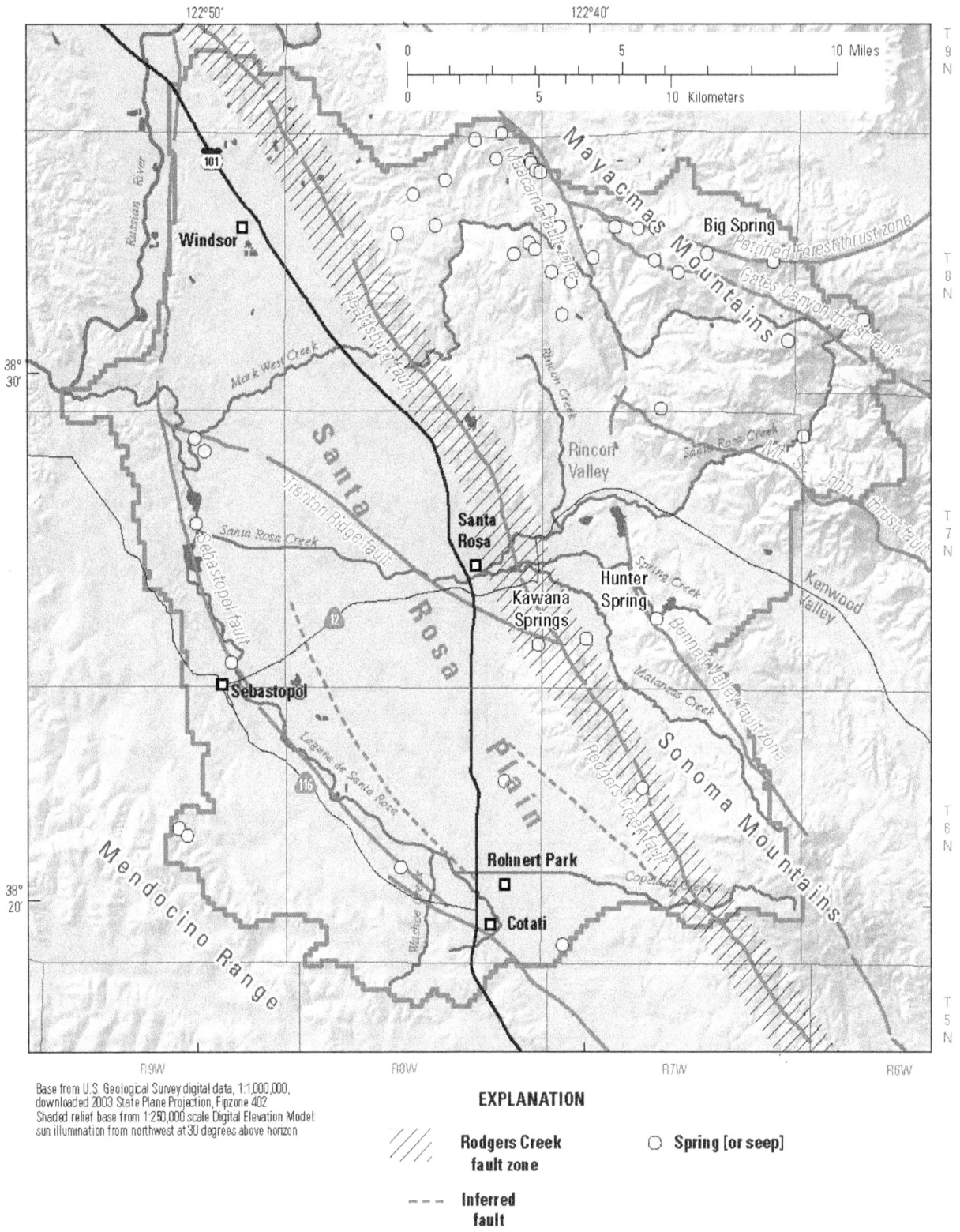

Figure 30. Current and historical springs in the Santa Rosa Plain watershed, Sonoma County, California.

to 23,400 acre-ft in water year 2002. The average total annual pumpage from all sources for water years 1975–2009 was about 47,400 acre-ft. The maximum total annual pumpage was about 78,500 acre-ft in water year 2008.

Groundwater Levels and Movement

Knowledge of groundwater levels and how they vary spatially and temporally is fundamental to understanding and managing water resources. The shape of the water-table and potentiometric surfaces, and combinations thereof, can reveal important hydrogeologic characteristics of the watershed, such as the locations of key areas of recharge and discharge, and geologic effects on groundwater flow. Declining long-term hydrographs can indicate that an aquifer is being pumped in excess of recharge.

Basin-wide Groundwater Levels

Groundwater-level data collected by the USGS, CDWR, SCWA, and the municipalities were used to construct groundwater-level maps of the study area for selected seasons and years to show changes in levels caused by changing patterns of water use and weather conditions for 1951, 1974, 1980, 1990, 2001, and 2007 (fig. 29A–I). In general, the water-level contours indicate that Santa Rosa Creek and tributaries gained water east of the Rodgers Creek fault zone, and Mark West and Santa Rosa creeks gained water in the western part of the SRP. The water-level contours are consistent with Copeland Creek gaining water. However, the water levels are generally 50–230 ft bls; therefore, groundwater could not have been entering the stream.

1951 (Predevelopment) Conditions

The earliest published groundwater-level data that have sufficient geographical distribution to allow an objective representation of groundwater-level altitudes in the study area date from the late 1940s to early 1950s (Cardwell, 1958). Cardwell (1958) depicted groundwater-level contours across the SRPW and adjacent areas for spring of 1951 (reproduced in modified form as figure 29A). The 1951 groundwater-level map was based on data collected before water-resources utilization in the study area had reached an amount that significantly altered water balance for the basin (Cardwell, 1958) and, thus, is an approximation of pre-development conditions. The 1951 groundwater-level map was based on data from about 450 wells that ranged in depth between 24 and 1,048 ft (Cardwell, 1958). Most of these wells were probably perforated through more than one hydrogeologic unit; therefore, the measured water level was a composite of the water levels from each perforated unit. For the spring 1951 data, the depth to water beneath the valley floor in most places was between 5 and 20 ft bls (Cardwell, 1958). As stated earlier, the spring 1951 groundwater-level contour map shows that groundwater

moved toward, and discharged into, stream channels in most of the SRP, likely sustaining baseflow (fig. 29A). In addition, the contour map shows that, on a larger scale, groundwater flowed from the Mayacmas and Sonoma Mountains in the UPL storage unit westward toward the Laguna de Santa Rosa on the western edge of the SRP and eastward from the highlands in the WG storage unit toward the Laguna de Santa Rosa.

Cardwell (1958) identified flowing wells in two parts of the study area: (1) a 10-mi by 2-mi wide area along the Laguna de Santa Rosa and (2) in the VAL storage unit in Bennett Valley (fig. 29A). The flowing wells along and west of the Laguna de Santa Rosa were indicative of groundwater upflow in this area and were consistent with the Laguna de Santa Rosa being a focal point of pre-development groundwater discharge. Most of the flowing wells identified in the eastern part of the study area were located close to mapped fault strands of the Rodgers Creek fault zone. The flowing wells also were within regions identified by Youngs and others (1983) as "known warm water zones." This observation is consistent with deep circulation of groundwater in some parts of the fault zones along the eastern side of SRPW (Cardwell, 1958).

Cardwell (1958) identified two groundwater-flow divides. The first was located between Santa Rosa Creek and Mark West Creek in the northern part of the basin (fig. 29A). The second divide was located about 1 mi southeast of Cotati, where the SRP narrows and terminates to the south (fig. 29A). Groundwater moved southwest and northwest, away from both divides (fig. 29A). The component of groundwater that moved southwest from the southern divide flowed out of the SRPW into Petaluma Valley.

Groundwater-level altitudes beneath the Laguna de Santa Rosa in 1951 were the lowest of anywhere in the study area (fig. 29A). The Laguna de Santa Rosa is a broad area of natural groundwater discharge; this discharge supplied, and still supplies, water for high rates of ET from marshy areas and baseflow to the lower reaches of the Laguna de Santa Rosa and Mark West Creek.

Groundwater levels in the Coast Ranges of California respond to precipitation. In many areas, groundwater levels can be positively correlated with the cumulative precipitation over several years (Faye, 1973; Farrar and Metzger, 2003). Precipitation records collected at Santa Rosa from 1906 to 2010 were obtained from the California Data Exchange Center (*http://cdec.water.ca.gov/*, accessed June 2011) to evaluate the representativeness of the 1951 water levels (fig. 33). The total precipitation for Santa Rosa for water year 1951 (October 1950 to September 1951) was 32.1 in., compared to a mean of 29.8 in. for the period 1906–2010 (fig. 33). Therefore, the amount of precipitation received in the wet season (November to March), prior to the time water levels were measured in 1951, was about 8 percent above the long-term mean. However, precipitation was less than the mean for all the years from 1944 to 1950. The cumulative precipitation deficit for

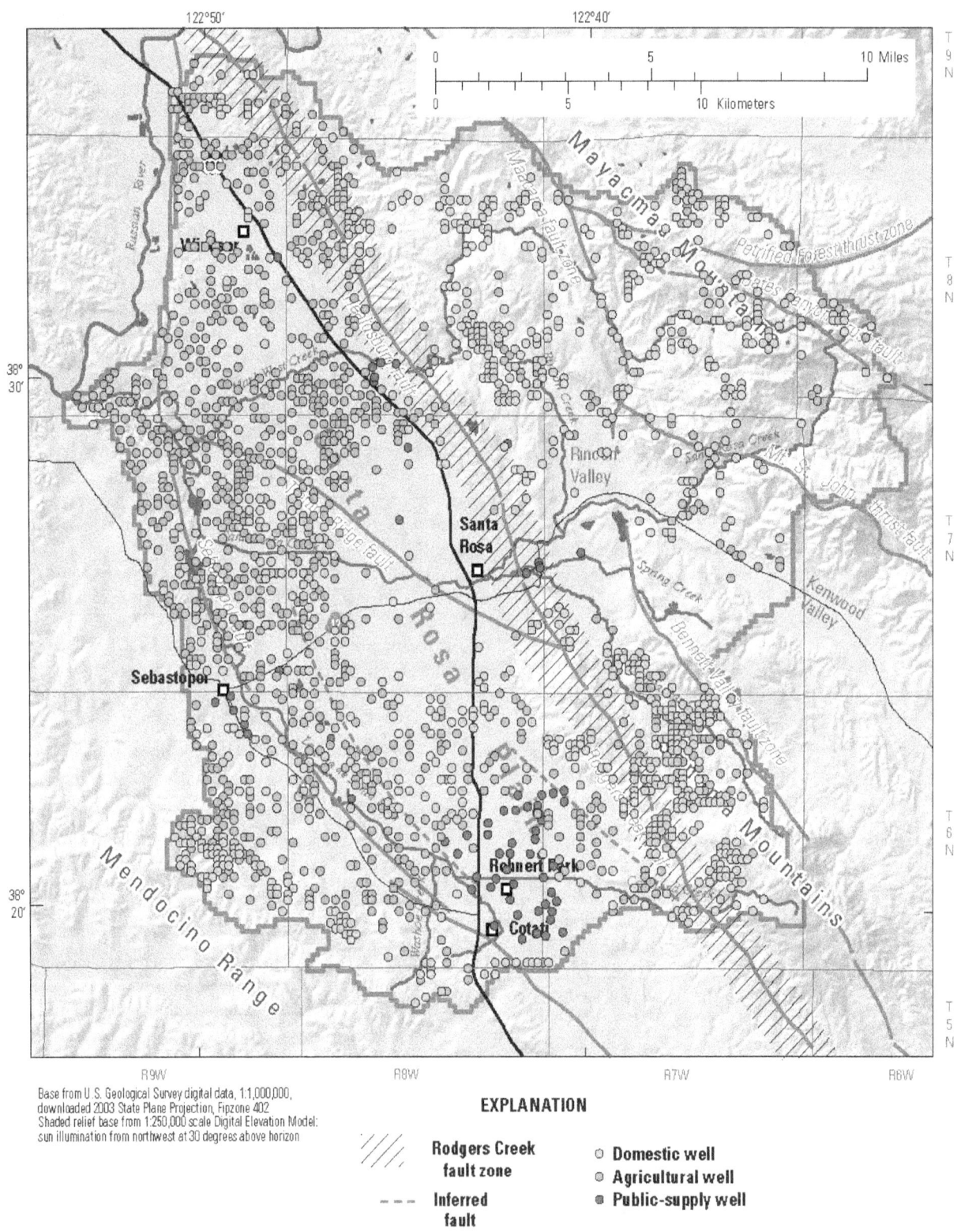

Figure 31.   Location of public-supply wells in the Santa Rosa Plain watershed, Sonoma County, California.

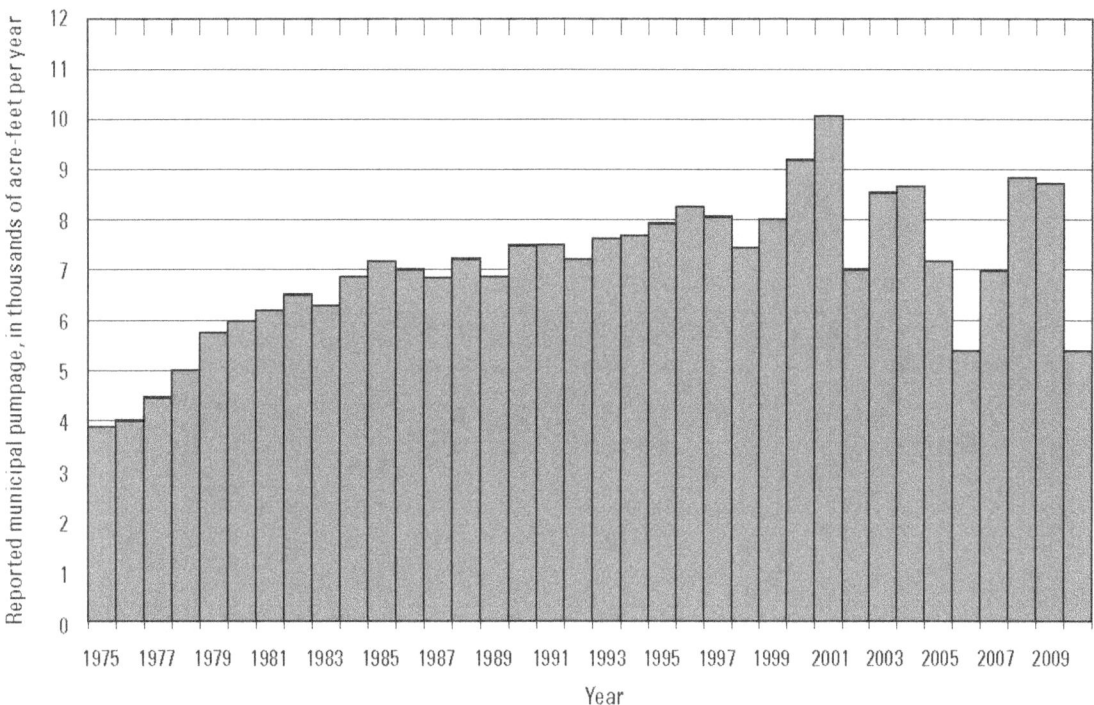

Figure 32. Total annual pumping from public-supply wells in the Santa Rosa Plain watershed, Sonoma County, California, 1975–2009.

that seven-year period was over 41 in., which is equivalent to about 1.4 years of mean precipitation. The 7-year dry period that preceded the 1951 water year probably caused ground-water levels measured in April 1951 to be lower than average, but the overall effect on groundwater levels in the study area is uncertain.

1974 Conditions

The spring 1974 groundwater-level contour map is broadly similar to the 1951 map (figs. 29*A* and *B*). The contours show the dominant direction of groundwater flow was from the east side of the plain toward the Laguna de Santa Rosa. However, the well-defined trough in contours near the Laguna de Santa Rosa, evident in the 1951 map, was not well defined for 1974 because of the fewer number of data points available (fig. 29*B*). Instead, there was a water-level depression in the southern part of the SRP near Rohnert Park-Cotati. Within this area, the data indicated that groundwater levels declined about 60 ft between 1951 and 1974 as a result of pumping. There was another water-level depression located between Mark West Creek and Santa Rosa Creek. This depression could have been caused by agricultural pumping as there are no municipal wells in the area. There were no depressions resulting from municipal pumping in the vicinities of Santa Rosa or Sebastopol. The spring 1974 water-level contours also indicated that Santa Rosa Creek could have lost water in the valleys and SRP (fig. 29*B*). The cause of Santa Rosa Creek becoming a losing stream is unclear, but it could have been caused by groundwater pumping.

1980 Conditions

The spring 1980 groundwater-level contour map indicates pumping depressions in the Rohnert Park-Cotati area and north of Sebastopol (fig. 29*C*). The same general distribution of heads as in 1974 is apparent, but the Laguna de Santa Rosa and Mark West Creek groundwater discharge area is better defined by the 60-ft contours along the west side of the map. The pumping depression in the Rohnert Park-Cotati area, evident in the 1974 map, is defined by the spring 1980 data as a more complex area of drawdown that is 20 ft deeper in places than in 1974 (fig. 29*B* and *C*). The spring 1980 water-level contours indicate that Santa Rosa Creek gained water east of the Rodgers Creek fault zone and that Mark West and Santa Rosa creeks gained water in the western end of the SRP (fig. 29*C*).

The groundwater-level contour map for fall 1980 is in similar shape and distribution to the spring 1980 map (fig. 29*D*); however, the fall data were measured at the end of a pumping and irrigation season that left groundwater levels generally 5 to 20 ft lower than in spring in most of the SRPW. In the discharge area along the Laguna de Santa Rosa, the fall levels were generally less than 10 ft lower than in spring. The pumping depression in the southern part of the area enlarged and deepened between spring and fall. Comparing the groundwater levels of spring 1980 to those of spring 1951, declines of over 40 ft were measured in the Rohnert Park-Cotati area, and the area of decline covered about one quarter of the SRPW. The area of declining groundwater

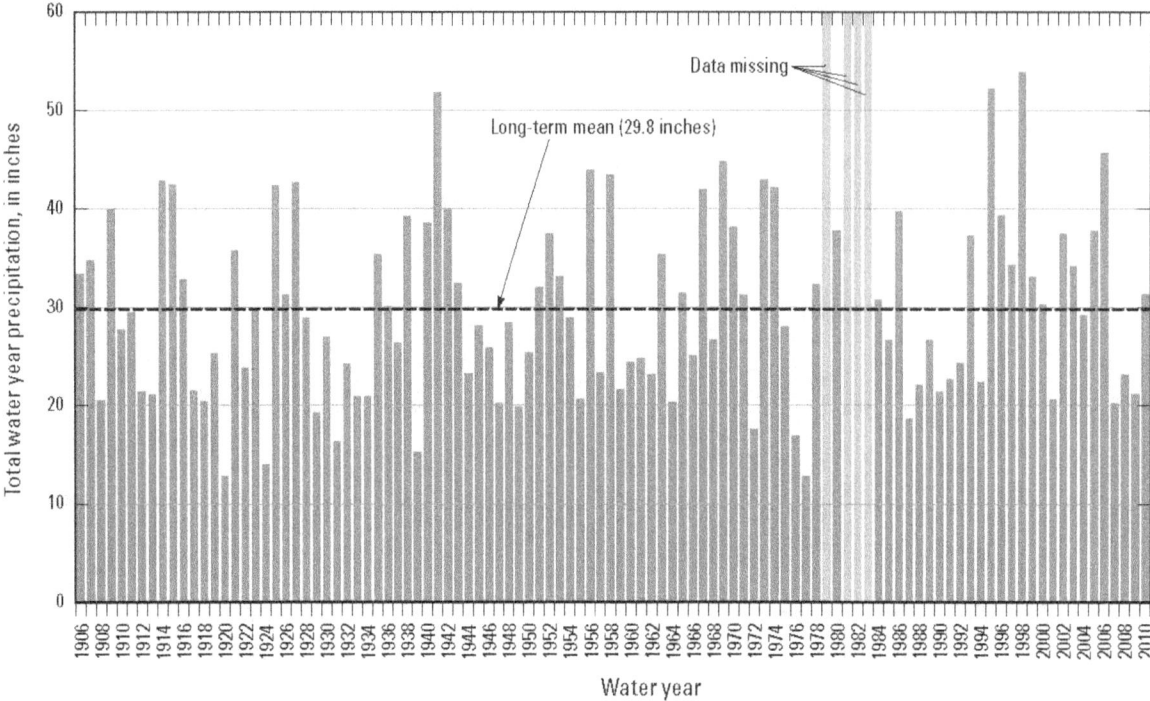

Figure 33.   Total annual precipitation for water-years 1906–2010, Santa Rosa Plain watershed, Sonoma County, California.

levels expanded, despite the slightly above average amounts of precipitation received in 1978 and the 124 percent of the mean received in 1980 (fig. 33). A smaller area of decline can also be defined in the northern region near Windsor. The fall 1980 water-level contours also indicate that Santa Rosa Creek gained water east of the Rodgers Creek fault zone (fig. 29*D*).

1990 Conditions

The groundwater-level contour map for spring 1990 is very similar in shape and position to the spring 1980 map; however, the pumping depression in the southern part of the area is more restricted areally and appeared to have shifted slightly southward (fig. 29*E*). In addition, groundwater levels beneath the Laguna de Santa Rosa area were higher than in 1980. The changes between spring 1980 and spring 1990 indicate there was some reduction in pumping in the Rohnert Park-Cotati area or that recharge during the 1989–90 rainfall season was greater than average. An increase in recharge seems unlikely because only 21.4 in. of precipitation was received in water–year 1990, about 70 percent of the mean, and the prior 3 years were all well below average, with a total precipitation deficit of about 23.7 in., or nearly 80 percent of the annual mean. The pumping depression expanded between spring and fall 1990 and deepened to include a large area where groundwater levels were below sea level (figs. 29*E–F*). The 1990 water-level contours indicate that Santa Rosa Creek gained water east of the Rodgers Creek fault zone (figs. 29*E–F*).

2001 Conditions

The spring 2001 groundwater-level contour map was based on measurements collected in mid-May, well past the rainy season and, presumably, past the peak groundwater levels; therefore, it is not presented. The fall 2001 groundwater-level contour map indicated that groundwater levels were generally lower than in fall 1990, as shown by the position of the 80-ft contour in the central part of the area (fig. 29*G*). The fall 2001 groundwater-level contours indicated that Santa Rosa Creek could be losing water east of the Rodgers Creek fault zone (fig. 29*G*). Water-year 2001 had relatively low precipitation (about 20 in.), which could explain Santa Rosa Creek losing water; however, the magnitude and duration were not equal to the 1976–77 or 1986–92 droughts (fig. 19*C*).

2007 Conditions

A comparison of the contours between spring and fall 2007 revealed a similarity in the general shape and distribution of contours. Groundwater-level contours for spring 2007 indicated that water levels were higher than the spring 2001 levels and the southern pumping depression had disappeared (fig. 29*H*), indicating that long-term reductions in pumping in the Rohnert Park-Cotati area allowed recovery of groundwater levels to altitudes typical of the early 1970s. The 2007 water-level contours indicate that Santa Rosa Creek was gaining water east of the Rodgers Creek fault zone (fig. 29*H–I*).

Water-level Changes in Monitoring Wells

Water-level data collected by the USGS, CDWR, SCWA, and others were used to construct water-level hydrographs for selected wells in the SRPW (fig. 34). Water-level hydrographs display the temporal variation in water levels at a particular location and can be used to refine the conceptualization of the hydrologic system and to calibrate a groundwater-flow model. Information on well construction and geology of the representative aquifer for the hydrograph wells is presented in table 6.

Windsor Basin Storage Unit

Well 7N/8W-24L1 is east of Santa Rosa in the Rodgers Creek fault zone in the southeastern part of the WB storage unit (fig. 34). Well 24L1 is perforated between 160 and 200 ft bls in the Petaluma Formation. Water levels in well 24L1 were fairly steady, with only 10–20 ft seasonal change in water levels (fig. 35A).

Wells 8N/8W-20Q1 and 29B1 are southeast of Windsor in the east-central part of the WB storage unit (fig. 34). Well 20Q1 is 312 ft deep and is perforated in the Glen Ellen and Petaluma Formations (table 6). Well 29B1 is 64 ft deep and is perforated in the Glen Ellen Formation (table 6). The water levels in well 20Q1 were variable and showed declines in the late 1970s and in the early 1990s (fig. 35B). These declines could be responses to the 1976–77 and 1987–92 droughts (fig. 19C). The water levels in well 29B1 were fairly steady but showed seasonal variability (fig. 35C).

Well 8N/9W-13A2 is just east of Windsor in the northeastern part of the WB storage unit (fig. 34). Well 13A2 is 109 ft deep and is perforated in the Glen Ellen Formation. The water levels varied over time; however, they generally increased about 30 ft between the late 1980s and the late 1990s (fig. 35D). The variability in the water levels could be a response to climate in addition to seasonal pumping fluctuations. For example, the water-level low in 1977 corresponds to the drought in 1977 (fig. 19C). However, effects of the 1987–92 drought is not evident in the water-level record.

Well 8N/9W-22R1 is southwest of Windsor in the western part of the WB storage unit (fig. 34) and is perforated between 122 and 142 ft bls in the Wilson Grove Formation. The water levels were fairly steady, but increased about a net 10 ft between the early 1980s and 2010 (fig. 35E).

Well 8N/9W-36P1 is south of Windsor in the southwestern part of the WB storage unit (fig. 34) and is perforated between 711 and 1,010 ft bls in the Petaluma Formation. Overall, water levels were steady; however, the data between 1978 and 1990 are sparse (fig. 35F). The water-level decline between 1976 and 1978 could be a response to the 1976–77 drought.

Table 6. Construction information of hydrograph wells, Santa Rosa Plain watershed, Sonoma County, California.

[**Abbreviations**: ft, feet; LSD, land surface datum; P, Petaluma Formation; QGE, Quaternary deposits and Glen Ellen Formation, undivided; SV, Sonoma Volcanics; WG, Wilson Grove Formation]

State well identifier	LSD (ft)	Well depth (ft)	Top perforation (ft)	Bottom perforation (ft)	Formation
Hydrograph wells located in the Windsor Basin storage unit					
7N/8W-24L01	185	200	160	200	P
8N/8W-20Q01	141	312	56	310	QGE and P
8N/8W-29B01	143	64	52	64	QGE
8N/9W-13A02	124	109	87	109	QGE
8N/9W-22R01	72	142	122	142	WG
8N/9W-36P01	94	1,010	711	1,010	P
Hydrograph wells located in the Cotati Basin storage unit					
6N/8W-15J03	93	166	65	90	QGE
6N/8W-23H01	100	1,500	300	1,500	P
6N/8W-25C01	105	[1]507	144	[1]490	P
6N/8W-26A01	101	462	288	462	P
6N/8W-26L01	99	94	54	94	QGE
6N/8W-27H01	95	82	62	82	QGE
7N/8W-21J01	124	360	148	360	P
7N/8W-35K01	128	206	185	205	P
Hydrograph wells located in the Valley storage unit					
7N/7W-06H02	303	100	60	80	QGE
7N/7W-09P01	381	296	286	296	SV
7N/7W-19B01	204	85	45	85	QGE
Hydrograph wells located in the Wilson Grove storage unit					
6N/9W-02C01	161	600	332	600	WG
7N/9W-15K01	71	70	59	69	WG
7N/9W-35D02	120	167	55	167	WG
Hydrograph wells located in the Uplands storage unit					
6N/7W-03D01	535	254	234	253	P
6N/7W-03M01	484	61	41	61	P
Muliple-level monitoring site					
6N/8W-07A04	78	80	60	80	QGE
6N/8W-07A05	78	257	237	257	P
6N/8W-07A06	78	570	550	570	WG

[1]Bottom 131 ft cemented in some time between 1985 and 2002. New depth is 376 ft.

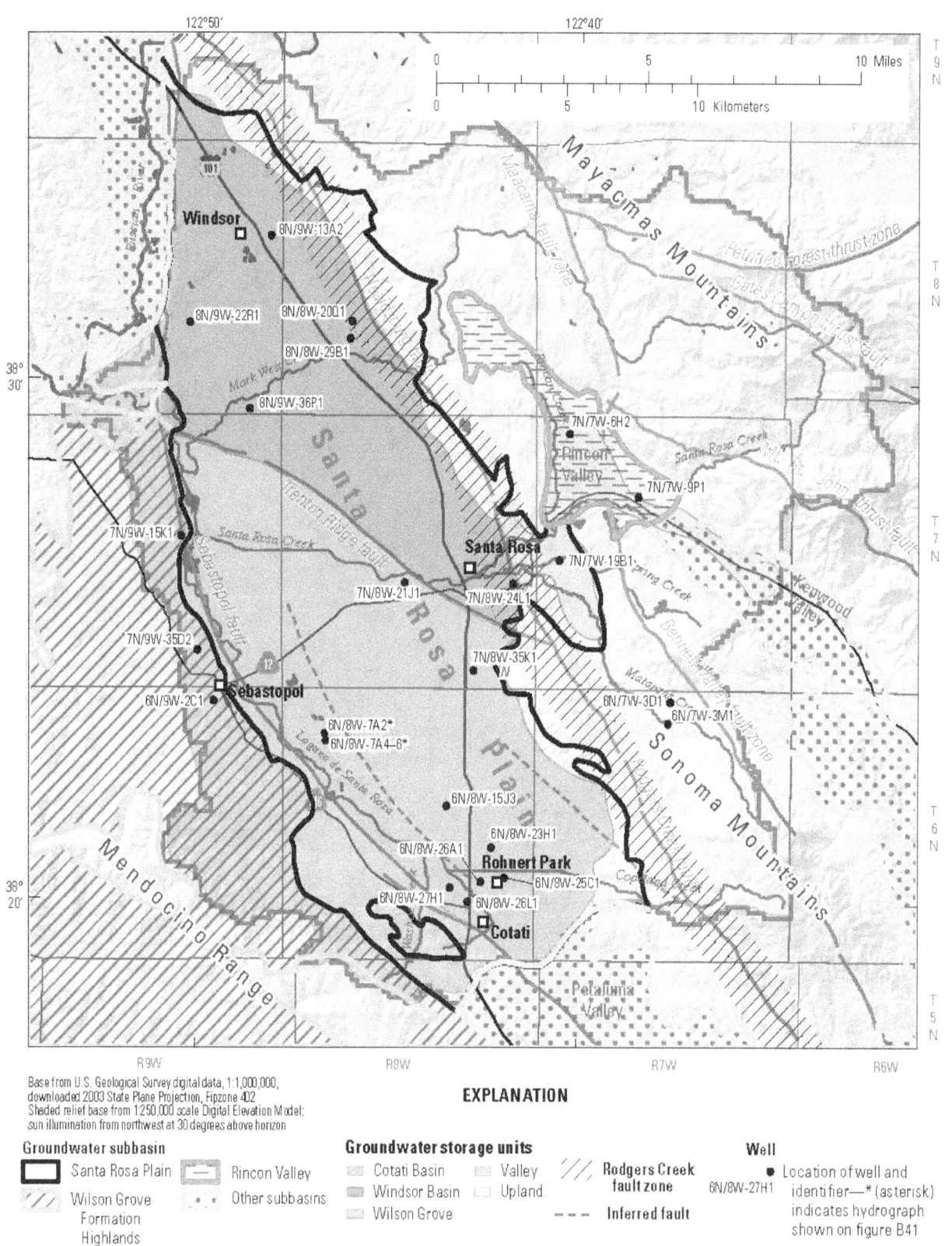

Figure 34. Locations of selected hydrograph wells, Santa Rosa Plain watershed, Sonoma County, California.

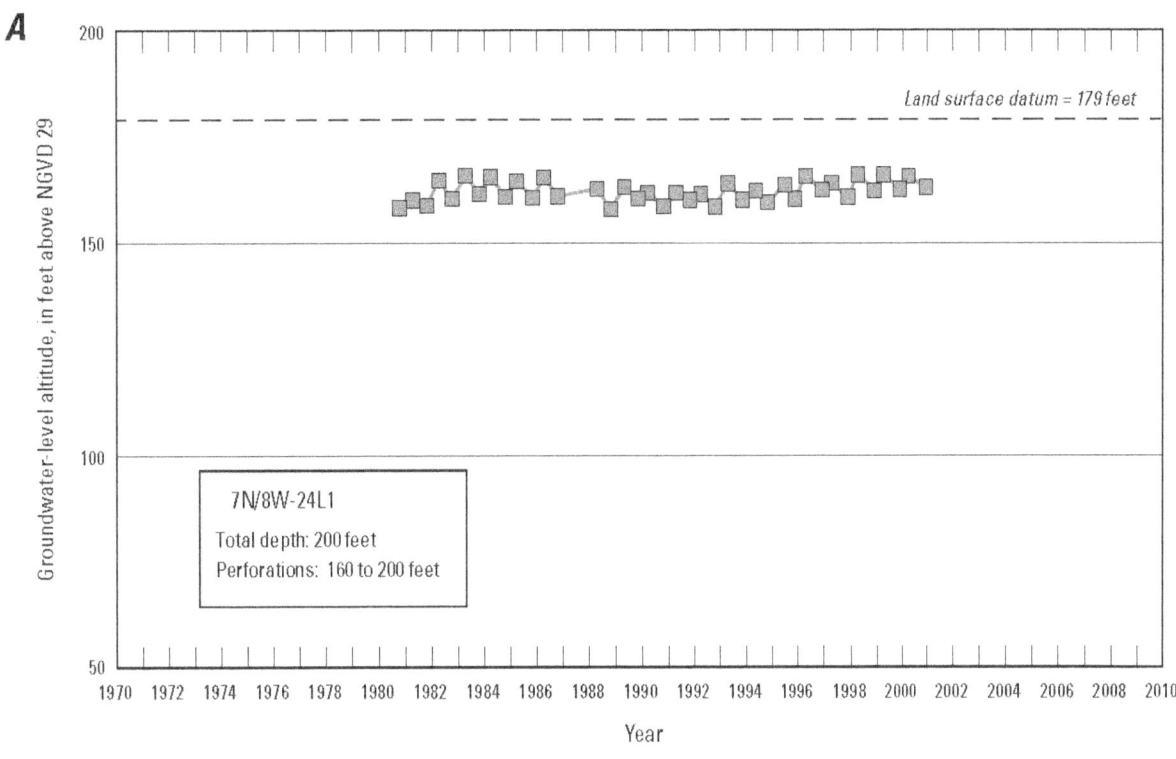

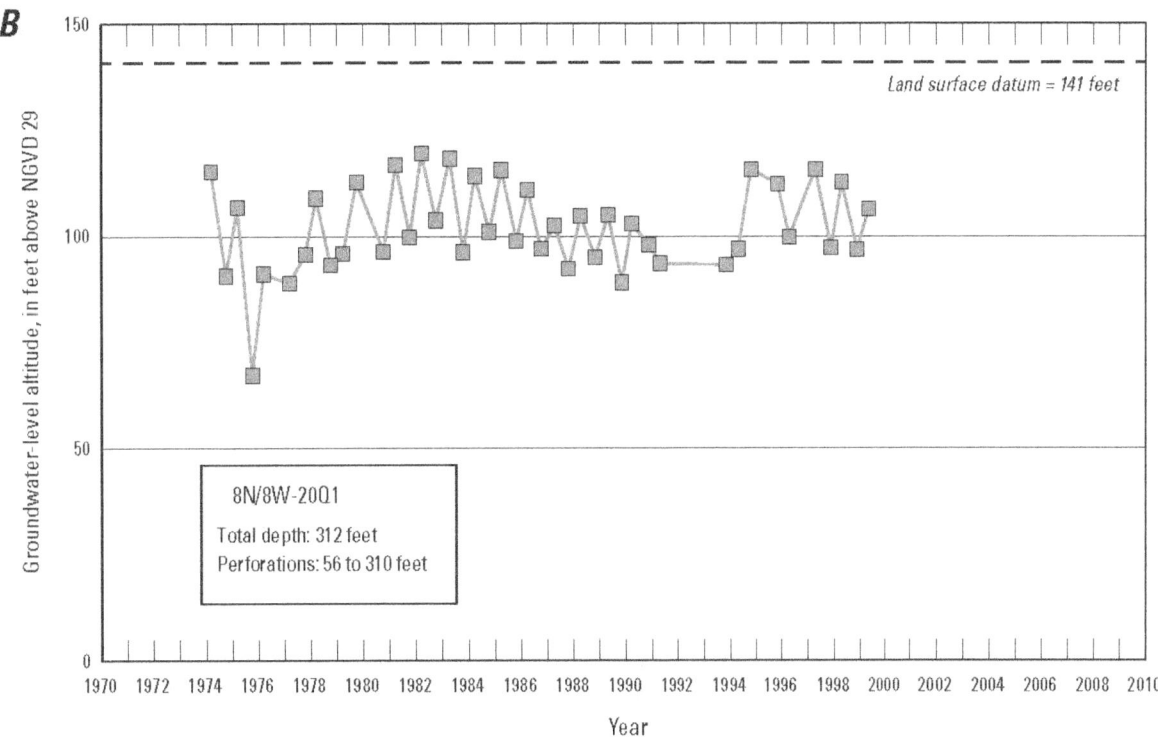

Figure 35. Water-level data for selected wells in the Windsor Basin storage unit, 1974–2010, Santa Rosa Plain watershed, Sonoma County, California: *A*, 7N/8W-24L1; *B*, 8N/8W-20Q1; *C*, 8N/8W-29B1; *D*, 8N/9W-13A2; *E*, 8N/9W-22R1; and *F*, 8N/9W-36P1.

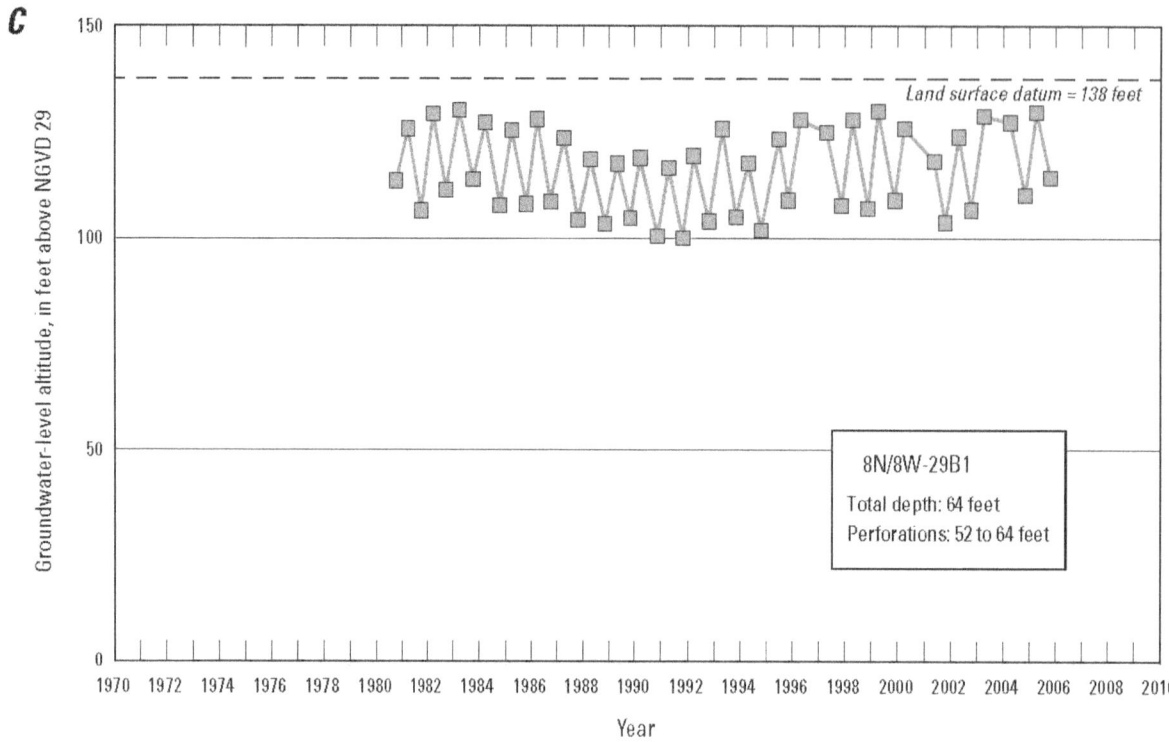

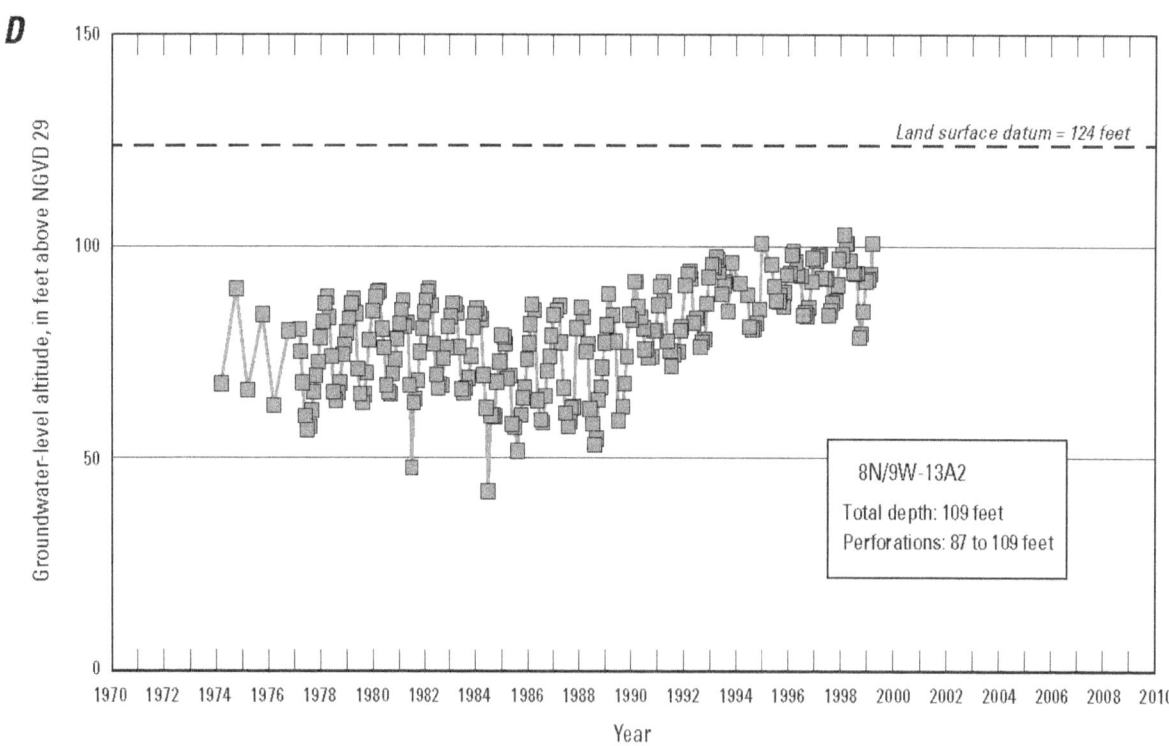

Figure 35. Continued.

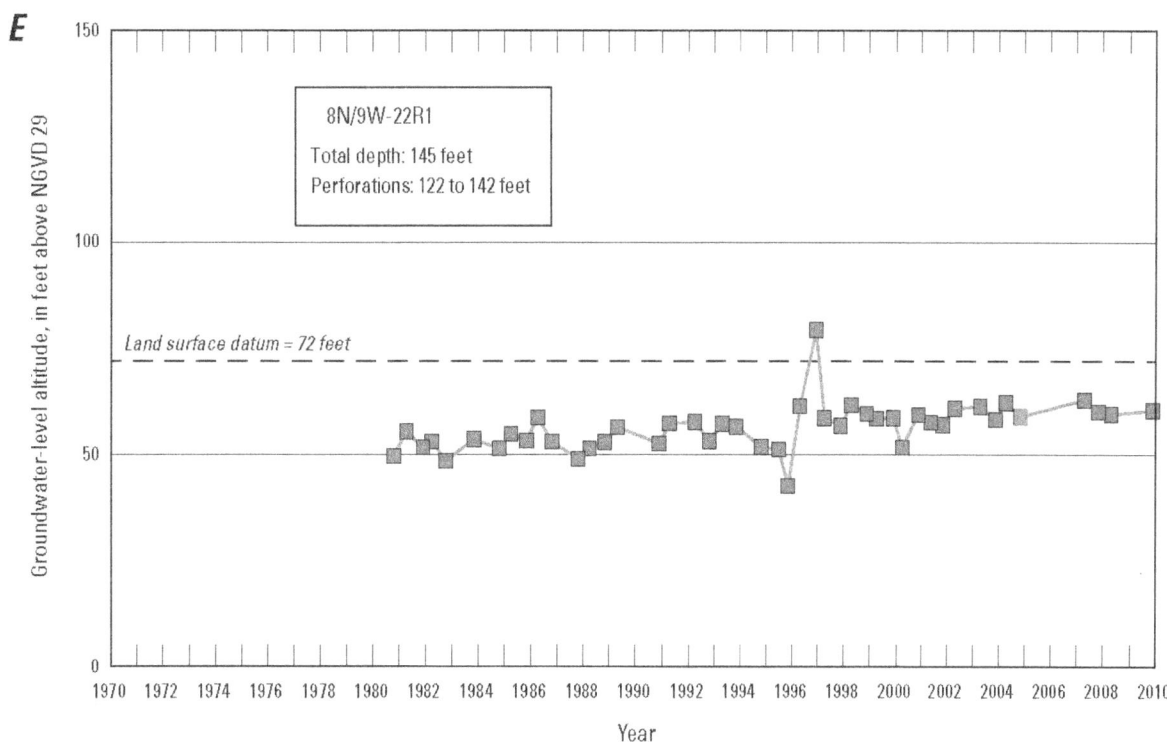

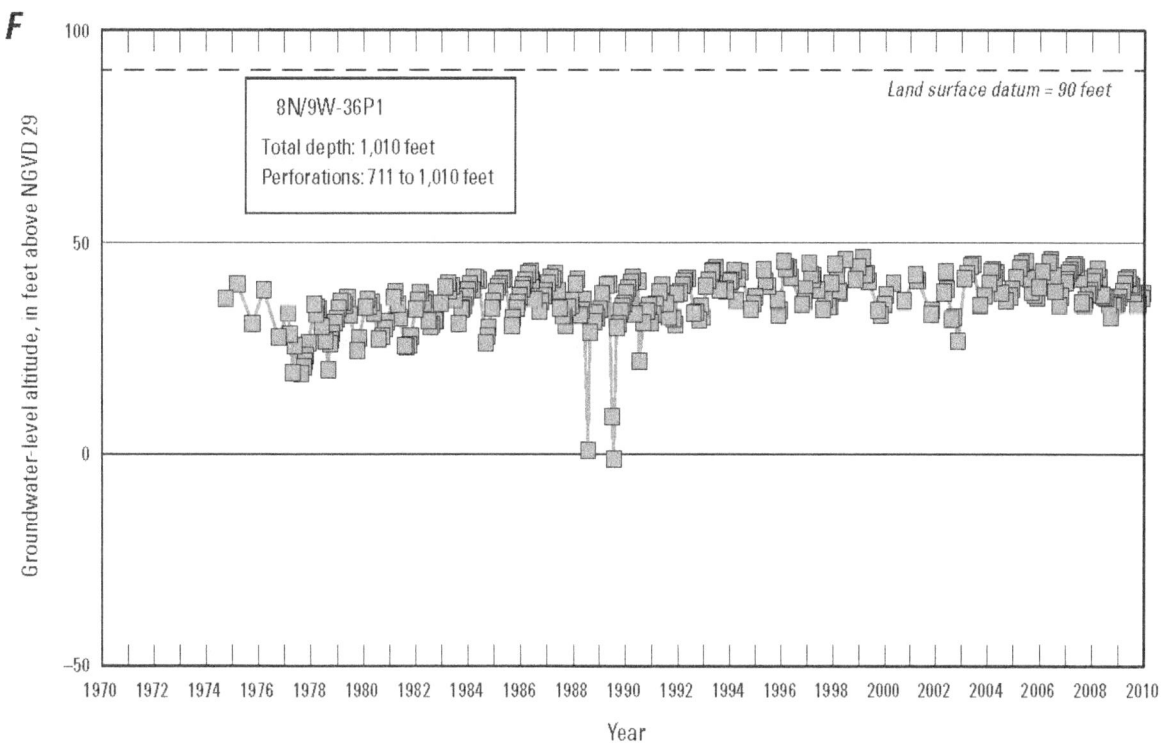

Figure 35.   Continued.

Cotati Basin Storage Unit

Wells 6N/8W-23H1, 25C1, and 26A1 are deep production wells located within 1 mi of each other in the Rohnert Park and Cotati area of the CB storage unit (fig. 34). Well 23H1 is perforated from 300 to 1,500 ft bls, well 25C1 is perforated from 144 to 376 ft bls, and well 26A1 is perforated from 288 to 462 ft bls (table 6). All of three of these wells are perforated in the Petaluma Formation (table 6). Water levels in well 23H1 showed a net increase of about 40 ft between 1977 and 2006; however, there was a water-level decline of at least 100–150 ft through the mid-1990s that inversely corresponded to the pumping pattern of Rohnert Park production wells, such that the water levels recovered as pumping declined (figs. 36A and 37). Water levels in well 25C1 showed a net increase of 40–60 ft between 1977 and 2006; however, the water levels declined through the 1990s in response to pumping and recovered once pumping decreased (figs. 36B and 37). There was no net change in measured water levels in well 26A1 between 1974 and 2005; however, the water levels declined through the 1990s in response to pumping and recovered once pumping decreased (figs. 36C and 37). The two deeper wells (23H1 and 26A1) showed similar water-level declines of about 100–150 ft, while the water-level decline in well 25C1 was about 30–40 ft. These patterns could indicate that the deeper wells are in confined zones with lower storativity than the shallower well and that the hydraulic connection between the deeper confined and shallow unconfined zones is impeded by low permeability deposits.

Wells 6N/8W-15J3, 26L1, and 27H1 are shallow wells located in the Rohnert Park/Cotati area (fig. 34), and all are perforated in the Glen Ellen Formation. Well 15J3 is perforated between 65 and 90 ft bls, well 26L1 is perforated between 54 and 94 ft bls, and well 27H1 is perforated between 62 and 82 ft bls (table 6). The water levels in well 15J3 showed a net decline of about 5–10 ft between 1970 and 2005; however, water levels declined about 35 ft between 1970 and the mid 1990s and recovered about 25 ft by 2005 (fig. 36D). Water levels in well 26L1 showed no net change between 1972 and 2004, although there was seasonal variability (fig. 36E). The water levels in well 27H1 showed a net increase of about 30 ft between 1989 and 2010 (fig. 36F). The water-level decline in well 15J3 and recoveries in wells 15J3 and 27H1 could be in response to Rohnert Park pumping (fig. 37).

Wells 7N/8W-21J1 and 35K1 are in the Santa Rosa area south of the Trenton Ridge fault in the northeastern part of the CB storage unit, and both are perforated in the Petaluma Formation (fig. 34, table 6). Well 21J1 is perforated between 148 and 360 ft bls; well 35K1 is perforated between 185 and 205 ft bls. The water levels in these wells were steady or increased by a small amount. The water levels in well 21J1 showed no net change between 1990 and 2011, but did show seasonal variability (fig. 36G). Water levels in well 35K1 showed a net increase of 10 ft between 1989 and 2011 (fig. 36H).

Valley Storage Unit

Wells 7N/7W-19B1 and 7N/7W-06H2 and 09P1 are in the VAL storage unit (fig. 34). Well 7N/7W-19B1 is in the Bennett Valley area (fig. 34) and is perforated between 45 and 85 ft bls in the Glen Ellen Formation (table 6). The water levels in well 19B1 were fairly steady, but showed a net decline of about 10 ft between 1980 and 2010 (fig. 38A)

Well 7N/7W-06H2 is in the Rincon Valley (fig. 34) and is perforated between 60 and 80 ft bls in the Glen Ellen Formation (table 6). The water levels showed seasonal variability, declined around 2001, and slightly recovered around 2006 (fig. 38B). Overall, there was a net decline in water level of about 10 ft between 1989 and 2011.

Well 7N/7W-09P1 is in the northern part of Kenwood Valley (fig. 34) and is perforated between 286 and 296 ft bls in the Sonoma Volcanics (table 6). Overall, the water levels showed a net increase of about 40 ft between 1989 and 2011 (fig. 38C).

Wilson Grove Storage Unit

Wells 6N/9W-02C1, 7N/9W-15K1, and 35D2 are in the WG storage unit (fig. 34) and all are perforated in the Wilson Grove Formation. Well 2C1 is a production well that is perforated between 332 and 600 ft bls, well 15K1 is perforated between 59 and 69 ft bls, and well 35D2 is perforated between 55 and 167 ft bls (table 6). The water levels in well 2C1 showed 20–40 ft of seasonal variability (80 ft maximum), which could be in response to pumping, and no long-term trend (fig. 39A). The water levels in well 15K1 showed a net increase of about 5 ft between 1980 and 2011 (fig. 39B). The water levels in well 35D2 showed no net change between 1970 and 2005 (fig. 39C).

Uplands Storage Unit

Wells 6N/7W-03D1 and 3M1 are in UPL storage unit in the Sonoma Mountains, and both are perforated in the Petaluma Formation (fig. 34, table 6). Well 3D1 is perforated between 234 and 253 ft bls, and well 3M1 is perforated between 41 and 61 ft bls (table 6). The water levels in well 3D1 showed a long-term decline of about 30 ft (fig. 40A). The water levels in well 3M1 were steady (fig. 40B). The wells are about 0.5 mi apart; however, the water levels in well 3D1 were more variable, and generally higher, than in well 3M1. Well 3M1 is next to Matanzas Creek, which could explain the steady nature of its water levels. In addition, the variability in the well 3D1 water levels could be due to pumping or climate variability.

Multiple-Level Monitoring Site

In 1977, SCWA constructed a multi-level monitoring site (6N/8W-07A4-6) on the west side of the basin in the CB storage unit (fig. 34). Well 7A4 is perforated between 60 and 80 ft bls in the Glen Ellen Formation, well 7A5 is perforated between 237 and 257 ft bls in the Petaluma Formation, and

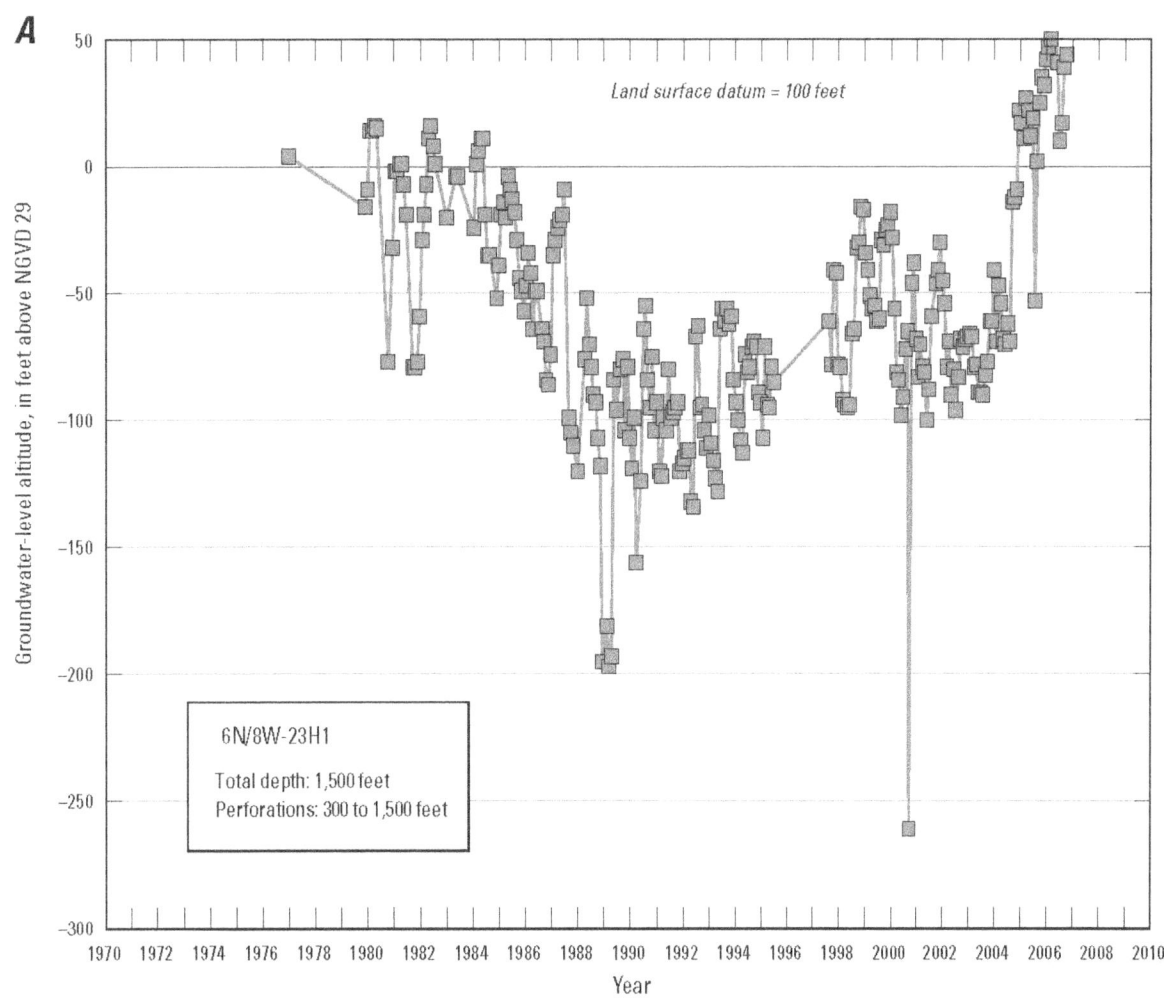

Figure 36. Water-level data for selected wells in the Cotati Basin storage unit, 1974–2010, Santa Rosa Plain watershed, Sonoma County, California: *A*, 6N/8W-23H1, *B*, 6N/8W-25C1, *C*, 6N/8W-26A1, *D*, 6N/8W-15J3, *E*, 6N/8W-26L1, *F*, 6N/8W-27H1, *G*, 7N/8W-21J1, and *H*, 7N/8W-35K1.

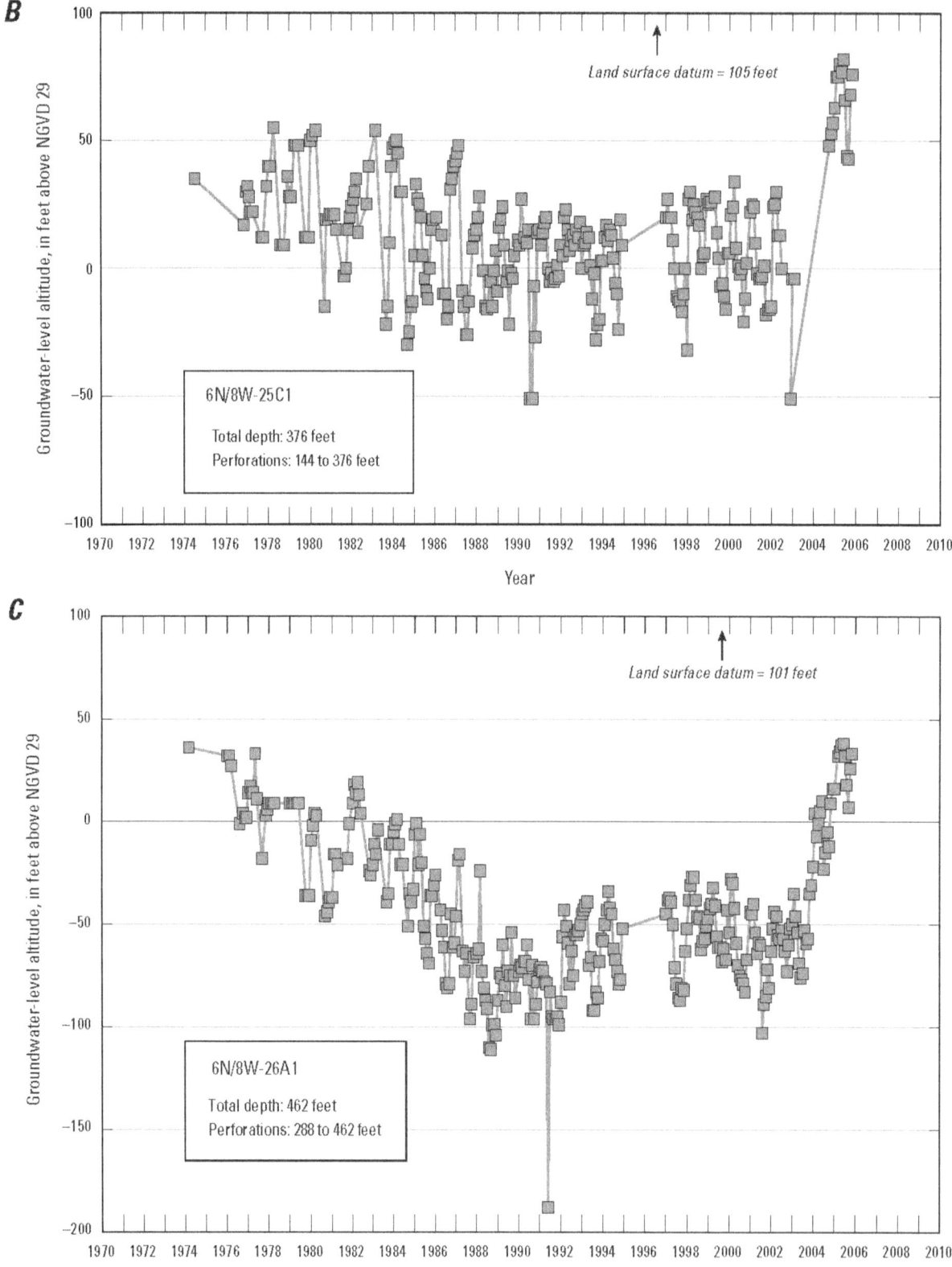

Figure 36.   Continued.

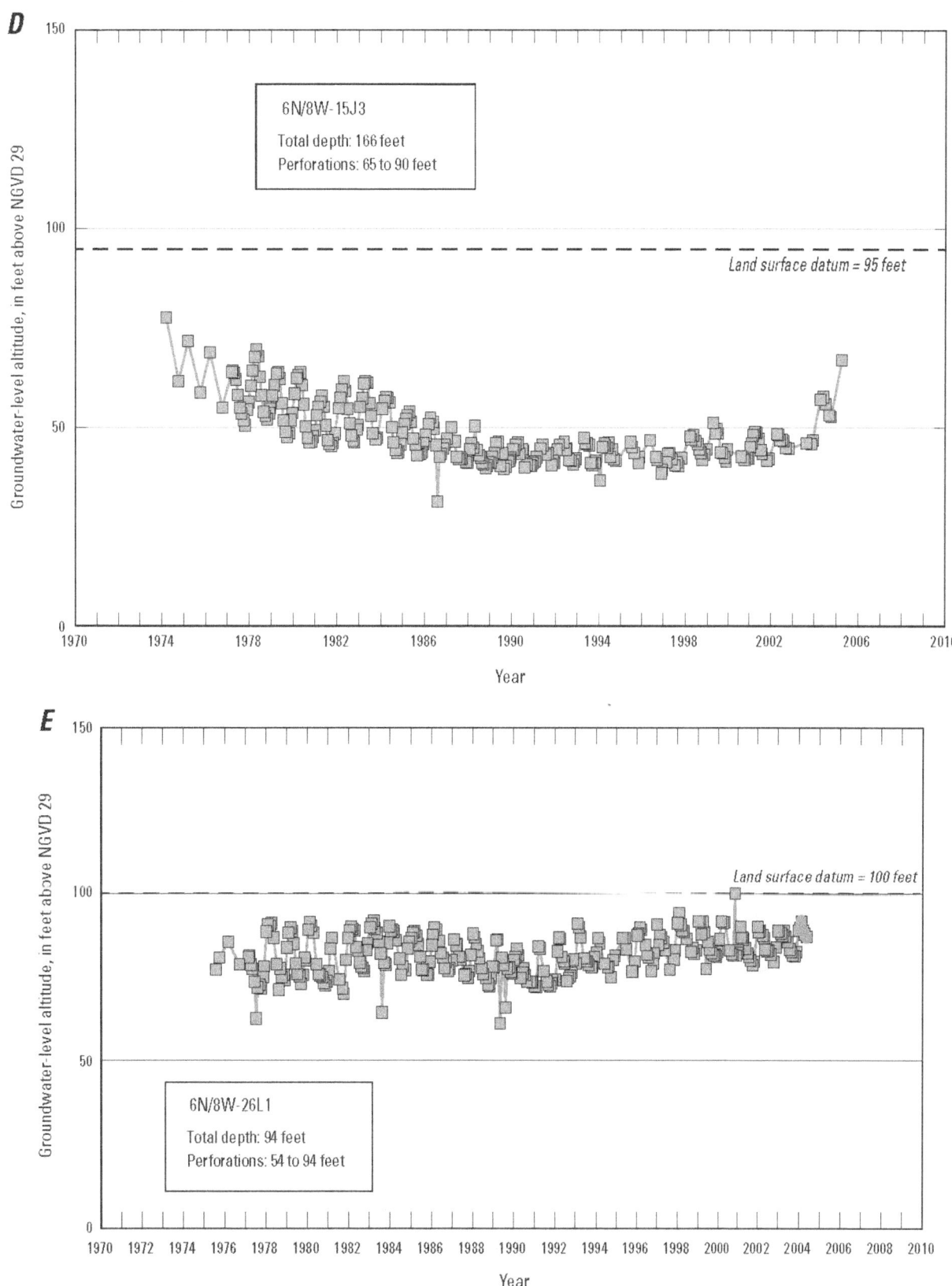

D

6N/8W-15J3

Total depth: 166 feet
Perforations: 65 to 90 feet

Land surface datum = 95 feet

E

Land surface datum = 100 feet

6N/8W-26L1

Total depth: 94 feet
Perforations: 54 to 94 feet

Figure 36.   Continued.

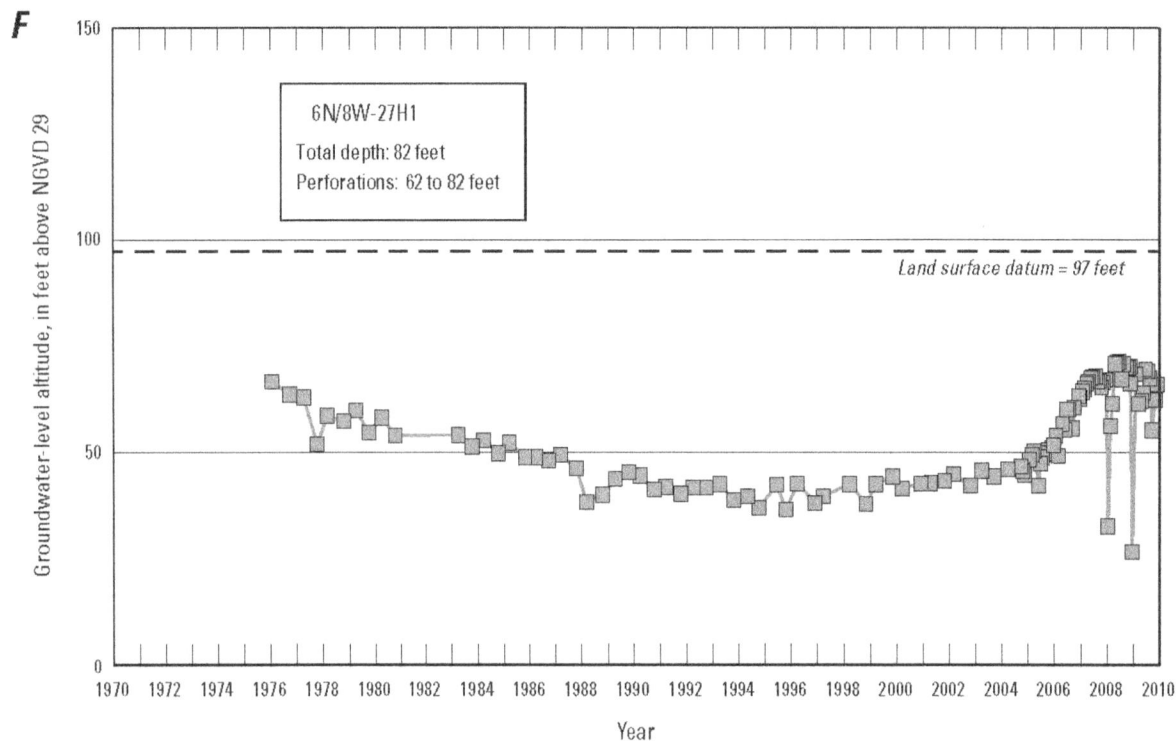

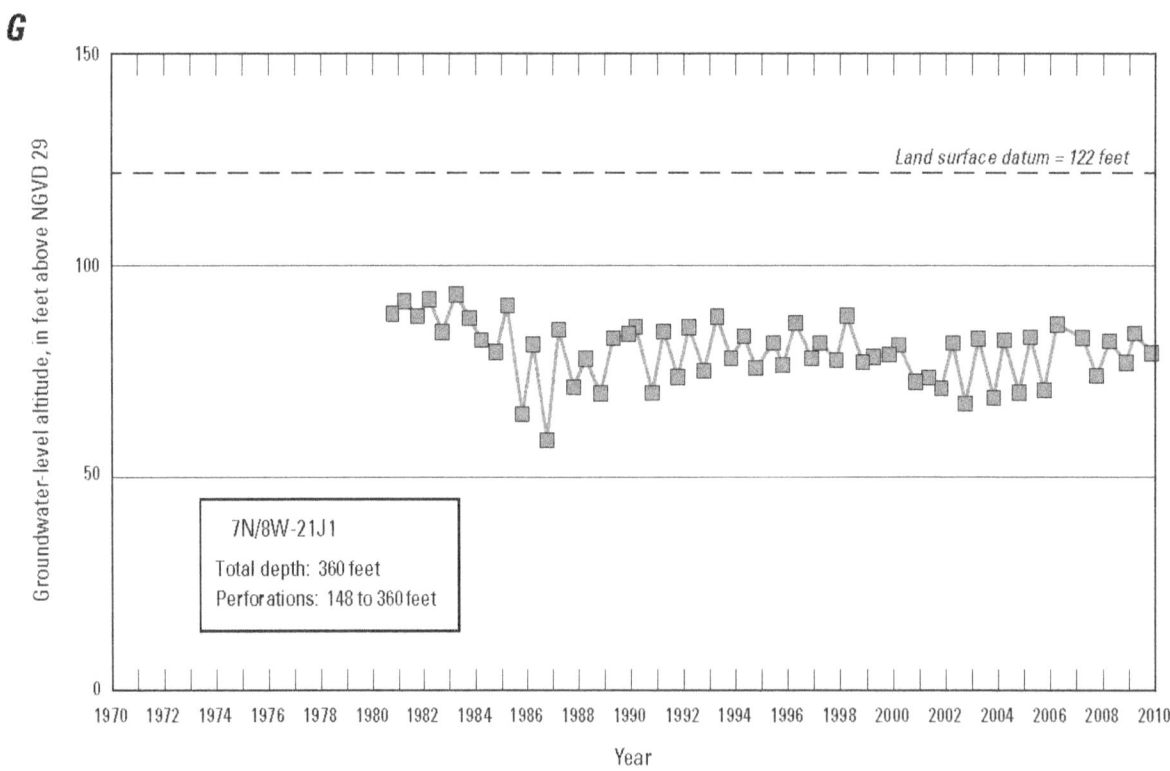

Figure 36.   Continued.

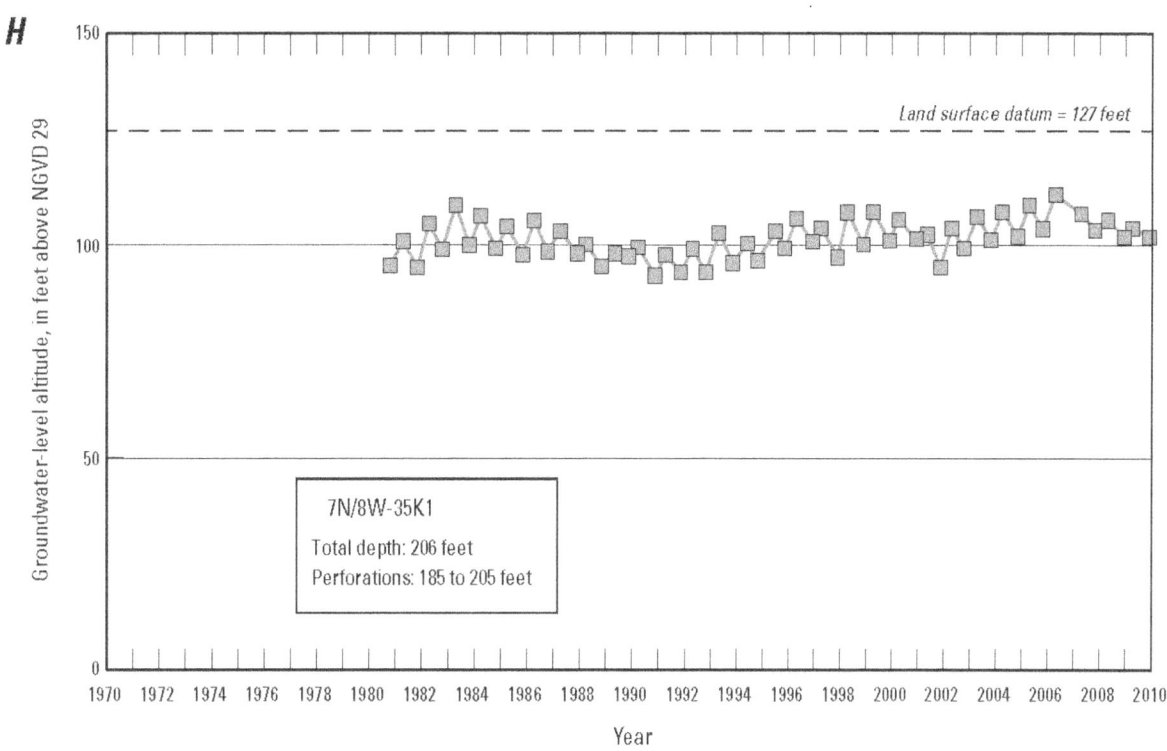

H

Land surface datum = 127 feet

Groundwater-level altitude, in feet above NGVD 29

7N/8W-35K1

Total depth: 206 feet
Perforations: 185 to 205 feet

Year

Figure 36. Continued.

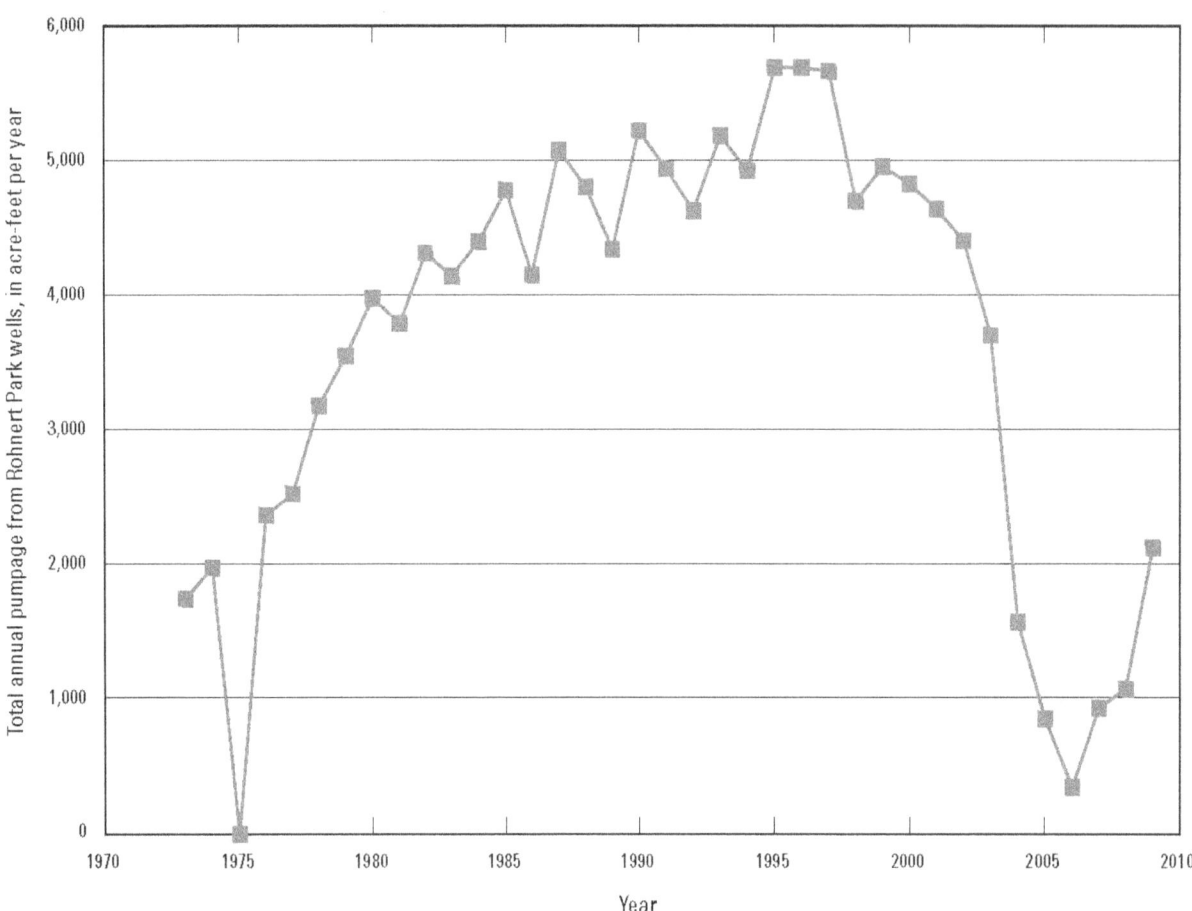

Figure 37. Total annual pumping from city of Rohnert Park production wells, Santa Rosa Plain watershed, Sonoma County, California, 1973–2009.

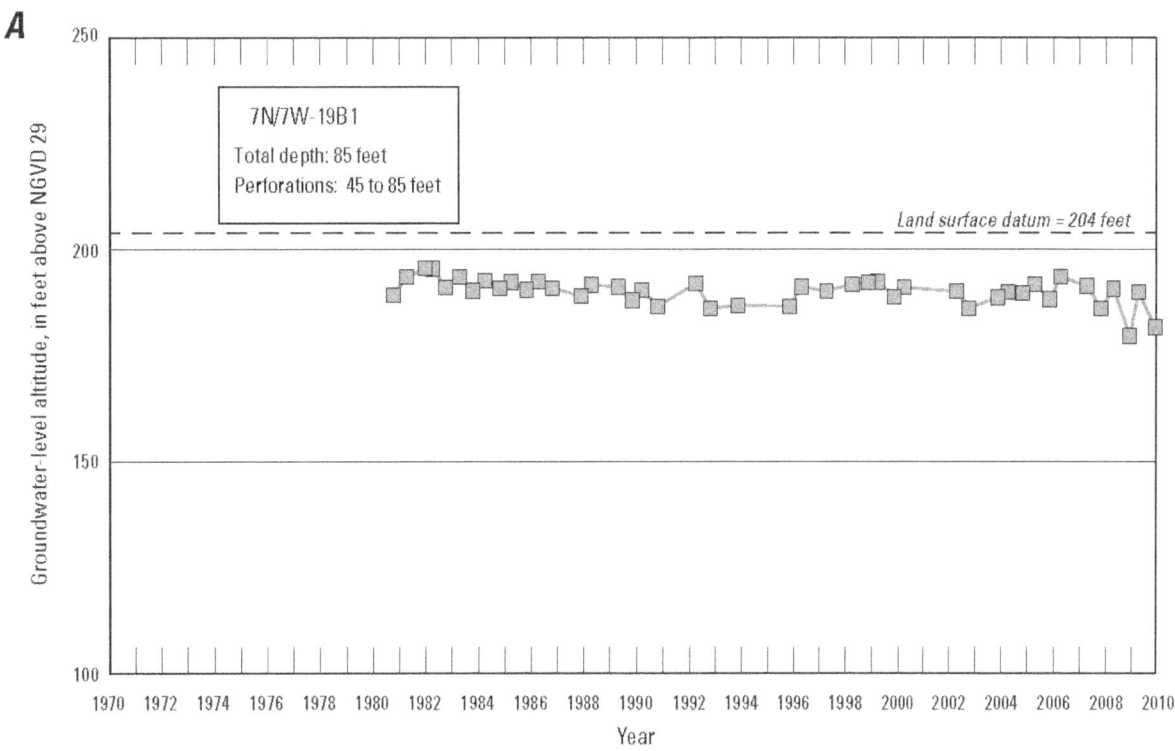

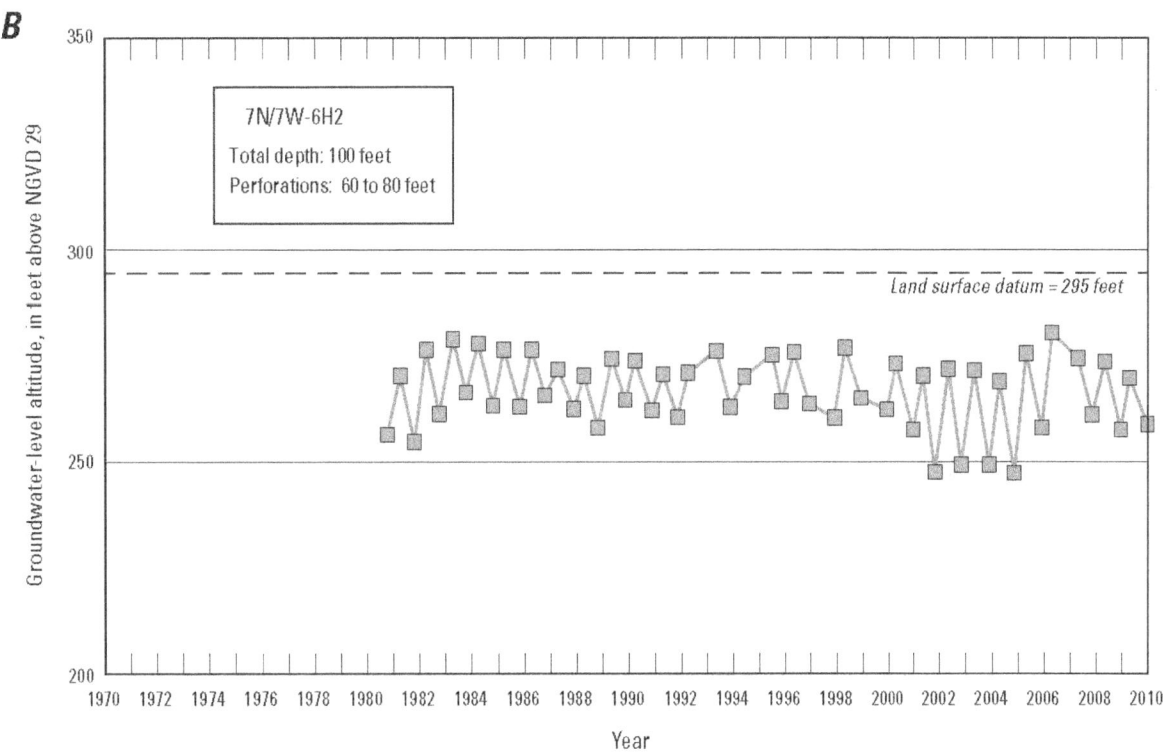

Figure 38.   Water-level data for selected wells in the Valley storage unit, 1974–2010, Santa Rosa Plain watershed, Sonoma County, California: *A*, 7N/7W-19B1, *B*, 7N/7W-06H2, and *C*, 7N/7W-09P1.

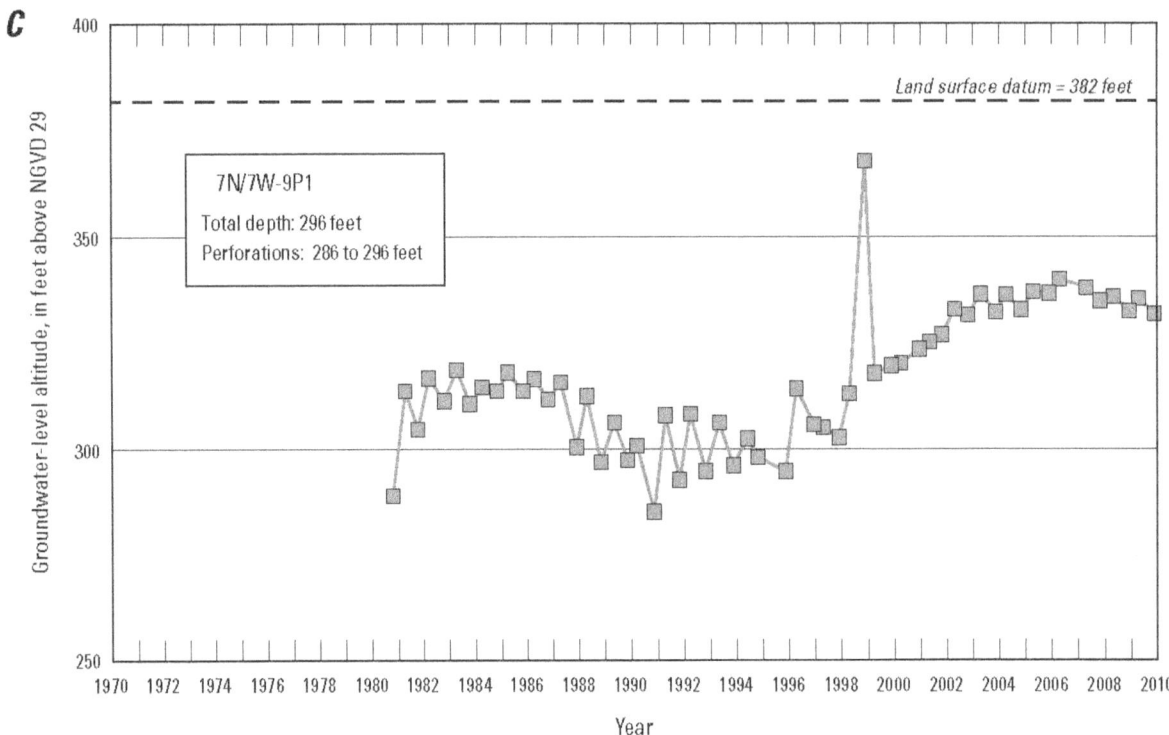

Figure 38. Continued.

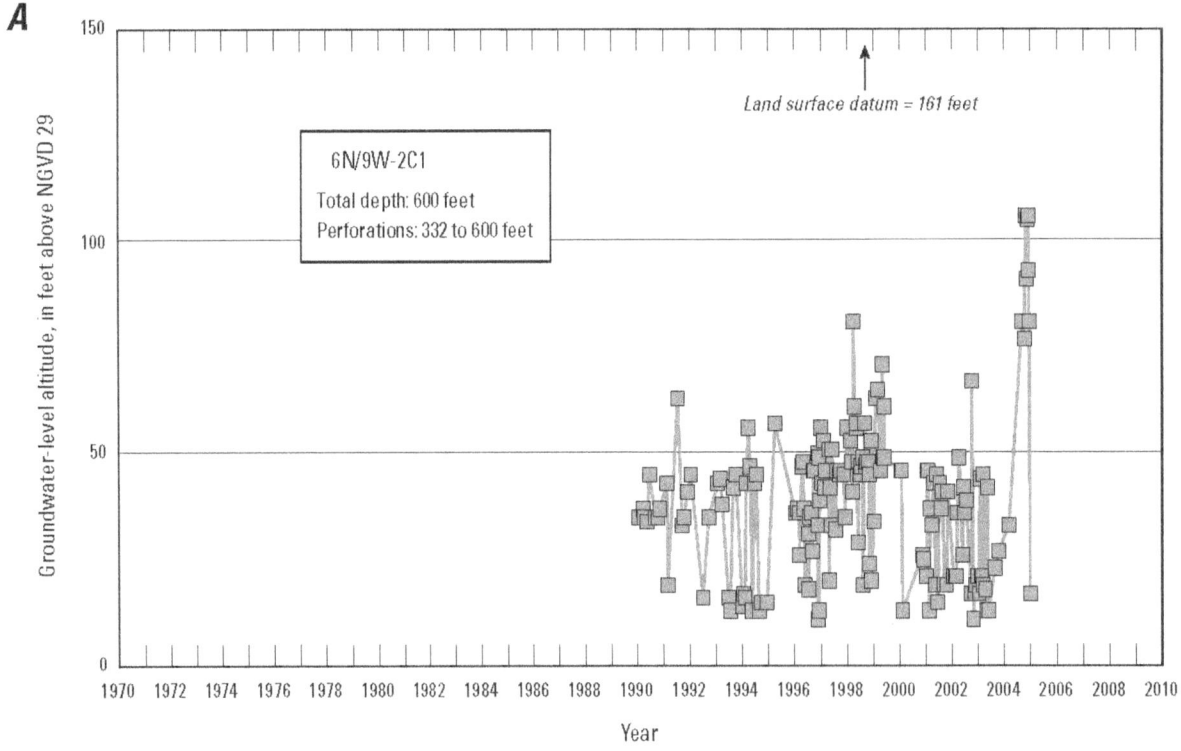

Figure 39. Water-level data for selected wells in the Wilson Grove storage unit, 1974–2010, Santa Rosa Plain watershed, Sonoma County, California: *A*, 6N/9W-02C1, *B*, 7N/9W-15K1, and *C*, 7N/9W-35D2.

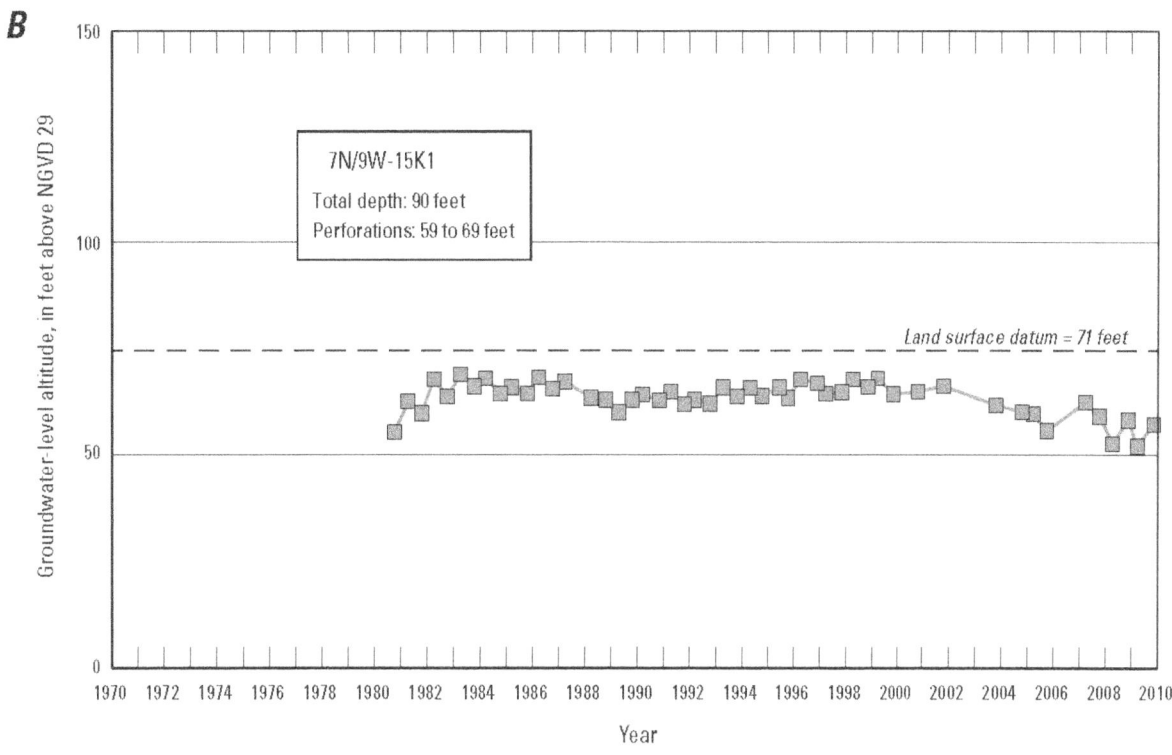

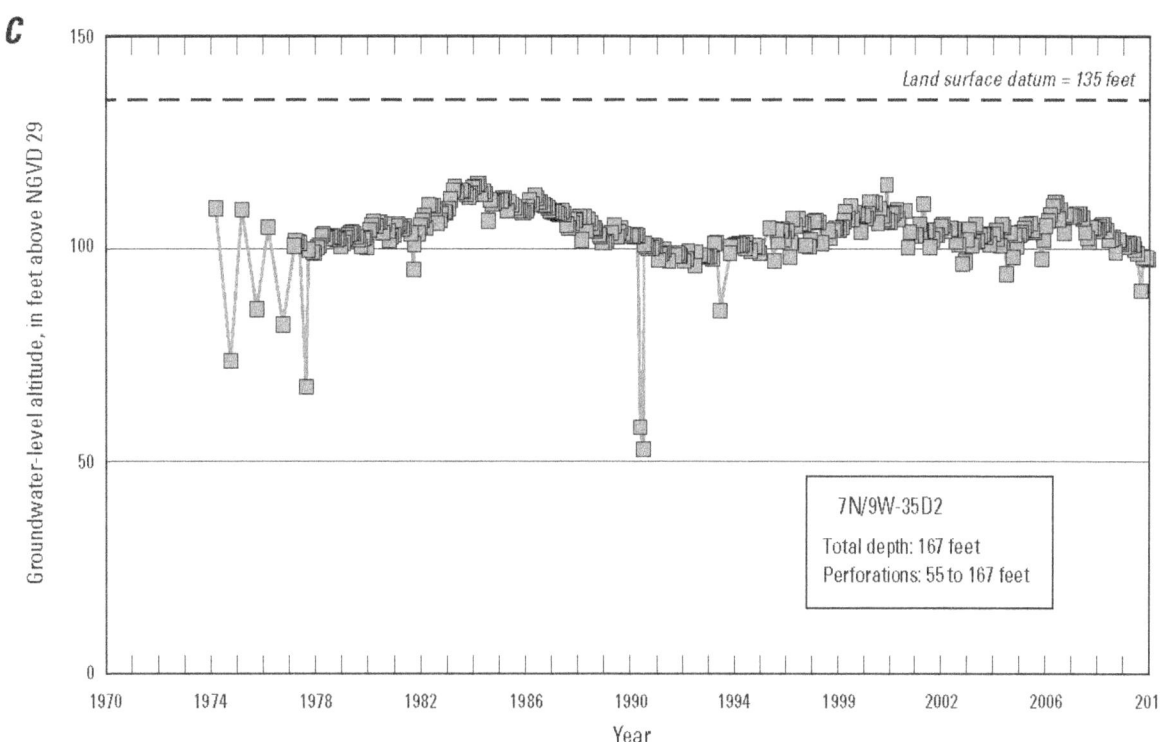

Figure 39. Continued.

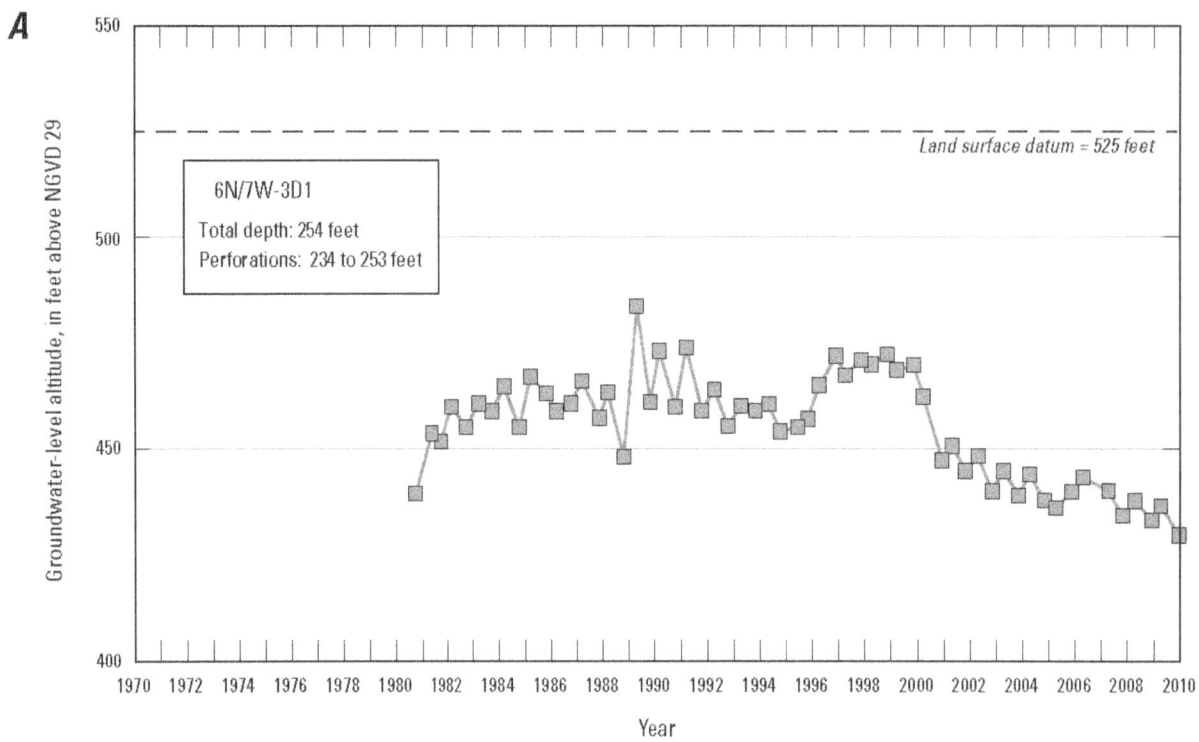

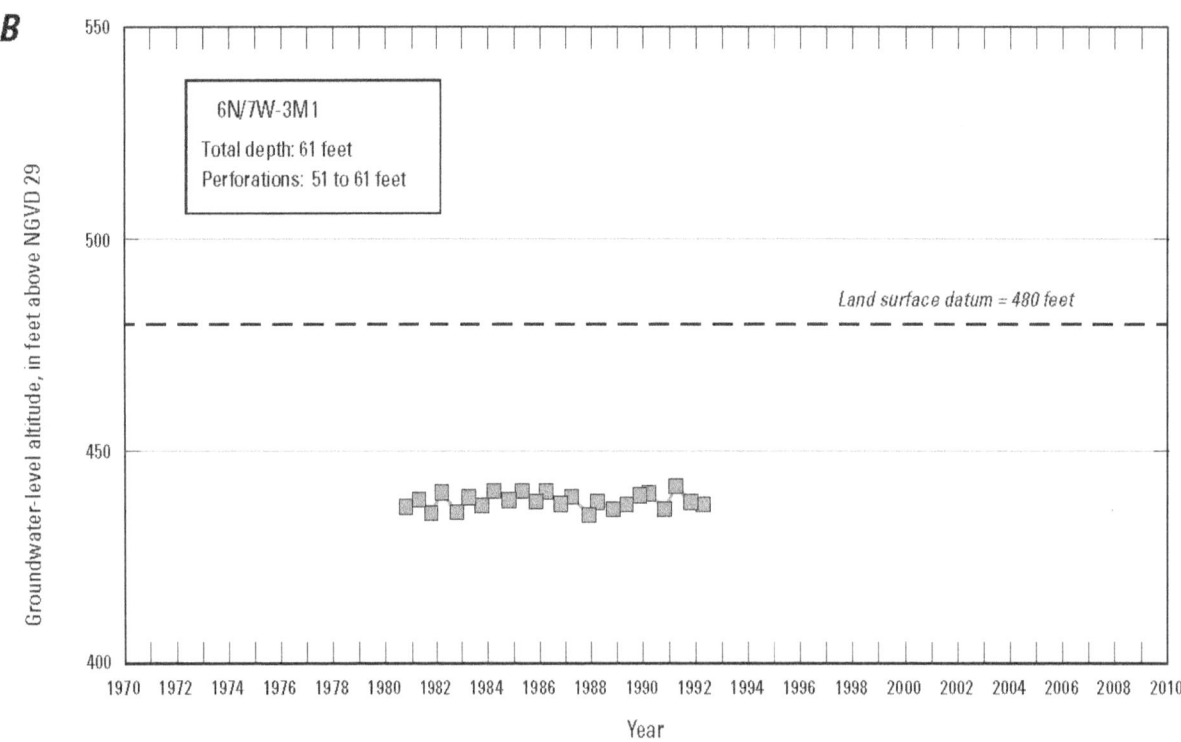

Figure 40.   Water-level data for selected wells in the Uplands storage unit, 1974–2010, Santa Rosa Plain watershed, Sonoma County, California: *A*, 6N/7W-03D1 and *B*, 6N/7W-03M1.

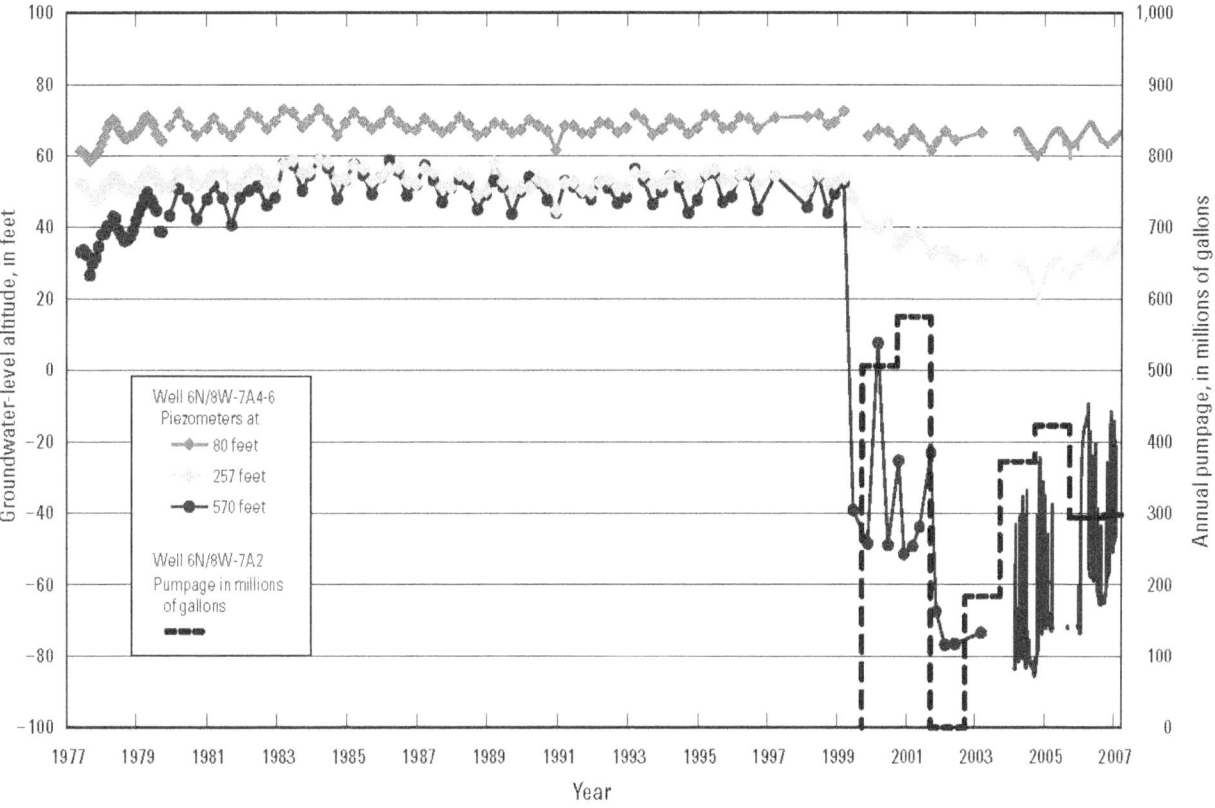

Figure 41. Water-level data for 6N/8W-07A4-6, 1977–2007, and daily pumpage data for 6N/8W-07A2, 2000–2007, Santa Rosa Plain, Sonoma County, California.

well 7A6 is perforated between 550 and 570 ft bls in the Wilson Grove Formation (table 6). Well 6N/8W-07A2, a SCWA production well installed in late 1999, is about 360 ft to the northwest of the multi-level monitoring site (fig. 34). The production well is perforated from 650 to 800 ft bls in the Wilson Grove Formation. Comparison of the water-level hydrographs of the piezometers to the annual pumpage from the production well showed that pumping from well 7A2 affects water-levels in the piezometers differently (fig. 41). The data indicated that the water levels in the shallowest piezometer respond the least to pumping from well 7A2 (fig. 41). This could be the result of low permeability deposits in the Glen Ellen and Petaluma Formations, the unconfined storage properties of the shallow aquifer, or both. The two deeper piezometers, however, particularly the 570-ft piezometer, have strong hydraulic communication with the pumping well, indicating that the 570-ft piezometer and the pumping well are perforated in the same hydrogeologic unit, which is under confined conditions. The data also indicated a downward vertical gradient between well 7A4 and the deeper wells; prior to pumping, the vertical gradients between 7A4 and 7A5 and between 7A4 and 7A6 were about 0.11 and 0.09, respectively. After pumping started, the vertical gradients increased to about 0.28 and 0.31 between the same respective wells.

Summary

The principal water-bearing strata in the study area include the Tertiary and Quaternary sediments and sedimentary rocks underlying the SRP floor to depths of at least 1,500 ft, as well as permeable zones in Tertiary volcanic rocks in contact with the sedimentary section along the valley margin and interfingered with it beneath the valley floor. The Tertiary material includes the Glen Ellen, Wilson Grove, and Petaluma Formations, and the Sonoma Volcanics. Geophysical investigations have identified two deep basins beneath SRP: the Windsor Basin to the north is about 1.2 mi deep, and the Cotati Basin to the south is about 2 mi deep. Geologic materials within the depth range perforated by wells are most commonly fine-grained or poorly-sorted sediments that have low hydraulic conductivities and specific yields that result in low to modest well yields and large drawdowns. In most parts of the study area, shallow groundwater flow is unconfined, and flow is confined at depth. The presence of fine-grained material—either as interbeds within coarser-grained materials or as thick, but not laterally extensive, beds—imparts anisotropic hydraulic characteristics typical of layered systems.

The available surface-water records indicate that a high percentage of streamflow is generated as overland flow, with a relatively fast response time to the larger storm events. The streamflow is highly seasonal, with peak flows limited to winter. Summer flows are low (less than 5 ft³/s). In general, December, January, and February are the months with the highest streamflow, while August and September are the driest months. Year to year variability in streamflow is also high, in terms of total annual flow, peak annual flow, and average summer flows. The primary source of the summer flow is likely baseflow from groundwater discharge, although irrigation and urban-area runoff also can contribute. Many of the streams are intermittent, with no flow during the summer; however, during drier than normal periods, normally perennial streams can become dry by late summer and early fall.

The 1951 (predevelopment) water-level contour map shows that groundwater flowed from the Mayacmas and Sonoma Mountains in the UPL storage unit westward toward the Laguna de Santa Rosa on the western edge of the SRP and eastward from the highlands in the WG storage unit toward the Laguna de Santa Rosa. The groundwater-level contour map also shows that groundwater moved toward, and discharged into, the stream channels, thereby, likely sustaining baseflow.

In the WB storage unit, water levels were fairly steady with some seasonal variability, and some hydrographs showed the effects of the 1976–77 and 1986–92 droughts. In the CB storage unit, water levels from wells in the area near Rohnert Park recovered to 1974 conditions by 2007 as a result of a large reduction in municipal pumping that began in 2000. In the VAL storage unit, water levels were fairly steady with some seasonal variability in the Bennett and Rincon valleys; however, water levels increased in the Kenwood Valley. In the WG storage unit, water levels showed seasonal variability but, generally, had no long-term trends. In the UPL storage unit, the water levels in the deeper well showed a long-term decline, while the water levels in the shallow well were steady. Sparse vertical profile data indicated that shallower wells have limited hydraulic communication with deeper aquifer material; this could be caused by the extensive clays in the Petaluma Formation and indicate confined aquifer conditions at depth.

References Cited

Allen, J.R., 2003, Stratigraphy and tectonics of Neogene strata, northern San Francisco Bay Area: San Jose, Calif., San Jose State University, unpublished Master's thesis, 183 p.

Bailey, E.H., 1966, Geology of Northern California: California Department of Mines and Geology Bulletin 190, 508 p.

Bailey, E.H., Irwin, W.P., Jones, D.L., 1964, Franciscan and related rocks and their significance in the geology of western California: California Division of Mines and Geology, Bulletin 183, 177 p., 2 plates.

Bezore, S.P., Koehler, R.D., and Witter, R.C., 2003, Geologic map of the Two Rock 7.5' quadrangle, Sonoma County, California: A digital database: California Geological Survey Preliminary Geologic Map, scale 1:24,000, accessed April, 2008 at URL: *http://www.consrv.ca.gov/CGS/rghm/rgm/ preliminary_geologic_maps.htm*

Blake, M.C., Jr., Graymer, R.W., and Stamski, R.E., 2002, Geologic Map and Map Database of Western Sonoma, Northernmost Marin, and Southernmost Mendocino Counties, California: U.S. Geological Survey Miscellaneous Field Studies Map MF-2402, 42 p., scale 1:100,000 *http:// pubs.usgs.gov/mf/2002/2402/*

Blake, M.C., Jr., Howell, D.G., and Jayko, A.S., 1984, Tectonostratigraphic terranes of the San Francisco Bay Region, *in* Blake, M.C., ed., 1984, Franciscan Geology of Northern California: Pacific Section, Society of Economic Paleontologists and Mineralogists, v. 43, p. 5–22.

California Department of Water Resources – Central District, 1979, Meeting water demands in the city of Rohnert Park, 127 p.

California Department of Water Resource, 1994, California water plan update: Bulletin 160-93, October 1994, v. 2, 315 p.

California Department of Water Resources, 2003, California's Groundwater: Bulletin 188 – Update 2003, 246 p. at URL *http://www.water.ca.gov/groundwater/bulletin118/ update2003.cfm*

California Interagency Watershed Mapping Committee, 2004, California Interagency Watershed Map of 1999 (Calwater 2.2.1): Sacramento, Calif., State of California, California Resources Agency, accessed September 7, 2007, at URL *http://atlas.ca.gov/download.html#/casil/inlandWaters*

Cardwell, G.T., 1958, Geology and Ground Water in the Santa Rosa and Petaluma areas, Sonoma County, California: U.S. Geological Survey Water-Supply Paper 1427, 273 p. and 5 plates.

City of Rohnert Park, 2007, 2005 Urban Water Management Plan (Prepared by Winzler and Kelly Consulting Engineers with Luhdorff and Scalmanini Consulting Engineers, Brown and Caldwell, and Maddaus Water Management): Rohnert Park, California, 108 p.

City of Santa Rosa, 2012, Geysers Project: accessed February 1, 2012, at URL *http://ci.santa-rosa.ca.us/departments/utilities/irwp/geysers/Pages/default.aspx*

Clahan, K.B., Bezore, S.P., Koehler, R.D., and Witter, R.C., 2003, Geologic map of the Cotati 7.5' quadrangle, Sonoma County, California: A digital database: California Geological Survey Preliminary Geologic Map, scale 1:24,000, accessed April, 2008 at URL: *http://www.consrv.ca.gov/ CGS/rghm/rgm/preliminary_geologic_maps.htm*

Cummings, John, 2004, Draining and Filling the Laguna de Santa Rosa, accessed February, 2011 at URL: *http://libweb. sonoma.edu/regional/papers/cummings/cummings3.pdf*

Dawson, Arthur, and Sloop, Christina, 2010, Final Report, Laguna de Santa Rosa historical hydrology project headwaters pilot study: Laguna de Santa Rosa Foundation and Sonoma Ecology Center, Santa Rosa, CA: accessed May, 2010 at URL: *http://www.lagunadesantarosa.org/ knowledgebase/*

Delattre, M.P., Koehler, R.D., and Gutierrez, C.I, 2008, Geologic map of the Sebastopol 7.5'quadrangle, Sonoma County California: A digital database: California Geological Survey Preliminary Geologic Map, scale 1:24,000, accessed April, 2008 at URL: *http://www.consrv.ca.gov/CGS/rghm/ rgm/preliminary_geologic_maps.htm.*

Dickinson, W.R., 1981, Plate tectonics and the continental margin of California, *in* Ernst, W.G., ed., The geotectonic development of California: New Jersey, Prentice-Hall, p. 1–28.

Driscoll, F.G., 1986, Groundwater and Wells: St. Paul Minnesota, Johnson Division, 1,089 p.

Farrar, C.D. and Metzger, L.F., 2003, Ground-water resources in the Lower Milliken–Sarco-Tulucay Creeks Area, southeastern Napa County, California, 2000–2002: U.S. Geological Survey Water Resources Investigations Report 03–4229, 94 p.

Faye, R.E., 1973, Ground-water hydrology of northern Napa Valley, California: U.S. Geological Survey Water-Resources Investigations 13–73, 64 p.

Federal Emergency Management Agency, 2002, Flood insurance study, Sonoma County, California.

Ford, R.S., 1975, Evaluation of ground water resources. Sonoma County, volume 1: geologic and hydrologic data: California Department of Water Resources Bulletin 118-4, 177 p., 1 plate.

Fox, K.F., Jr., 1983, Tectonic setting of late Miocene, Pliocene, and Pleistocene, rocks in part of the Coast Ranges north of San Francisco, California: U.S. Geological Survey Professional Paper 1239, 92 p.

Gealey, W.K., 1951, Geology of the Healdsburg quadrangle, California: California Division of Mines and Geology Bulletin 161, scale 1:62,500.

Graymer, R.W., Brabb, E.E., Jones, D.J., Barnes, J., Nicholson, R.S., and Stamski, R.E., 2007, Geologic map and map database of eastern Sonoma and western Napa counties, California: U.S. Geological Survey Scientific Investigations Map 2956, scale 1:100,000, accessed November 2009 at *http://pubs.usgs.gov/sim/2007/2956/.*

Graymer, R.W., Bryant, William, McCabe, C.A., Hecker, Suzanne, and Prentice, C.S., 2006, Map of Quaternary-active faults in the San Francisco Bay region: U.S. Geological Survey Scientific Investigations Map 2919, scale 1:275,000, accessed November 2009 at *http://pubs.usgs. gov/sim/2006/2919/.*

Gutierrez, C. Bryant, W., Saucedo, G., and Willis, C., 2010, Geologic Map of California: California Geological Survey, scale 1: 750,000, 1 sheet.

Herbst, C.M., Jacinto, D.M., and McGuire, R.A., 1982, Evaluation of ground water resources, Sonoma County, volume 2: Santa Rosa Plain: California Department of Water Resources Bulletin 118-4, 107 p., 1 plate.

Helsel, D.R., and Hirsch, R.M., 1992, Statistical methods in water resources: U.S. Geological Survey, Techniques of Water-Resources Investigations Book 4, Chapter A3, 522 p.

Hevesi, J.A., Woolfenden, L.R., Niswonger, R.G., Regan, R.S., and Nishikawa, Tracy, 2011, Decoupled application of the integrated hydrologic model, GSFLOW, to estimate agricultural irrigation in the Santa Rosa Plain, California, *in* Maxwell, R.M., Poeter, E.P., Hill, M.C., and Zheng, Chunmiao, eds., MODFLOW and More 2011: Integrated hydrologic modeling–Conference proceedings, June 5–8, 2011: Golden, CO, International Groundwater Modeling Center, p. 115–119.

Heyenkamp, M.R., Goodwin, L.B., Mozley, P.S., and Haneberg, W.C., 1999, Controls on fault-zone architecture in poorly lithified sediments, Rio Grande Rift, New Mexico, implications for fault-zone permeability and fluid flow, *in* Haneberg, W.C., Mozley, P.S., Moore, C.J., and Goodwin, L.B., eds., Faults and subsurface fluid flow in the shallow crust: Washington, D.C., American Geophysical Union, p. 27–50.

Hirsch, R.M., 1982, A comparison of four streamflow record extension techniques: Water Resources Research, v. 18, no. 4., p. 1081–1088.

Holland, P., Rubin, R., Wakabayashi, J., and Allen, J., 2009, Geology along the Rodgers Creek and Tolay faults: Late Cenozoic stratigraphy and tectonics, Franciscan mélange and high grade metamorphism, and Quaternary landslide complexes: Association of Engineering Geologists, San Francisco Section Field Trip guide, August 2009, 42 p.

Hopson, C.A., Mattinson, J.M., Pessagno, E.A., Jr., and Luyendyk, B.P., 2008, California Coast Range ophiolite: Composite Middle and Late Jurassic oceanic lithosphere: *in* Wright, J.E., and Shervais, J.W., eds. Ophiolites, Arcs, and Batholiths: A Tribute to Cliff Hopson: Geological Society of America Special Paper 483, p. 1–101.

Ingersoll, R. V., 1990, Nomenclature of upper Mesozoic strata of the Sacramento Valley of California: review and recommendations, *in* Ingersoll, R. V., and Nilsen, T. H., eds., Sacramento Valley symposium and guidebook: Pacific Section, Society of Economic Paleontologists and Mineralogists, Book 65, p. 1–3.

Kadir, T.N. and McGuire, R.A., 1987, Santa Rosa Plain ground water model: California Department of Water Resources Central District, 318 p.

Kharaka, Y.K., Thordsen, J.J., Evans, W.C., and Kennedy, B.M., 1999, Geochemistry and hydromechanical interaction of fluids associated with the San Andreas fault system, California, *in* Haneberg, W.C., Mozley, P.S., Moore, C.J., and Goodwin, L.B., eds., Faults and subsurface fluid flow in the shallow crust: Washington, D.C., American Geophysical Union, p. 129–148.

Kunkel, Fred, and Upson, J.E., 1960, Geology and ground water in Napa and Sonoma Valleys, Napa and Sonoma Counties, California: U.S. Geological Survey Water-Supply Paper 1495, 252 p.

Langenheim, V.E., Graymer, R. W., Jachens, R.C., McLaughlin, R.J., Wagner, D.L., and Sweetkind, D.S., 2010, Geophysical framework of the Northern San Francisco Bay region, California: Geosphere, v. 6, no. 5, p. 594–620.

Langenheim, V.E., McLaughlin, R.J., McPhee, D.K., Roberts, C.W., McCabe, C.A., and Elmira Wan, 2008, Chapter B, Geophysical framework of the Santa Rosa 7.5' quadrangle, Sonoma County, California: U.S. Geological Survey Open-File Report 2008–1009, 51 p., three sheets, scale 1:24,000, at URL: *http://pubs.usgs.gov/of/2008/1009/*.

Langenheim, V.E., Roberts, C.W., McCabe, C.A., McPhee, D.K., Tilden, J.E., and Jachens, R.C., 2006, Preliminary isostatic gravity map of the Sonoma Volcanic Field and vicinity, Sonoma and Napa Counties, California: U.S. Geological Survey Open-File Report 2006–1056, scale 1:100,000, at URL: *http://pubs.usgs.gov/of/2006/1056/*.

McLaughlin, R.J., Langenheim, V.E., Sarna-Wojcicki, A.M., Fleck, R.J., McPhee, D.K., Roberts, C.W., McCabe, C.A., and Wan, Elmira, 2008, Geologic and geophysical framework of the Santa Rosa 7.5' quadrangle, Sonoma County, California: U.S. Geological Survey Open- File Report 2008–1009, 51 p., three sheets, scale 1:24,000, at URL: *http://pubs.usgs.gov/of/2008/1009/*.

McLaughlin, R.J., and Ohlin, H.N., 1984, Tectonostratigraphic framework of the Geysers-Clear Lake region, California, in M. C, Blake, Jr., ed., Franciscan geology of northern California: S.E.P.M., Pacific section, v. 43, p. 221–254.

McLaughlin, R.J., Sarna-Wojcicki, A.M., Fleck, R.J., Langenheim, V.E., and Jachens, R.C., 2005, Framework geology and structure of the Sonoma Volcanics and associated sedimentary deposits, of the right-stepped Rodgers Creek-Maacama fault system and concealed basins beneath Santa Rosa plain *in* Stevens, C. and Cooper, J., editors, Late Neogene transition from transform to subduction margin east of the San Andreas fault in the wine country of the northern San Francisco Bay Area, California, Fieldtrip Guidebook and Volume Prepared for the Joint Meeting of the Cordilleran Section-GSA and Pacific Section-AAPG, April 29–May 1, 2005, San Jose, California, Fieldtrip 10, Pacific Section SEPM, p. 29–81.

McLaughlin, R.J., Sarna-Wojcicki, A.M., Fleck, R.J., Wright, W.H., Levin, V.R.G., and Valin, Z.C., 2004, Geology, tephrochronology, radiometric ages, and cross sections of the Mark West Springs 7.5' quadrangle, Sonoma and Napa Counties, California: U.S. Geological Survey Scientific Investigations Map 2858, scale 1:24,000, 2 sheets and pamphlet, at URL: *http://pubs.usgs.gov/sim/2004/2858/*.

McPhee, D.K., Langenheim, V.E., Hartzell, S., McLaughlin, R.J., Aagaard, B.T., Jachens, R.C., and McCabe, C., 2007, Basin structure beneath the Santa Rosa Plain, Northern California: Implications for damage caused by the 1969 Santa Rosa and 1906 San Francisco Earthquakes: Bulletin of the Seismological Society of America, v. 97, p. 1449–1457.

Morse, R.R. and Bailey, T.L., 1935, Geological observations in the Petaluma district, California: Geological Society of America Bulletin, v. 46, p. 1437–1456.

Nelson, E.P., Kullman, A.J., Gardner, M.H., Batzle, M., 1999, Fault-fracture networks and related fluid flow and sealing, Brushy Canyon Formation, West Texas, *in* Haneberg, W.C., Mozley, P.S., Moore, C.J., and Goodwin, L.B., eds., Faults and subsurface fluid flow in the shallow crust: Washington, D.C., American Geophysical Union, p. 69–82.

Nettleton, L.L., 1976, Gravity and Magnetics in Oil Prospecting: New York, McGraw-Hill, 464 p.

Norris, R.M. and Webb, R.W., 1976, Geology of California: New York, John Wiley and Sons, 365 p.

Page, R.W., 1986, Geology of the fresh ground-water basin of the Central Valley, California, with texture maps and sections: U.S. Geological Survey Professional Paper 1401-C, 54 p. and 5 plates.

Powell, C.L., II, Allen, J.R., and Holland, P.J., 2004, Invertebrate paleontology of the Wilson Grove Formation (late Miocene to late Pliocene), Sonoma and Marin Counties, California, with some observations on its stratigraphy, thickness, and structure: U.S. Geological Survey Open-File Report 2004–1017, 105 p., at URL: *http://pubs.usgs.gov/of/2004/1017/*.

Simley, J.D. and Carswell Jr., W.J., 2009, The National Map – Hydrography: U.S. Geological Survey Fact Sheet 2009-3054, 4 p.

Sloop, C., Honton, J., Creager, C., Chen, L., Andrews, E.S., Bozkurt, S., 2007, The altered Laguna, A conceptual model of watershed stewardship: Laguna de Santa Rosa Foundation, Santa Rosa, CA, 48 p., at URL *http://www.lagunafoundation.org/knowledgebase/?q=node/182.*

Sowers, J.M., Noller, J.S., and Lettis, W.R., 1998, Quaternary geology and liquefaction susceptibility, Napa, California, 1:100,000–A digital database: U.S. Geological Survey Open-File Report 98-460: U.S. Geological Survey Open-File Report 98–460.

Swartz, R.J. and Hauge, Carl, 2003, California's groundwater: California Department of Water Resources Bulletin 118—Update 2003, 246 p.

Sweetkind, D.S., Taylor, E.M., McCabe, C.A., Langenheim, V.E., and McLaughlin, R.J., 2010, Three-dimensional geologic modeling of the Santa Rosa Plain, California: Geosphere, v. 6, p. 237–274.

Tamanyu, Shiro, 1999, How do fracture-vein systems form in a geothermal reservoir? Examples from Northern Honshu, Japan, *in* Haneberg, W.C., Mozley, P.S., Moore, C.J., and Goodwin, L.B., eds., Faults and subsurface fluid flow in the shallow crust: Washington, D.C., American Geophysical Union, p. 185–206.

U.S. Army Corps. Of Engineers, 2002, Santa Rosa Creek ecosystem restoration feasibility study; Sonoma County, California: Hydrologic Engineering Office. August 5, 2002.

U.S. Geological Survey, 2011, National Water Information System data available on the World Wide Web (Water Data for the Nation), accessed May 13, 2011 at URL: *http://wdr.water.usgs.gov/wy2009/pdfs/11466800.2009.pdf.*

Wagner, D.L. and Bortugno, E.J., 1982, Geologic map of the Santa Rosa Quadrangle, California Division of Mines and Geology Regional Geologic Map Series Map 2A, scale 1:250,000.

Wagner, D.L., and Gutierrez, C.I., 2010, Geologic Map of the Napa 30'x 60' Quadrangle, California: California Geological Survey Preliminary Geologic Map, scale 1:100,000, accessed June, 2013 at URL: *http://www.conservation.ca.gov/cgs/rghm/rgm/Pages/northern_region_quads.aspx.*

Wagner, D.L. Fleck, R.J., Sarna-Wojcicki, A.M., and Deino, Alan, 2005, Golden Gate to southern Sonoma County, Rodgers Creek fault, Burdell Mountain, Donnell Ranch, and southern Sonoma Volcanics *in* Stevens, Calvin and Cooper, John, eds., Late Neogene transition from transform to subduction margin east of the San Andreas fault in the wine country of the northern San Francisco Bay Area, California—Field trip 10 Guidebook and Volume Prepared for the Joint Meeting of the Cordilleran Section—Geological Society of America and Pacific Section—American Association of Petroleum Geologists, San Jose, California, April 29–May 1, 2005: Pacific Section–Society for Sedimentary Geology, p. 1–28.

Wagner, D.L., Saucedo, G.J., Clahan, K.B., Fleck, R.J., Langenheim, V.E., McLaughlin, R.J., Sarna-Wojcicki, A.M., Allen, J.R., and Deino, A.L., 2011, Geology, geochronology, and paleogeography of the southern Sonoma volcanic field and adjacent areas, northern San Francisco Bay region, California: Geosphere, v. 7, p. 658–683.

Weaver, C.E., 1949, Geology of the Coast Ranges immediately north of the San Francisco Bay region, California: Geological Society of America Memoir, v. 35, 242 p.

Williams, R.A., Langenheim, V.E., McLaughlin, R.J., Odum, J.K., Worley, D.M., Stephenson, W.J., Kent, R.L., McCullough, S.M., Knepprath, N.E., and Leslie, S.R., 2008, Seismic reflection profiles image the Rodgers Creek fault and Trenton Ridge beneath urban Santa Rosa, California: Seismological Research Letters, v. 79, p. 317.

Woolfenden, L.R., Hevesi, J.A., Niswonger, R.G., and Nishikawa, Tracy, 2011, Modeling a complex hydrologic system with an integrated hydrologic model: preliminary results, *in* Maxwell, R.M., Poeter, E.P., Hill, M.C., and Zheng, Chunmiao, eds., MODFLOW and More 2011: Integrated hydrologic modeling–Conference proceedings, June 5–8, 2011: Golden, Colo., International Groundwater Modeling Center, p. 134–138.

Wong, I.G. and Bott, J.D., 1995, A new look back at the 1969 Santa Rosa, California earthquakes: Seismological Society of America Bulletin, v. 85, p. 334–341.

Wong, Teng-fong and Zhu, Wenlu, 1999, Brittle faulting and permeability evolution: hydromechanical measurement, microstructural observation, and network modeling, *in* Haneberg, W.C., Mozley, P.S., Moore, C.J., and Goodwin, L.B., eds., Faults and subsurface fluid flow in the shallow crust: Washington, D.C., American Geophysical Union, p. 83–100.

Youngs, L.G., Campion, L.F., Chapman, R.H., Higgins, C.T., Levias, E., Chase, G.W., and Bezore, S.P., 1983, Geothermal resources of the northern Sonoma Valley Area, California: California Division of Mines and Geology, Open-File Report 83–27, 106 p., 6 plates.

Chapter C. Groundwater Quality and Source and Age of Groundwater in the Santa Rosa Plain Watershed, Sonoma County, California

By Peter Martin, Loren F. Metzger, Jill N. Densmore, and Roy A. Schroeder

Introduction

Groundwater-quality and source and age of groundwater were characterized for the Santa Rosa Plain (SRP) and the surrounding area within the Santa Rosa Plain watershed (SRPW) by using analyses for selected physical properties, inorganic constituents, and stable and radioactive isotopes. This chapter describes (1) the areal, depth-dependent, and temporal variations in groundwater quality; (2) groundwater-quality constituents of particular concern for drinking water and irrigation; and (3) the recharge source and age of groundwater. Data used in this characterization were compiled from previous investigations; from databases maintained by the California Department of Public Health (CDPH), the California Department of Water Resources (CDWR), and various public-supply purveyors; and from analyses from surface-water sites and groundwater wells sampled either as part of this study or concurrently by private consultants. These data represent untreated water samples, which are not equivalent to analyses used for compliance to drinking-water quality standards. To place these results within a health-based context, however, these untreated water analyses were compared with federal (U.S. Environmental Protection Agency, or USEPA) and state (CDPH) drinking-water regulatory standards. As with *chapters A* and *B*, groundwater conditions in the SRP portion of the study area are the primary focus of this chapter. For the purposes of this chapter, specific surface-water sites, springs, and groundwater wells are identified by the abbreviations "SW," "SPR," and "W," respectively.

Previous Investigations

One of the earliest published geohydrologic investigations of the SRPW to include water quality was by Cardwell (1958). His study included a limited assessment of water quality based on analyses from 200 wells, of which 80 analyses were relatively complete for major ions. Cardwell (1958) characterized the quality of groundwater in the SRPW as generally satisfactory for most uses on the basis of dissolved solids and hardness concentrations in the range of 250 to 350 mg/L

and 60 to 160 mg/L, respectively. However, Cardwell (1958) noted several areas where concentrations of boron, iron, and manganese were high enough to classify water as being of unsatisfactory quality without treatment. Boron was cited as a constituent of particular concern on the western side of the Windsor Basin in samples from several irrigation wells perforated in the Glen Ellen Formation. Cardwell (1958) hypothesized that boron concentrations of up to 2,200 micrograms per liter (μg/L) in well samples from the area of township, range, and section (T/R/S) 8N/9W-23 could be attributable to deeply circulating groundwater rising along a fault zone. Iron and manganese were noted to be high (greater than the 1946 drinking water standard of 300 μg/L for both elements) in localized areas of the Wilson Grove Formation (formerly known as the Merced Formation; Fox, 1983). High concentrations of iron and manganese were especially prevalent in samples from wells perforated in thick sections of blue (unoxidized) sandstone and where the Wilson Grove Formation crops out on the western side of the SRP (Cardwell, 1958).

Water-quality conditions in the SRPW were described in detail in several reports by the CDWR. Ford (1975) described that groundwater in localized areas was of poor quality for agricultural and domestic use because of elevated concentrations of boron, sodium, iron, manganese, and dissolved solids. Herbst and others (1982), characterized groundwater quality by type based on major-ion composition and grouped water samples by geologic formations. Hardness was described by Herbst and others (1982) as one of the most widespread water-quality problems in the SRP as a result of an area of soft water (0 to 60 mg/L as calcium carbonate) northwest of the city of Santa Rosa and hard water (121 to 180 mg/L as calcium carbonate) to the west and southwest of the city of Santa Rosa. Areas of high-dissolved solids concentrations were identified in water from wells in the Rohnert Park and Cotati area (6N/7W-18 and 6N/8W-16 and -26), north of Santa Rosa (7N/8W-13C), and east of Sebastopol (7N/8W-29 and -30). As with previous reports, boron was cited as a constituent of particular concern. High concentrations of boron were associated with sodium-bicarbonate type water, particularly along the Rodgers Creek fault and along an area that Herd and Helley (1976) described as a possible branch of the Rodgers Creek fault in the southeast portion of 8N/9W. This hypothesized branch of the Rodgers Creek fault coincides with the western end of the shallow west-northwest-striking Trenton Ridge fault, which was described earlier in this report (*chapter B*). Boron also was reported as being widespread in the Windsor area in samples from relatively shallow depths of about 60 to 150 feet below land surface (ft bls), but at concentrations generally less than 500 µg/L (Herbst and others, 1982).

In 2000, the California State Water Resources Control Board (SWRCB), the U.S. Geological Survey (USGS), and the Lawrence Livermore National Laboratory (LLNL) initiated the Groundwater Ambient Monitoring Assessment (GAMA) Program to assess the quality of ambient groundwater in aquifers used for drinking-water supply and to establish a baseline groundwater-quality monitoring program. The GAMA Program partitioned California into 10 hydrogeologic provinces and 35 study units within these provinces (Belitz and others, 2003). Within each study unit, statistical and graphical methods to explain relations between water quality and various causative factors, such as well depth, groundwater age, oxidation-reduction (redox) status of the subsurface, and position along a conceptual flowpath (Belitz and others, 2003). One such study unit was the North San Francisco Bay (NSF) Study Unit, which includes the SRPW (Kulongoski and others, 2006; 2010). In 2004, samples of untreated water were collected from 89 wells in the NSF Study Unit, including 28 wells from the SRPW. As with other GAMA study units, analysis of samples collected from wells in the SRPW focused primarily on organic constituents, including volatile organic compounds (VOCs), pesticides and pesticide degradates, and organic wastewater indicators (Kulongoski and others, 2006). Fewer wells (five) within the SRPW were sampled for inorganic constituents (major and minor elements and nutrients). Stable and radioactive isotopes were used to characterize the sources or ages of groundwater, including samples collected from 25 wells for oxygen (^{18}O) and hydrogen (^2H), 27 wells for tritium (^3H), and 5 wells for carbon-14 (^{14}C) and carbon-13 (^{13}C) (Kulongoski and others, 2006). Results for the NSF Study Unit as a whole, showed that no anthropogenic constituents were detected at concentrations higher than regulatory thresholds, and few naturally occurring constituents were detected at concentrations above regulatory thresholds (Kulongoski and others, 2006 and 2010).

Historical Monitoring

Groundwater quality in the SRPW has been monitored since at least 1947 to ensure minimum quality standards are met for a variety of water demands, including domestic, agricultural, and industrial use. The monitoring of groundwater for these various purposes has primarily focused on physical properties and inorganic constituents, such as specific conductance, pH, temperature, dissolved solids, hardness, major ions, nutrients, and trace (minor) elements. In the last several decades, however, attention has shifted to include organic constituents

such as VOCs, pesticides, organic wastewater indicators, and other constituents of potential concern associated with anthropogenic activities. For this study, analyses completed on 2 springs and 162 wells by various agencies were compiled to help describe the groundwater quality of the SRPW (fig. 1, table 1, and appendix A).

Water-quality sampling of springs in the SRPW consists of one analysis from each of two springs located in the Mayacmas Mountains (fig. 1A and appendix A). These two analyses from the mid-1970s include physical properties and inorganic constituents (major ions, selected trace elements, and nutrients).

Groundwater-quality data from wells in the SRPW includes one-time sampling for short-term studies and multiple samples collected over time from selected wells. The longest sustained water-quality monitoring effort in the SRPW has been done by the CDWR. Since the late 1940s, the CDWR has sampled and analyzed selected wells for major ions (calcium, magnesium, potassium, sodium, chloride, and sulfate), boron, nitrate, total dissolved solids, alkalinity, specific conductance, pH, and water temperature.

Methods of Sample Collection and Analysis

The methods for the collection and analysis of surface-water and groundwater samples for this study (collected between 2006 and 2010, referred to as recent samples) and for previous studies (primarily collected prior to 2006, referred to as historical samples) by the CDWR, the CDPH, and public-supply purveyors have varied in accordance with different sampling objectives and in response to improvements in sampling protocols and analytical techniques. The sampling protocols and analytical techniques for the recent and historical samples are briefly described in the following sections. The sampling protocols and analytical techniques for samples collected and analyzed as part of the NSF GAMA study are described in Kulongoski and others (2006). All non-USGS laboratories used as sources of data in this report have met the requirements of California Code of Regulations Title 22, Division 4, Chapter 19, on the certification of environmental laboratories, including successful participation in the California Environmental Laboratory Accreditation Program (ELAP).

Recent Sample Collection and Analysis

Fifteen surface-water samples (fig. 1A) were collected for the analysis of the stable isotopes of oxygen (^{18}O) and hydrogen (^2H) in the SRPW and from neighboring areas to the west of the study area (SW14 and SW18) to help determine the source of groundwater to the SRPW. Specific conductance and water temperature were measured at the time of sample collection. For ^{18}O and ^2H, unfiltered water samples were

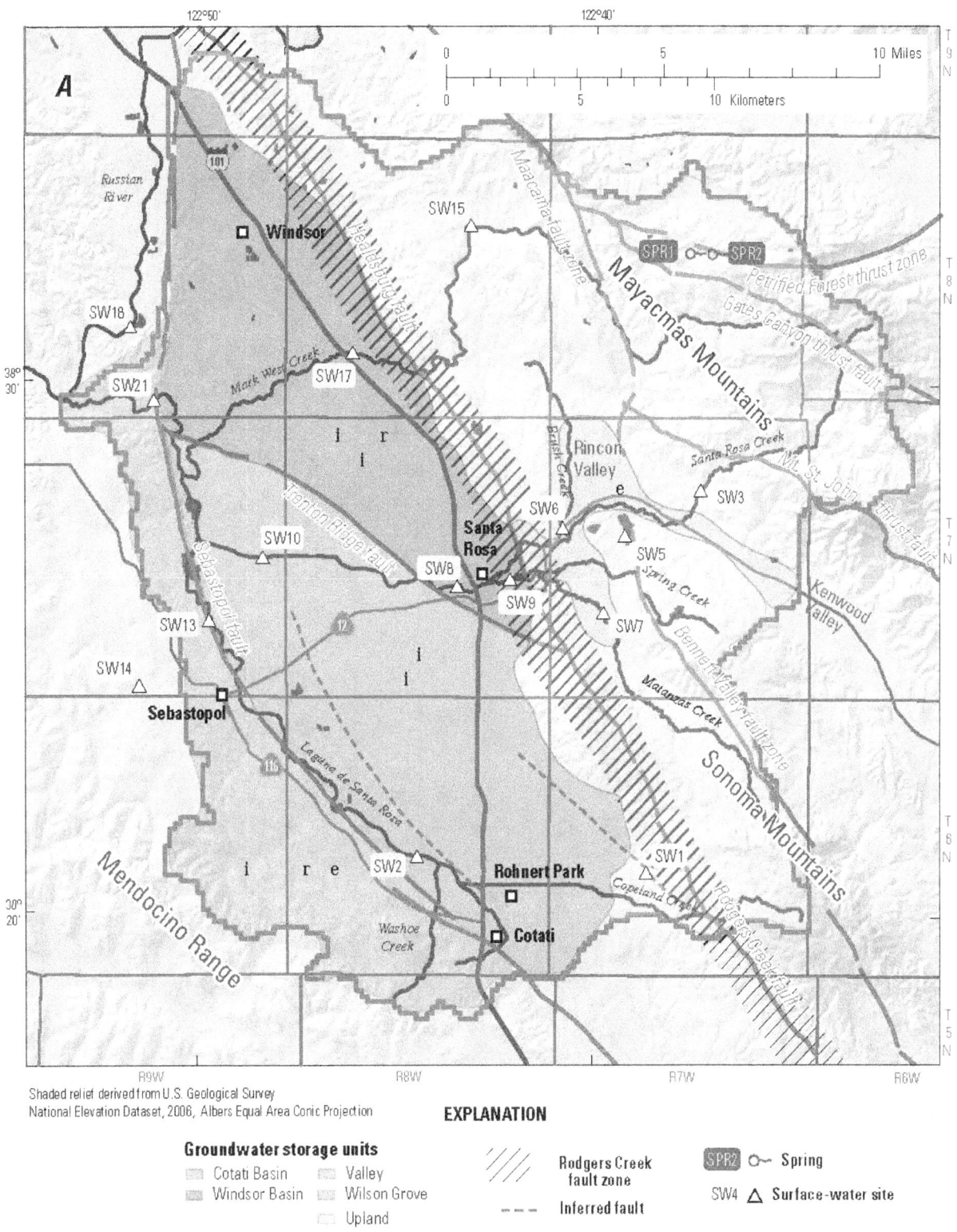

EXPLANATION

Groundwater storage units

Cotati Basin   Valley

Windsor Basin   Wilson Grove

Upland

///// Rodgers Creek fault zone

--- Inferred fault

SPR2 ○~ Spring

SW4 △ Surface-water site

Figure 1.   Locations of *A*, surface-water sites and springs and *B*, wells sampled in the Santa Rosa Plain watershed, Sonoma County, California, 1947–2010.

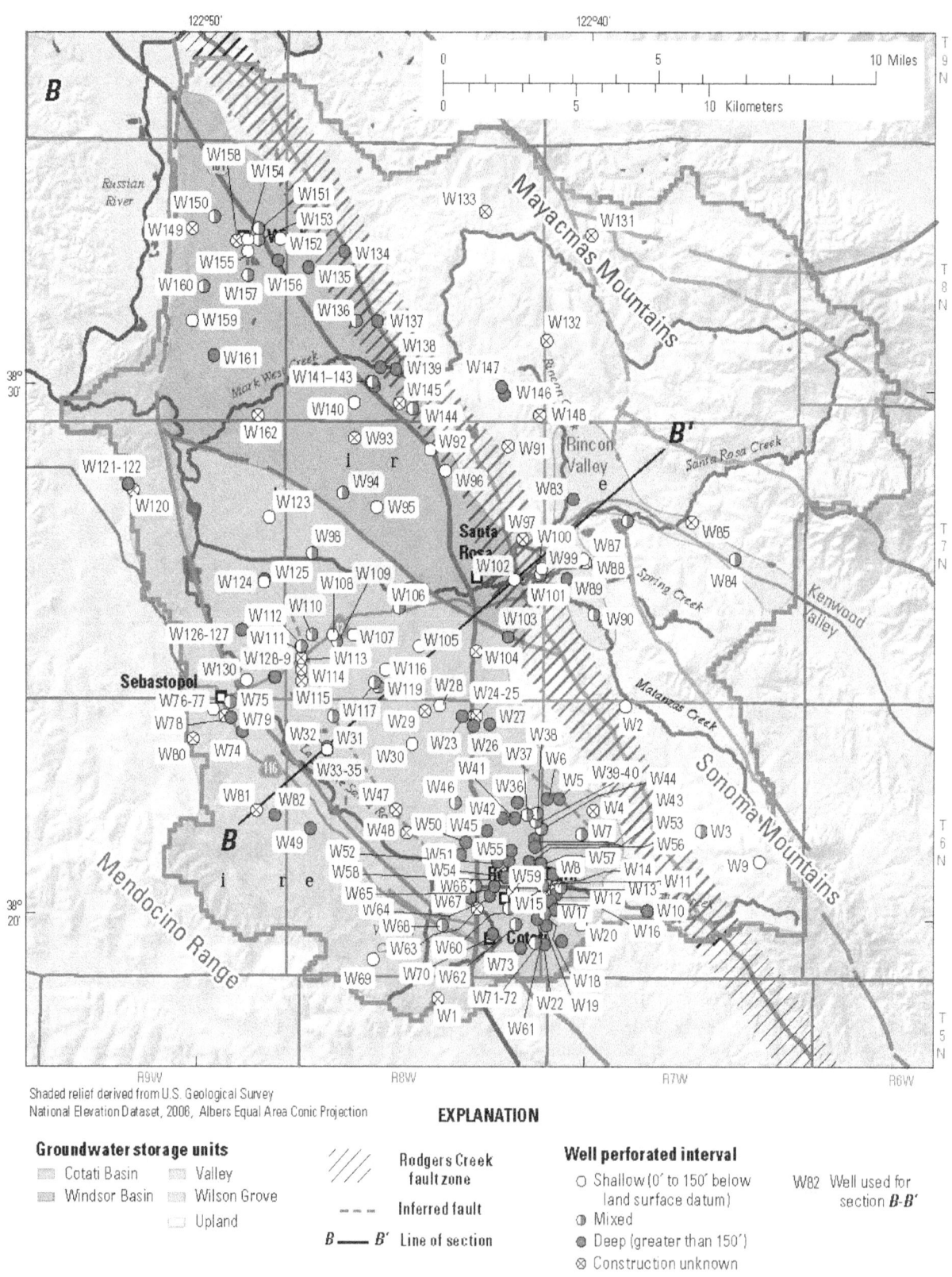

Shaded relief derived from U.S. Geological Survey
National Elevation Dataset, 2006, Albers Equal Area Conic Projection

EXPLANATION

Groundwater storage units

- Cotati Basin
- Windsor Basin
- Valley
- Wilson Grove
- Upland

/// Rodgers Creek
fault zone

- - - Inferred fault

B ——— B' Line of section

Well perforated interval

- ○ Shallow (0′ to 150′ below land surface datum)
- ◑ Mixed
- ● Deep (greater than 150′)
- ⊗ Construction unknown

W82 Well used for section **B-B′**

Figure 1. Continued.

Table 1. Construction information for selected wells used to obtain water-quality data in the Santa Rosa Plain watershed, Sonoma County, California.

[See figure 1 for locations (map numbers) of wells, springs, and surface-water sites. U.S. Geological Survey (USGS) site identification number is the unique number for each site in USGS National Water Information System (NWIS) database. Depths in feet below land surface (fbls). Depth Category based on average of perforated interval. Land surface altitude, where available, in feet above sea level which refers to the North American Vertical Datum of 1988 (NAVD of 1988). **Abbreviations:** CDPH, California Department of Public Health; CDWR, California Department of Water Resources; GAMA, Groundwater Ambient Monitoring and Assessment; owner, data provided by well owner; Unk, unknown; USGS-GAMA, Groundwater Ambient Monitoring and Assessment Program (USGS and California State Water Resources Control Board); <, actual value is less than value shown; —, no data]

Map number	Groundwater storage unit	State well number	USGS site identification number	GAMA identification number	Depth drilled (fbls)	Depth cased (fbls)	Depth of top perforation (fbls)	Depth of bottom perforation (fbls)	Depth category	Land surface altitude[1]	Period of data included in this report	Water quality data source
W1	Wilson Grove	5N/8W-03G1	3818271224434601	NSFVP-04	397	397	—	—	Unk	211	2004	USGS-GAMA
W2	Uplands	6N/7W-05A1	—	—	120	[2]120	—	—	Shallow	—	1960	CDWR
W3	Uplands	—	—	—	306	306	127	306	Mixed	—	1996–2002	CDPH
W4	Cotati	6N/7W-17E1	—	—	650	[2]650	—	—	Unk	—	1957–1963	CDWR
W5	Cotati	—	—	—	600	600	323	580	Deep	—	1987	CDPH
W6	Cotati	—	—	—	602	560	297	540	Deep	158	1988	CDPH
W7	Cotati	6N/7W-18R1	382138122393901	—	250	250	18	240	Mixed	—	1957–2007	CDWR
W8	Cotati	—	—	—	370	364	242	344	Deep	—	1980–2007	CDPH, CDWR
W9	Uplands	6N/7W-23H1	382038122354001	NSFVOL-10	140	136	76	136	Shallow	1,704	2004–2006	CDPH, USGS-GAMA
W10	Cotati	6N/7W-28L1	382009122382601	NSFVP-01	921	864	441	862	Deep	331	1984–2009	CDPH, CDWR, USGS-GAMA
W11	Cotati	—	—	—	325	325	100	—	Unk	—	1974	CDPH, CDWR
W12	Cotati	—	—	—	430	415	225	395	Deep	—	2005	CDPH
W13	Cotati	—	—	—	417	400	100	390	Mixed	—	2005–2008	CDPH
W14	Cotati	6N/7W-30D1	—	—	—	—	—	—	Unk	—	1957–1961	CDWR
W15	Cotati	—	—	—	470	420	120	380	Mixed	109	1982–2009	CDPH, CDWR, owner
W16	Cotati	—	—	—	526	401	238	398	Deep	—	2000–2009	CDPH
W17	Cotati	—	—	—	689	498	220	480	Deep	—	2000–2006	CDPH
W18	Cotati	6N/8W-02E1	382349122424201	—	726	387	130	380	Mixed	114	1989–2002	CDPH
W19	Cotati	—	—	—	715	705	210	695	Deep	—	1991	CDPH
W20	Cotati	6N/7W-30R1	381955122395901	—	150	150	120	150	Shallow	178	1976–2002	CDWR
W21	Cotati	6N/7W-31C1	381934122403601	NSFVP-06	716	700	170	680	Deep	144	1998–2007	CDPH, USGS-GAMA
W22	Cotati	—	—	—	700	603	160	590	Deep	—	1998–2007	CDPH
W23	Cotati	6N/8W-02E1	—	—	172	172	167	172	Deep	114	1974–1984	CDWR
W24	Cotati	—	—	—	—	—	—	—	Unk	—	1994	CDPH
W25	Cotati	—	—	—	—	—	—	—	Unk	—	1994	CDPH
W26	Cotati	6N/8W-02J2	382337122422902	NSFVP-23	530	530	369	530	Deep	115	2004	USGS-GAMA
W27	Cotati	—	—	—	900	900	274	880	Deep	108	2006	CDPH
W28	Cotati	6N/8W-03B1	382402122434401	—	60	60	50	60	Shallow	109	1957–2007	CDWR
W29	Cotati	—	—	—	—	—	—	—	Unk	—	1996	CDPH
W30	Cotati	6N/8W-04R1	382315122443001	NSFVP-03	120	120	55	85	Shallow	95	2004	USGS-GAMA
W31	Cotati	6N/8W-05E1	—	—	638	638	20	638	Mixed	[3]90	1951	CDWR
W32	Cotati	6N/8W-07A2	382307122463801	NSFVP-12	815	808	650	800	Deep	82	1982–2009	CDPH, CDWR, USGS-GAMA, owner

Table 1.   Construction information for selected wells used to obtain water-quality data in the Santa Rosa Plain watershed, Sonoma County, California.—Continued

[See figure 1 for locations (map numbers) of wells, springs, and surface-water sites. U.S. Geological Survey (USGS) site identification number is the unique number for each site in USGS National Water Information System (NWIS) database. Depths in feet below land surface (fbls). Depth Category based on average of perforated interval. Land surface altitude, where available, in feet above sea level which refers to the North American Vertical Datum of 1988 (NAVD of 1988). **Abbreviations:** CDPH, California Department of Public Health; CDWR, California Department of Water Resources; GAMA, Groundwater Ambient Monitoring and Assessment; owner, data provided by well owner; Unk, unknown; USGS-GAMA, Groundwater Ambient Monitoring and Assessment Program (USGS and California State Water Resources Control Board); <, actual value is less than value shown; —, no data]

Map number	Groundwater storage unit	State well number	USGS site identification number	GAMA identification number	Depth drilled (fbls)	Depth cased (fbls)	Depth of top perforation (fbls)	Depth of bottom perforation (fbls)	Depth category	Land surface altitude[1]	Period of data included in this report	Water quality data source
W33	Cotati	6N/8W-07A4	382308122463903	—	900	80	60	80	Shallow	81	2007	USGS
W34	Cotati	6N/8W-07A5	382308122463902	—	900	257	237	257	Deep	81	2007	USGS
W35	Cotati	6N/8W-07A6	382308122463901	—	900	570	550	570	Deep	81	2007	USGS
W36	Cotati	—	—	—	615	502	302	462	Deep	103	1988–1989	CDPH, CDWR
W37	Cotati	—	—	—	520	510	120	490	Mixed	—	1998–2007	CDPH, CDWR
W38	Cotati	—	—	—	600	510	110	510	Mixed	—	1988–2009	CDPH
W39	Cotati	—	—	—	411	411	38	411	Mixed	—	1996	CDPH
W40	Cotati	—	—	—	606	445	300	440	Deep	—	2002–2008	CDPH
W41	Cotati	—	—	—	580	580	224	490	Deep	107	1976–2009	CDPH, CDWR, owner
W42	Cotati	—	—	—	624	562	298	522	Deep	—	1980–2007	CDPH, CDWR
W43	Cotati	—	—	—	1,328	1,290	275	1,275	Deep	113	1987–2005	CDPH, CDWR
W44	Cotati	6N/8W-13R3	382141122410901	NSFVP-05	600	502	130	450	Mixed	115	1988–2009	CDPH, USGS-GAMA
W45	Cotati	—	—	—	1,512	1,501	351	1,491	Deep	100	1981–2009	CDPH, CDWR
W46	Cotati	—	—	—	240	240	120	240	Mixed	—	1991	CDWR
W47	Cotati	6N/8W-16G1[4]	—	—	—	—	—	—	Unk	—	1959	CDWR
W48	Cotati	6N/8W-16R1	—	—	1,204	1,204	—	1,204	Unk	—	1958–1961	CDWR
W49	Wilson Grove	—	—	—	310	300	204	300	Deep	—	1996	CDPH
W50	Cotati	—	—	—	721	715	175	675	Deep	—	1999–2008	CDPH
W51	Cotati	—	—	—	610	590	190	580	Deep	—	1988–1997	CDPH
W52	Cotati	—	—	—	1,504	1,500	300	1,500	Deep	103	1987–2008	CDPH, CDWR
W53	Cotati	—	—	—	614	614	260	594	Deep	—	1998–2007	CDPH
W54	Cotati	—	—	—	496	496	224	494	Deep	106	1981–2008	CDPH, CDWR
W55	Cotati	—	—	—	725	424	161	421	Deep	—	2000–2009	CDPH
W56	Cotati	—	—	—	510	510	200	450	Deep	113	2002–2008	CDPH
W57	Cotati	—	—	—	460	405	190	—	Deep	—	1980–2007	CDPH, CDWR, owner
W58	Cotati	—	—	—	630	458	265	458	Deep	104	1982–2009	CDPH, CDWR, owner
W59	Cotati	—	—	—	503	503	—	—	Unk	108	1987–2008	CDPH, CDWR
W60	Cotati	—	—	—	500	500	118	478	Mixed	106	1982–2009	CDPH, CDWR, owner
W61	Cotati	—	—	—	710	688	156	666	Deep	—	2007	CDPH
W62	Cotati	—	—	—	610	500	60	425	Mixed	109	1988–2000	CDPH, CDWR
W63	Cotati	6N/8W-26[4]	—	—	—	—	—	—	Unk	—	1956	CDWR
W64	Cotati	—	—	—	462	462	288	462	Deep	100	1982–2009	CDPH, CDWR, owner
W65	Cotati	—	—	—	473	473	160	463	Deep	103	1982–2009	CDPH, CDWR, owner

Table 1. Construction information for selected wells used to obtain water-quality data in the Santa Rosa Plain watershed, Sonoma County, California.—Continued

[See figure 1 for locations (map numbers) of wells, springs, and surface-water sites. U.S. Geological Survey (USGS) site identification number is the unique number for each site in USGS National Water Information System (NWIS) database. Depths in feet below land surface (fbls). Depth Category based on average of perforated interval. Land surface altitude, where available, in feet above sea level which refers to the North American Vertical Datum of 1988 (NAVD of 1988). **Abbreviations**: CDPH, California Department of Public Health; CDWR, California Department of Water Resources; GAMA, Groundwater Ambient Monitoring and Assessment; owner, data provided by well owner; Unk, unknown; USGS-GAMA, Groundwater Ambient Monitoring and Assessment Program (USGS and California State Water Resources Control Board); <, actual value is less than value shown; —, no data]

Map number	Groundwater storage unit	State well number	USGS site identification number	GAMA identification number	Depth drilled (fbls)	Depth cased (fbls)	Depth of top perforation (fbls)	Depth of bottom perforation (fbls)	Depth category	Land surface altitude[1]	Period of data included in this report	Water quality data source
W66	Cotati	—	—		603	475	128	460	Mixed	102	1998–2007	CDPH, CDWR
W67	Cotati	6N/8W-26C2	382021122425701	NSFVP-07	795	685	295	670	Deep	95	1988–2009	CDPH / USGS
W68	Cotati	—	—		350	337	135	337	Mixed	—	1996	CDPH
W69	Wilson Grove	6N/8W-33M1	381915122452701	NSFWilson Grove-12	261	261	100	128	Shallow	203	2004	CDPH, USGS-GAMA
W70	Cotati	—	—		660	552	370	400	Deep	100	1957–1987	CDPH, CDWR
W71	Cotati	—	—		510	500	125	490	Mixed	120	1999–2008	CDPH, CDWR
W72	Cotati	—	—		642	642	520	—	Deep	—	1996–2005	CDPH
W73	Cotati	—	—		515	500	220	485	Deep	—	1976–2008	CDPH, CDWR, owner
W74	Wilson Grove	—	—		1,015	572	172	552	Deep	—	1974–2009	CDPH, CDWR
W75	Wilson Grove	6N/9W-02B1	382400122491201	NSFWilson GroveFP-01	647	528	138	528	Mixed	83	2004	CDPH, CDWR, USGS-GAMA
W76	Wilson Grove	6N/9W-02C1	382352122493301	NSFWilson Grove-08	600	600	332	600	Deep	115	1960–2004	CDPH, CDWR, USGS-GAMA
W77A	Wilson Grove	6N/9W-02C2	382352122493801		660	650	100	[6]120	Shallow	116	2006	USGS
W77B	Wilson Grove	6N/9W-02C2	382352122493801		660	650	[6]340	[6]360	Deep	116	2006	USGS
W77C	Wilson Grove	6N/9W-02C2	382352122493801		660	650	[6]640	650	Deep	116	2006	USGS
W78	Wilson Grove	6N/9W-02G1			552	[2]552			Unk	—	1957–1964	CDWR
W79	Wilson Grove	—			776	530	237	468	Deep	83	1975–2008	CDPH, CDWR
W80	Wilson Grove	6N/9W-03R84			280	[2]280	—	—	Unk	—	1973	CDWR
W81	Wilson Grove	6N/9W-13F1[4]			220	[2]220	—	—	Unk	—	1952	CDWR
W82	Wilson Grove	6N/9W-13J2	382153122480301	NSFWilson Grove-04	452	432	432	452	Deep	197	1997	CDPH, USGS-GAMA
W83	Valley	7N/7W-07K1	382754122402501	NSFVP-25	265	265	160	265	Deep	236	2004	USGS-GAMA
W84	Valley	7N/7W-14P1			828	812	111	812	Mixed	—	1961–1963	CDWR
W85	Valley	7N/7W-15C1	382737122372501		—		—	—	Unk	378	1957–2007	CDWR
W86	Valley	7N/7W-17A2	382730122390201		1,018	1,008	84	1,008	Mixed	300	2007	USGS
W87	Valley	—			160	160	—	—	Unk	211	1982–2005	CDPH, CDWR
W88	Valley	7N/7W-18R2	382647122400702	NSFVP-16	206	206	50	100	Shallow	205	1983–2005	CDPH, USGS-GAMA
W89A	Valley	7N/7W-19G2	382623122403301		1,074	870	250	[6]290	Deep	215	2010	owner, USGS
W89B	Valley	7N/7W-19G2	382623122403301		1,074	870	[6]500	[6]620	Deep	215	2010	owner, USGS
W89C	Valley	7N/7W-19G2	382623122403301		1,074	870	[6]750	860	Deep	215	2010	owner, USGS
W90	Valley	7N/7W-29D1	382548122395501		588	588	40	—	Mixed	243	1957–1984	CDWR

Table 1.   Construction information for selected wells used to obtain water-quality data in the Santa Rosa Plain watershed, Sonoma County, California.—Continued

[See figure 1 for locations (map numbers) of wells, springs, and surface-water sites. U.S. Geological Survey (USGS) site identification number is the unique number for each site in USGS National Water Information System (NWIS) database. Depths in feet below land surface (fbls). Depth Category based on average of perforated interval. Land surface altitude, where available, in feet above sea level which refers to the North American Vertical Datum of 1988 (NAVD of 1988). **Abbreviations:** CDPH, California Department of Public Health; CDWR, California Department of Water Resources; GAMA, Groundwater Ambient Monitoring and Assessment; owner, data provided by well owner; Unk, unknown; USGS-GAMA, Groundwater Ambient Monitoring and Assessment Program (USGS and California State Water Resources Control Board); <, actual value is less than value shown; —, no data]

Map number	Groundwater storage unit	State well number	USGS site identification number	GAMA identification number	Depth drilled (fbls)	Depth cased (fbls)	Depth of top perforation (fbls)	Depth of bottom perforation (fbls)	Depth category	Land surface altitude[1]	Period of data included in this report	Water quality data source
W91	Uplands	7N/8W-01M1	3828491224220501	NSFVOL-01	323	—	—	—	Unk	329	1983–2005	CDPH, CDWR, USGS-GAMA
W92	Windsor	7N/8W-03L1	3828521224440301	—	—	135	17	95	Shallow	145	1957–1984	CDWR
W93	Windsor	7N/8W-05G1	3828551224460501	—	—	—	—	—	Unk	142	1957–1984	CDWR
W94	Windsor	7N/8W-08L1	3827591224462001	NSFVP-02	350	350	65	341	Mixed	138	2004	USGS-GAMA
W95	Windsor	7N/8W-09N1	3827401224452601	—	63	63	35	63	Shallow	125	1975–2003	CDWR
W96	Windsor	7N/8W-10A1	3828251224341101	NSFVP-24	85	85	65	85	Shallow	149	2004	USGS-GAMA
W97	Valley	7N/8W-13[4]	—	—	—	—	—	—	Unk	—	1957	CDWR
W98	Cotati	7N/8W-18Q1	—	—	811	[2]811	50	—	Mixed	—	1947–1969	CDWR
W99	Valley	—	—	—	1,200	915	[2]120	910	Mixed	198	1958–2009	CDPH, CDWR
W100	Valley	—	—	—	291	290	50	288	Mixed	198	1983–2005	CDPH, CDWR
W101	Valley	—	—	—	1,000	940	280	—	Deep	—	1950–2009	CDPH, CDWR
W102A	Windsor	7N/8W-24F1	3826221224149001	—	1,020	200	130	180	Mixed	196	2009	owner, USGS
W102B	Windsor	7N/8W-24F1	3826221224149001	—	1,020	—	[6]786	[6]806	Deep	196	2009	owner, USGS
W103	Cotati	—	—	—	315	296	165	292	Deep	[3]174	1996	CDPH
W104	Cotati	—	—	—	—	—	—	—	Unk	—	1997	CDPH
W105	Cotati	7N/8W-27N2	382509122441701	—	65	65	45	65	Shallow	119	1974–1992	CDWR
W106	Cotati	—	—	—	256	256	116	256	Mixed	—	2001	CDPH
W107	Cotati	7N/8W-29K1	382513122455101	—	67	67	47	67	Shallow	99	1976–2007	CDWR
W108	Cotati	7N/8W-29M2	3825181224630001	—	—	98	78	98	Shallow	95	1975–1992	CDWR
W109	Cotati	7N/8W-19P1	382510122462701	NSFVPFP-02	250	231	61	231	Mixed	95	2004	USGS-GAMA
W110	Cotati	7N/8W-30K1	—	—	290	290	105	290	Mixed	[3]95	1976–2005	CDWR
W111	Cotati	7N/8W-30P1	382459122472101	—	—	196	—	—	Unk	86	1963–1979	CDWR
W112	Cotati	7N/8W-30P6	—	—	158	158	65	156	Mixed	—	1981–1996	CDWR
W113	Cotati	7N/8W-31C1	—	—	320	[2]320	—	—	Unk	—	1960	CDWR
W114	Cotati	7N/8W-31F1	—	—	780	[2]780	—	—	Unk	—	1957	CDWR
W115	Cotati	7N/8W-31L1	—	—	780	[2]780	—	—	Unk	—	1961	CDWR
W116	Cotati	7N/8W-33F1	—	—	110	[2]110	—	—	Shallow	—	1961	CDWR
W117	Cotati	7N/8W-33M1	—	—	452	452	140	452	Mixed	—	1957–1963	CDWR
W118	Cotati	—	—	—	503	503	143	503	Mixed	—	2005–2008	CDPH
W119	Cotati	—	—	—	220	220	75	220	Mixed	—	2005	CDPH
W120	Wilson Grove	7N/9W-09[4]	—	—	—	—	—	—	Unk	—	1951	CDWR
W121	Wilson Grove	7N/9W-09F1	382804122514701	—	—	186	—	—	Unk	283	1957–1991	CDWR
W122	Wilson Grove	7N/9W-09F3	382807122515401	—	280	280	240	280	Deep	275	1978–2007	CDWR

Table 1. Construction information for selected wells used to obtain water-quality data in the Santa Rosa Plain watershed, Sonoma County, California.—Continued

[See figure 1 for locations (map numbers) of wells, springs, and surface-water sites. U.S. Geological Survey (USGS) site identification number is the unique number for each site in USGS National Water Information System (NWIS) database. Depths in feet below land surface (fbls). Land surface altitude, where available, in feet above sea level which refers to the North American Vertical Datum of 1988 (NAVD of 1988). Depth Category based on average of perforated interval. **Abbreviations:** CDPH, California Department of Public Health; CDWR, California Department of Water Resources; GAMA, Groundwater Ambient Monitoring and Assessment; owner, data provided by well owner; Unk, unknown; USGS-GAMA, Groundwater Ambient Monitoring and Assessment Program (USGS and California State Water Resources Control Board); <, actual value is less than value shown; —, no data]

Map number	Groundwater storage unit	State well number	USGS site identification number	GAMA identification number	Depth drilled (fbls)	Depth cased (fbls)	Depth of top perforation (fbls)	Depth of bottom perforation (fbls)	Depth category	Land surface altitude[1]	Period of data included in this report	Water quality data source
W123	Cotati	7N/9W-12K1	382751122482501	NSFVP-15	135	60	40	60	Shallow	105	2004	USGS-GAMA
W124	Cotati	7N/9W-24G1	382619122482101	NSFVP-14	550	550	507	547	Deep	95	2004	USGS-GAMA
W125	Cotati	—	—	—	250	20	—	—	Shallow	—	1997	CDPH
W126	Cotati	—	—	—	1,521	830	410	805	Deep	94	1999	CDPH, CDWR
W127	Cotati	—	—	—	790	770	310	750	Deep	—	2003	CDPH
W128	Cotati	—	—	—	1,506	1,065	425	1,045	Deep	76	1982–2004	CDPH, CDWR
W129	Cotati	7N/9W-36K2	382429122480501	NSFVPFP-01	1,074	1,040	410	1,020	Deep	79	2002–2005	CDPH, USGS-GAMA
W130	Cotati	7N/9W-36M1	382419122484301	—	—	[2]288	—	—	Shallow	73	1957–1972	CDWR
W131	Uplands	8N/7W-07R1	383257122400501	—	380	380	—	—	Unk	790	2004	USGS-GAMA
W132	Uplands	8N/7W-30E1	383053122410801	NSFVOL-02	—	—	—	—	Unk	1,155	2004	USGS-GAMA
W133	Uplands	8N/8W-11G1	383319122424301	NSFVOL-09	—	—	—	—	Unk	525	2004	USGS-GAMA
W134	Windsor	8N/8W-17F1	383231122462201	NSFVP-13	355	355	155	355	Deep	221	1975–1991	CDWR
W135	Windsor	—	—	—	440	432	240	432	Deep	—	1990–1993	CDPH
W136	Windsor	8N/8W-20Q1	383111122455701	—	312	312	56	310	Mixed	142	1950–2007	CDWR
W137	Uplands	8N/8W-21N1	383115122452701	—	406	394	294	394	Deep	223	1976–1991	CDWR
W138	Windsor	—	—	—	512	352	180	342	Deep	153	1981–2009	CDPH, CDWR
W139	Windsor	—	—	—	570	565	210	520	Deep	—	2009	CDPH, owner
W140	Windsor	8N/8W-32K1	382937122460601	NSFVP-20	123	110	70	110	Shallow	141	2004	USGS-GAMA
W141	Windsor	—	—	—	380	380	90	283	Mixed	153	1984	CDPH, CDWR
W142	Windsor	—	—	—	295	282	165	270	Deep	—	2003–2004	CDPH, owner
W143	Windsor	—	—	—	690	545	158	541	Deep	—	2002–09	CDPH, owner
W144	Windsor	—	—	—	220	220	80	220	Mixed	—	1994–2001	CDPH
W145	Windsor	8N/8W-33K1	382938122445401	—	400	400	—	—	Unk	158	1975–1984	CDWR
W146	Uplands	—	—	—	558	558	364	558	Deep	—	1980	CDPH
W147	Uplands	—	—	—	470	464	323	464	Deep	—	1996	CDPH
W148	Uplands	8N/8W-36R1[4]	—	—	—	—	—	—	Unk	—	1952	CDWR
W149	Windsor	8N/9W-10R1	—	—	400	[2]400	—	—	Unk	—	1952	CDWR
W150	Windsor	—	—	—	637	558	105	192	Mixed	—	1984	CDPH, CDWR
W151	Windsor	8N/9W-12P1	383254122483301	—	187	187	67	187	Mixed	113	1975–1984	CDWR
W152	Windsor	8N/9W-13A2	383243122480501	—	109	109	87	109	Shallow	123	1974–1993	CDWR
W153	Windsor	8N/9W-13C1	—	—	202	202	96	200	Mixed	—	1952	CDWR
W154	Windsor	8N/9W-13D2	—	—	100	[2]100	—	—	Shallow	—	1951	CDWR
W155	Windsor	8N/9W-13E1	—	—	130	[2]130	—	—	Shallow	—	1951	CDWR

Table 1. Construction information for selected wells used to obtain water-quality data in the Santa Rosa Plain watershed, Sonoma County, California.—Continued

[See figure 1 for locations (map numbers) of wells, springs, and surface-water sites. U.S. Geological Survey (USGS) site identification number is the unique number for each site in USGS National Water Information System (NWIS) database. Depths in feet below land surface (fbls). Depth Category based on average of perforated interval. Land surface altitude, where available, in feet above sea level which refers to the North American Vertical Datum of 1988 (NAVD of 1988). **Abbreviations**: CDPH, California Department of Public Health; CDWR, California Department of Water Resources; GAMA, Groundwater Ambient Monitoring and Assessment; owner, data provided by well owner; Unk, unknown; USGS-GAMA, Groundwater Ambient Monitoring and Assessment Program (USGS and California State Water Resources Control Board); <, actual value is less than value shown; —, no data]

Map number	Groundwater storage unit	State well number	USGS site identification number	GAMA identification number	Depth drilled (fbls)	Depth cased (fbls)	Depth of top perforation (fbls)	Depth of bottom perforation (fbls)	Depth category	Land surface altitude[1]	Period of data included in this report	Water quality data source
W156	Windsor	8N/9W-13J2	383220122480501	NSFVP-09	400	400	260	400	Deep	118	1973–2008	CDPH, CDWR, USGS-GAMA
W157	Windsor	8N/9W-13N1	—	—	208	188	14	184	Mixed	—	1951	CDWR
W158	Windsor	8N/9W-14A1	—	—	185	[2]185	—	—	Unk	—	1951	CDWR
W159	Windsor	8N/9W-22R1	—	—	145	142	122	142	Shallow	—	2001–2005	CDWR
W160	Windsor	8N/9W-23D1	—	—	370	[3]370	35	165	Mixed	—	1959–1964	CDWR
W161	Windsor	8N/9W-26L1	383038122494301	—	265	265	246	265	Deep	118	1977–1984	CDWR
W162	Windsor	8N/9W-36P1	—	—	1,048	[2]1,048	—	—	Unk	—	1958–1984	CDWR

[1]Land-surface altitude listed where available.

[2]Information uncertain.

[3]Well used on section; land-surface altitude estimated from TOPO.

[4]Unofficial or incomplete State well number based on approximate location.

[5]Represents sample depth using packers at this site.

[6]Missing footnote?

collected in 60-milliliter (mL) glass bottles by immersing the bottle beneath the water surface until filled. Samples and measurements were collected from the center of flow with the exception of Spring Lake (SW5), where they were collected from the end of a pier, and the Laguna de Santa Rosa (SW2 and SW13) and the Russian River (SW18), where deep water necessitated collection from stream banks.

Groundwater samples were collected from six wells (W33–35, W77, W89, and W102; fig. 1B) by the USGS, private consulting firms (PES Environmental, Inc., or PES, for the city of Sebastopol; and The Environment, Community, and Opportunity Network, or ECON, for the city of Santa Rosa), or a combination of both (table 2). All samples from W33–35 were collected by USGS personnel. Samples from W77 were collected concurrently by the USGS and PES from three depth intervals (100–120, 340–360, and 640–650 ft bls) and by only PES from seven additional depth intervals. All samples from W89 and W102 were collected by ECON.

The samples from these wells were collected by using submersible pumps that were temporarily installed in monitor wells (W33-35, W89A, and W102A) and were put between inflatable packers in long-screened wells (W77, W89B, and W89C) and in an open borehole (W102B; table 2). Inflatable packers were used to sample 10 depth intervals in W77 (at 10 to 20 ft intervals between 100–650 ft bls); 2 depth intervals in W89 (500–620 and 750–860 ft bls), and 1 depth interval in the open borehole of W102 (W102B; 786–806 ft bls) prior to the well casing installation (table 2). Prior to sample collection, each well or open borehole was pumped continuously to purge at least three casing volumes of water. Samples were collected after specific conductance, pH, and temperature stabilized (three successive measurements within 5-percent of one another) to ensure representative samples.

Samples from wells W33–35, W77, W89, and W102 were collected for the analysis of selected inorganic constituents, including major and minor (trace) elements, nutrients, chemical and physical properties (turbidity, dissolved oxygen, pH, specific conductance, temperature, and dissolved solids), and stable and radioactive isotopes, including ^{18}O, ^2H, tritium (^3H), carbon-14 (^{14}C), and carbon-13 (^{13}C). These samples were collected, treated, and preserved following or by using procedures similar to those outlined by the U.S. Geological Survey (2010). Analyses were done using a combination of USGS and consultant-designated laboratories. All samples collected by the USGS from W33–35 and 3 of the 10 depth intervals for W77 were analyzed by the USGS or USGS-contracted laboratories. Samples collected from seven additional depth intervals for W77 by PES were analyzed for arsenic by Analytical Sciences of Petaluma, which is an ELAP-certified laboratory (Carl Michelsen, PES, written commun., 2006). Samples collected from W89 and W102 by ECON were analyzed for major ions, trace elements, and nutrients by the city of Santa Rosa Utilities Department, Laguna Environment Lab, an ELAP-certified laboratory (Andy Rogers, ECON, written commun., 2009 and 2010), and were analyzed for ^{18}O, ^2H, ^3H, ^{14}C, and ^{13}C by the USGS Reston Stable Isotope Laboratory

(RSIL) or USGS-contracted laboratories. Procedures for on-site measurements and laboratory analysis of samples by the non-USGS laboratories followed standard methods listed in table 3. Procedures for the collection and analysis of samples by the USGS or USGS-contracted laboratories are described in greater detail in the following paragraphs.

Five laboratories managed or contracted by the USGS performed chemical analyses for the SRPW study. The USGS National Water-Quality Laboratory (NWQL), in Denver, Colorado, performed analyses for inorganic analytes (major ions, trace elements, and nutrients) in samples collected from W33–35 and the three previously described depth intervals of W77. The following analytical methods were employed by the NWQL: major ions were analyzed by inductively coupled plasma with atomic-emission spectrometry (ICP-AES; Fishman and Friedman, 1989; Fishman, 1993; the American Public Health Association, 1998); trace elements were analyzed by ICP-AES, inductively coupled plasma with mass spectrometry (ICP-MS), auto-segmented-flow/ion-selective electrode, colorimetry, and automated batch analyzer (Fishman and Friedman, 1989; Fishman, 1993; Struzeski and others, 1996; Garbarino, 1999; Garbarino and others, 2006); and nutrients were analyzed by colorimetry (Fishman, 1993; Patton and Truitt, 2000). The USGS Isotope Fractionation Project in Reston, Virginia (*currently* [2013] the Reston Stable Isotope Laboratory) analyzed surface and groundwater samples for ^{18}O and ^2H by using a hydrogen-water-equilibration technique (Coplen and others, 1991; K. Revesz and T. Coplen, U.S. Geological Survey internal standard operating procedure, written commun., 2004) and an automated version of the carbon dioxide equilibration technique of Epstein and Mayeda (1953; Revesz and Coplen, U.S. Geological Survey, internal standard operating procedure, written commun., 2004). The University of Miami Tritium Laboratory performed analysis on samples collected for tritium by using an electrolytic enrichment with gas proportional counting technique, as described by Ostlund and Dorsey (1975) and the University of Miami Tritium Laboratory (2010).

Samples for ^{14}C and ^{13}C were analyzed by two different USGS-contracted laboratories during the SRPW study: the University Waterloo Environmental Isotope Laboratory, Waterloo, Ontario, Canada, analyzed samples collected in 2006 and 2007, and the Woods Hole Oceanographic Institute's National Ocean Sciences Accelerator Mass Spectrometry Facility (NOSAMS), Woods Hole, Massachusetts, analyzed samples collected in 2009 and 2010. Both facilities used accelerator mass spectrometry techniques similar to the methodology described by Fifield (1999).

Historical Sample Collection and Analysis

Water-quality samples presented in this report from the CDWR (appendix A) were collected and analyzed following referenced methods of the American Public Health Association (2005) and the U.S. Environmental Protection Agency (1993,

Table 2.   Summary of discrete-depth samples from selected wells, Santa Rosa Plain Watershed, Sonoma County, California, 2006–2010.

[See figure 1 for locations (map numbers) of wells. U.S. Geological Survey (USGS) identification number is the unique number for each site in USGS National Water Information System (NWIS) database. Sample depth in feet below land surface. The five-digit number below the constituent name is the USGS parameter code used to uniquely identify a specific constituent or property; not listed where no USGS values are shown. Alkalinity values collected and analyzed by USGS represent field measurements with parameter code 39086; all others represent laboratory analyses. **Abbreviations:** CaCO$_3$, calcium carbonate; E, estimated value; ECON, The Environment, Community, and Opportunity Network; ft, foot; in., inch; mg/L, milligram per liter; ND, non detect; PES, PES Environmental, Inc.; °C, degree Celsius; µS/cm, microsiemen per centimeter at 25°C; µg/L, microgram per liter; (L), measured in laboratory; <, actual value is less than value shown; —, no data]

Map number	Groundwater storage unit	State well number	USGS identification number	Perforated or sampled depth (fbls)	Sample date (mm/dd/yyyy)	Collecting and analyzing agency	Method of sample collection
W33	Cotati	6N/8W-07A4	382308122463903	60–80	08/22/2007	USGS	Dedicated 2 in. well in multiple-well monitoring site.
W34	Cotati	6N/8W-07A5	382308122463902	237–257	08/22/2007	USGS	Dedicated 2 in. well in multiple-well monitoring site.
W35	Cotati	6N/8W-07A6	382308122463901	550–570	08/23/2007	USGS	Dedicated 2 in. well in multiple-well monitoring site.
W77	Wilson Grove	6N/9W-02C2	382352122493801	100–120	10/16/2006	USGS	Inflatable bladders (packers) used to isolate sampled interval within open borehole.
W77	Wilson Grove	—	—	200–210	10/16/2006	PES	Inflatable bladders (packers) used to isolate sampled interval within open borehole.
W77	Wilson Grove	—	—	240–250	10/16/2006	PES	Inflatable bladders (packers) used to isolate sampled interval within open borehole.
W77	Wilson Grove	—	—	280–300	10/13/2006	PES	Inflatable bladders (packers) used to isolate sampled interval within open borehole.
W77	Wilson Grove	—	—	340–360	10/16/2006	USGS	Inflatable bladders (packers) used to isolate sampled interval within open borehole.
W77	Wilson Grove	—	—	420–440	10/13/2006	PES	Inflatable bladders (packers) used to isolate sampled interval within open borehole.
W77	Wilson Grove	—	—	490–510	10/13/2006	PES	Inflatable bladders (packers) used to isolate sampled interval within open borehole.
W77	Wilson Grove	—	—	540–560	10/13/2006	PES	Inflatable bladders (packers) used to isolate sampled interval within open borehole.
W77	Wilson Grove	—	—	590–610	10/12/2006	PES	Inflatable bladders (packers) used to isolate sampled interval within open borehole.
W77	Wilson Grove	—	—	640–650	10/12/2006	USGS	Inflatable bladders (packers) used to isolate sampled interval within open borehole.
W89A	Valley	7N/7W-19G2	382623122403301	250–290	03/12/2010	ECON	Dedicated 2 in. well in borehole.
W89B	Valley	—	—	500–620	03/17/2010	ECON	Inflatable bladders (packers) used to isolate sampled interval within well screen.
W89C	Valley	—	—	750–860	03/11/2010	ECON	Inflatable bladders (packers) used to isolate sampled interval within well screen.
W102A	Windsor	7N/8W-24F1	382622122414901	130–180	08/24/2009	ECON	Permanent well.
W102B	Windsor	—	—	786–806	07/23/2009	ECON	Inflatable bladders (packers) used to isolate sampled interval within open borehole.

Table 3.   Minimum reporting levels and analytical method references for water-quality analyses performed by non-U.S. Geological Survey laboratories, Santa Rosa Plain Watershed, Sonoma County, California, 2006–2010.

[All standard methods (SM) are from "Standard Methods for the Examination of Water and Wastewater" (American Public Health Association, 2005). **Abbreviations**: CaCO$_3$, calcium carbonate; ECON, The Environment, Community, and Opportunity Network; EPA, U.S. Environmental Protection Agency; EPA 200.7: "Trace Metals by ICP" (U.S. Environmental Protection Agency, 1994); EPA 200.8: "Trace Metals by ICP" (U.S. Environmental Protection Agency, 1994); EPA 200.9: "Methods for the Determination of Metals in Environmental Samples, Supplement 1" (U.S. Environmental Protection Agency, 1994); EPA 245.1: "Methods for the Determination of Metals in Environmental Samples, Supplement 1" (U.S. Environmental Protection Agency, 1994); EPA 300.0: "Methods for the Determination of Inorganic Anions by Ion Chromatography" (U.S. Environmental Protection Agency, 1993); mg/L, milligram per liter; NTU, nephelometric turbidity unit; PES, PES Environmental, Inc.; µS/cm, microsiemen per centimeter at 25°C; °C, degree Celsius; µg/L, microgram per liter]

Property or constituent (unit of measurement)	Minimum reporting level	Reporting entity	Analytical method reference
Physical properties, major ions, and nutrients			
Dissolved oxygen (mg/L)	0.1	ECON, PES	SM 4500 O G: Electrometric Method.
pH (standard units)	0.1	ECON, PES	SM 4500 H$^+$ B: Electrometric Method.
Specific conductance (µS/cm)	10	ECON, PES	SM 2510 A-B: Conductivity.
Temperature (°C)	0.1	ECON, PES	SM 2550 B: Temperature.
Hardness, total (mg/L, as CaCO$_3$)	1.0	ECON	SM 2340 B: Total Hardness by Calculation.
Calcium, dissolved (mg/L)	0.1	ECON	EPA 200.7: Inductively Coupled Plasma—Atomic Emission Spectrometry.
Magnesium, dissolved (mg/L)	0.02	ECON	EPA 200.7: Inductively Coupled Plasma—Atomic Emission Spectrometry.
Potassium, dissolved (mg/L)	0.5	ECON	EPA 200.7: Inductively Coupled Plasma—Atomic Emission Spectrometry.
Sodium, dissolved (mg/L)	0.5	ECON	EPA 200.7: Inductively Coupled Plasma—Atomic Emission Spectrometry.
Chloride, dissolved (mg/L)	1.0	ECON	EPA 300.0: Ion Chromatography.
Fluoride, dissolved (mg/L)	0.1	ECON	EPA 300.0: Ion Chromatography.
Sulfate, dissolved (mg/L)	0.5	ECON	EPA 300.0: Ion Chromatography.
Nitrate as NO$_3$ (mg/L)	2.0	ECON	EPA 300.0: Ion Chromatography.
Nitrite, dissolved as N (mg/L)	0.4	ECON	SM 4500 NO2 B: Colorimetry.
Turbidity (NTU)	1	ECON, PES	SM 2130 B: Temperature.
Trace elements			
Aluminum, dissolved (µg/L)	10	ECON	EPA 200.8: Inductively Coupled Plasma—Mass Spectrometry.
Antimony, dissolved (µg/L)	2.0	ECON	EPA 200.8: Inductively Coupled Plasma—Mass Spectrometry.
Arsenic, dissolved (µg/L)	1.0	ECON	EPA 200.8: Inductively Coupled Plasma—Mass Spectrometry.
Arsenic, dissolved (µg/L)	1.0	PES	EPA 200.9: Stablized Temperature Graphite Furnance Atomic Absorption.
Barium, dissolved (µg/L)	1.0	ECON	EPA 200.8: Inductively Coupled Plasma—Mass Spectrometry.
Beryllium, dissolved (µg/L)	0.5	ECON	EPA 200.8: Inductively Coupled Plasma—Mass Spectrometry.
Boron, dissolved (µg/L)	50	ECON	EPA 200.7: Inductively Coupled Plasma—Atomic Emission Spectrometry.
Cadmium, dissolved (µg/L)	1.0	ECON	EPA 200.8: Inductively Coupled Plasma—Mass Spectrometry.
Chromium, dissolved (µg/L)	2.0	ECON	EPA 200.8: Inductively Coupled Plasma—Mass Spectrometry.
Copper, dissolved (µg/L)	2.0	ECON	EPA 200.8: Inductively Coupled Plasma—Mass Spectrometry.
Iron, dissolved (µg/L)	40	ECON	EPA 200.7: Inductively Coupled Plasma—Atomic Emission Spectrometry.
Lead, dissolved (µg/L)	1.0	ECON	EPA 200.8: Inductively Coupled Plasma—Mass Spectrometry.
Manganese, dissolved (µg/L)	1.0	ECON	EPA 200.8: Inductively Coupled Plasma—Mass Spectrometry.
Mercury, dissolved (µg/L)	0.2	ECON	EPA 245.1: Manual Cold-Vapor Atomic-Absorption Spectrometry.
Molybdenum, dissolved (µg/L)	2.0	ECON	EPA 200.8: Inductively Coupled Plasma—Mass Spectrometry.
Nickel, dissolved (µg/L)	2.0	ECON	EPA 200.8: Inductively Coupled Plasma—Mass Spectrometry.
Selenium, dissolved (µg/L)	2.0	ECON	EPA 200.8: Inductively Coupled Plasma—Mass Spectrometry.
Silver, dissolved (µg/L)	10	ECON	EPA 200.8: Inductively Coupled Plasma—Mass Spectrometry.
Thallium, dissolved (µg/L)	1.0	ECON	EPA 200.8: Inductively Coupled Plasma—Mass Spectrometry.
Vanadium, dissolved (µg/L)	2.0	ECON	EPA 200.8: Inductively Coupled Plasma—Mass Spectrometry.
Zinc, dissolved (µg/L)	20	ECON	EPA 200.8: Inductively Coupled Plasma—Mass Spectrometry.

1994). All CDWR samples were analyzed at the California Department of Water Resources Bryte Analytical Laboratory in West Sacramento, California (Bruce Agee, California Department of Water Resources, written commun., 2005).

Water-quality samples presented in this report from public-supply purveyors and the CDPH (appendix A) were generally collected, analyzed, or both by consultants or laboratories contracted by individual public-supply purveyors. Sampling and analysis were done in accordance with requirements of California Code of Regulations Title 22 (California Department of Public Health, 2011), by using methods referenced by the American Public Health Association (2005) and the U.S. Environmental Protection Agency (1993, 1994).

Well Construction of Sampled Wells

Groundwater wells in this study area span a wide range of completed depths and perforated (screened) intervals (table 1). Most are production (municipal, irrigation, and domestic) wells and, therefore, are typically perforated over long intervals (greater than or equal to 100 ft). Some wells, especially the older ones, could have been constructed in uncased open holes, but this information is unavailable. Perforated intervals for many of the wells having water-quality data are composed either of a single, continuous interval or multiple, individual intervals covering 50 percent or more over the completed hole depth. For the purposes of this report, the wells were categorized into four main groups on the basis of well construction: (1) shallow wells—entire perforated or open interval above 150 ft bls, (2) deep wells—entire perforated or open interval below 150 ft bls, (3) mixed depth wells—perforated or open interval extends above and below 150 ft bls, and (4) unknown depth wells—the depth of the perforated or open interval is unknown (table 1). A total of 167 wells were sampled: 24 were shallow, 69 were deep, 40 were mixed, and 34 were unknown-depth. Given the spatial distribution of sampled wells (fig. 1B) and the relative uniform distribution of sample depths, it is expected that the data are representative of the geochemistry of the SRPW. In the SRP, the shallow wells primarily sample the Glen Ellen Formation and the deep wells primarily sample the Petaluma Formation.

Chemical Character of Groundwater

Water-quality data compiled for this report (appendix A) were used to characterize the areal, vertical, and temporal variations in groundwater quality and to identify water-quality constituents of potential concern. These data include physical properties and inorganic constituents (major ions, trace elements, and nutrients) collected during 1947–2010, with an average sample date of about 1990. Wells with multiple chemical analyses are represented in selected summary tables and figures by a single value corresponding to the most recent analysis for that particular well.

Major-Ion Composition

The major-ion composition of groundwater is controlled by the natural chemistry of the recharge water, geochemical reactions in the subsurface, primarily dissolution and precipitation of minerals, and anthropogenic factors, such as the disposal of wastewater and irrigation return flows. The major-ion composition of groundwater was characterized for this report by using trilinear and Stiff diagrams. Dissolved-solids concentrations of the samples were also used to help characterize the water quality of the SRPW.

A trilinear diagram shows the relative contribution of major cations and anions, on a charge-equivalent basis, to the ionic content of the water (Piper, 1944). Percentage scales along the sides of the diagram indicate the percentage of total cations or anions, in milliequivalents per liter (meq/L), of each major ion. Cations are shown in the left triangle, anions are shown in the right triangle, and the central diamond integrates the data (fig. 2). Trilinear diagrams are useful in depicting the range in chemical composition, in grouping water types, and in determining if there is simple mixing between chemically different water (Hem, 1992). In this report, the dominant cation and anion species are used to describe the water type of a water sample when a single cation or anion composes more than 60 percent of the total cations or anions, respectively. Where no one cation or anion exceeds 60 percent, the sample is described as mixed and the first and second most abundant cations or anions are given for description purposes.

Stiff diagrams depict the concentrations of major ions in meq/L and indicate relative proportions of major ions (Stiff, 1951). Analyses with similarly shaped diagrams represent groundwater of similar chemical characteristics with respect to major ions. Changes in the width of the diagrams indicate differences in the concentration of dissolved constituents. Water that contains higher concentrations of major ions has a larger polygon than does the diagram for water with lower concentrations. All Stiff diagrams in this report are shown at the same scale in units of meq/L (fig. 3). The left side of the diagram shows the major cations: sodium plus potassium at the top, calcium in the middle, and magnesium at the bottom. The right side of the diagram shows major anions: chloride plus fluoride at the top, carbonate plus bicarbonate in the middle, and sulfate on the bottom.

The dissolved-solids concentration in water ordinarily is determined from the weight of the dry residue remaining after evaporation of an aliquot of the water sample and is referred to as residue on evaporation (ROE) (appendix A). The dissolved-solids concentration in water can also be computed if the concentrations of major ions are known. For this report, the computed dissolved-solids concentration represents the total concentration of all the major ions reported, but does not include an adjustment of bicarbonate to carbonate ions that commonly is made to make the computation comparable to the ROE value for the same sample. Because many of the samples were not analyzed for ROE, the dissolved-solids concentrations discussed in this report are the computed values presented in appendix A.

A

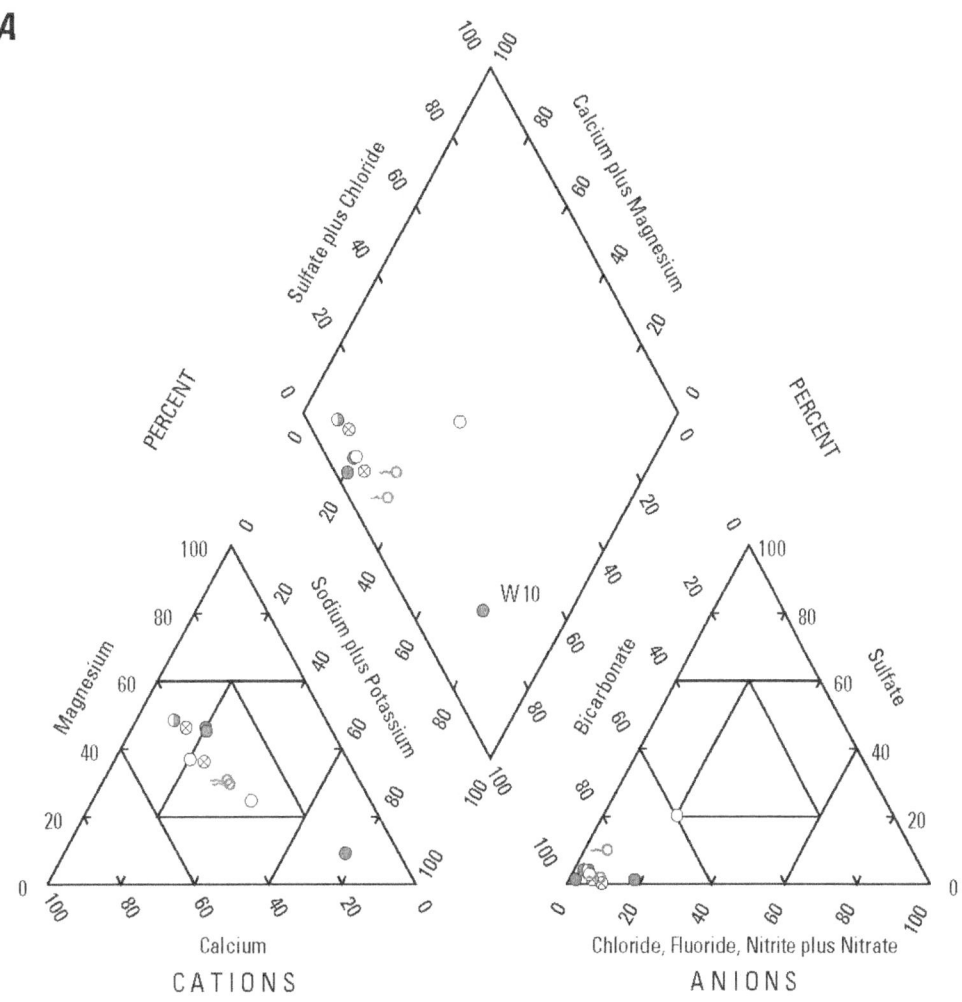

PERCENTAGE OF TOTAL MILLIEQUIVALENTS PER LITER

EXPLANATION

Uplands storage unit

Well perforated interval

○ Shallow (0' to 150' blsd)

◑ Mixed

● Deep (Greater than 150')

⊗ Construction unknown

⊸○ Spring

W 10 Well number

Figure 2. Trilinear diagrams of the most recent, complete sample from selected surface-water sites, springs, and wells in storage units in the Santa Rosa Plain watershed, Sonoma County, California, 1947–2010: *A*, Uplands; *B*, Valley; *C*, Windsor; *D*, Cotati; and *E*, Wilson Grove.

B

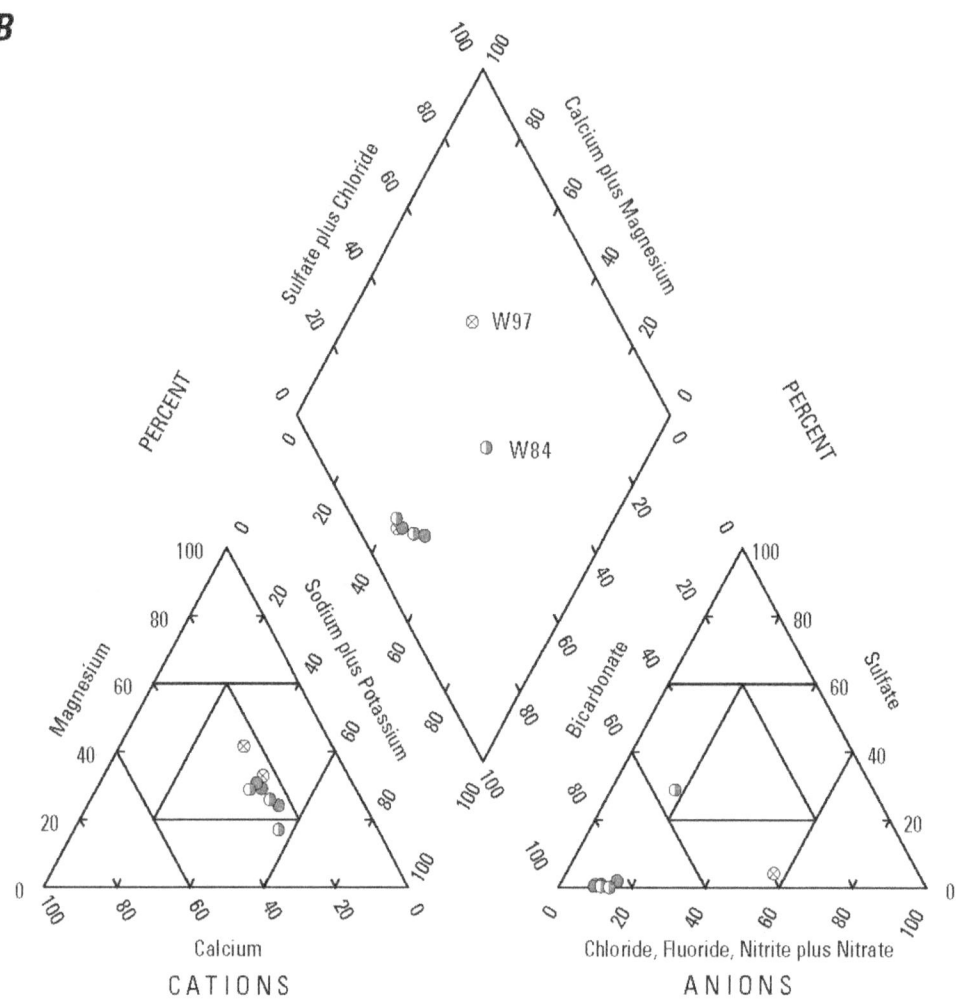

PERCENTAGE OF TOTAL MILLIEQUIVALENTS PER LITER

EXPLANATION

Valley storage unit

Well perforated interval
- ○ Shallow (0' to 150' blsd)
- ◑ Mixed
- ● Deep (Greater than 150')
- ⊗ Construction unknown

W84   Well number

Figure 2.   Continued.

C

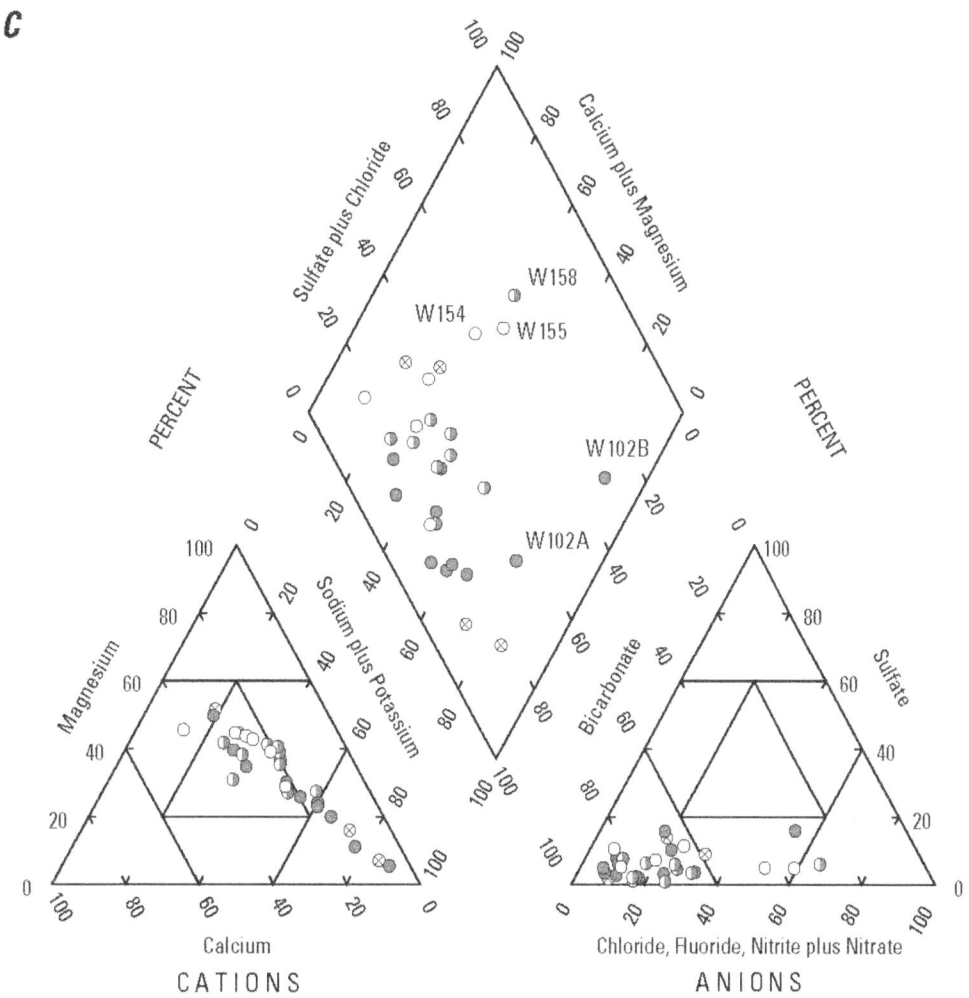

PERCENTAGE OF TOTAL MILLIEQUIVALENTS PER LITER

EXPLANATION

Windsor Basin storage unit

Well perforated interval
- ○ Shallow (0′ to 150′ blsd)
- ◑ Mixed
- ● Deep (Greater than 150′)
- ⊗ Construction unknown

W155   Well number

Figure 2.   Continued.

D

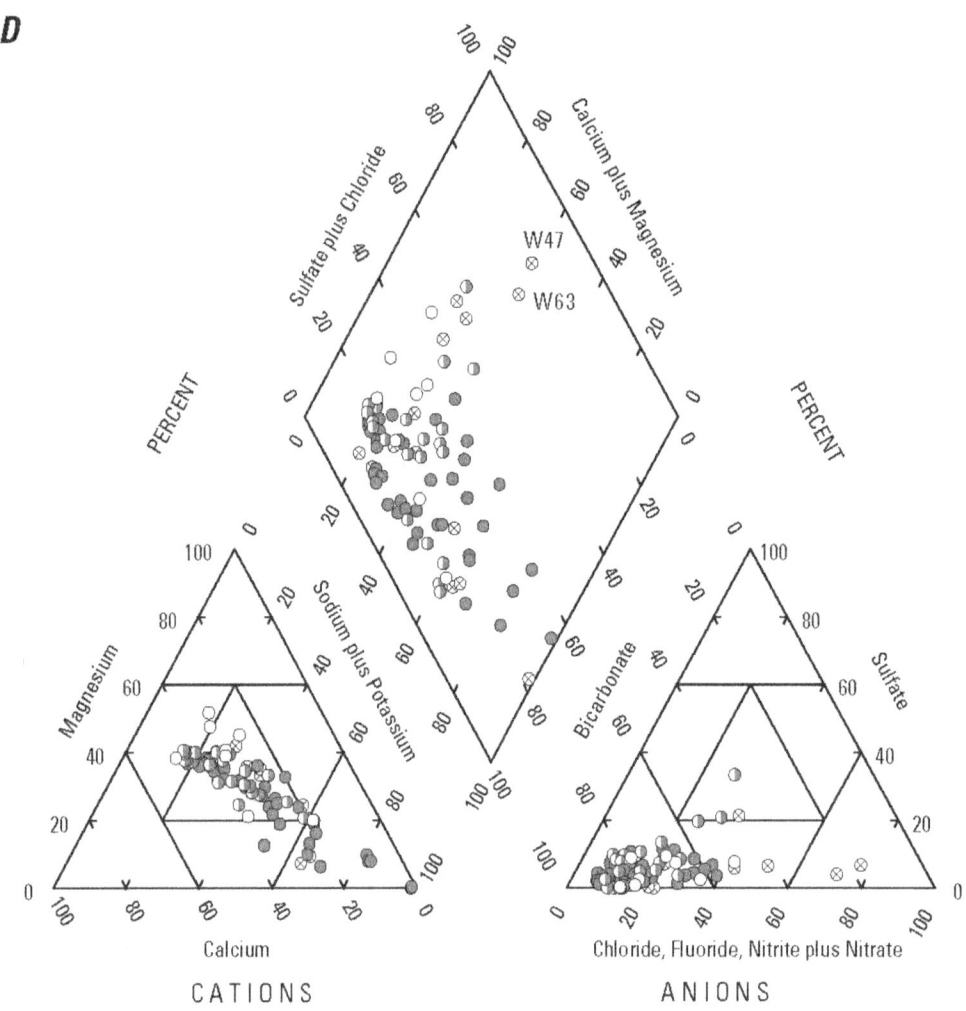

PERCENTAGE OF TOTAL MILLIEQUIVALENTS PER LITER

EXPLANATION

Cotati Basin storage unit

Well perforated interval

○ Shallow (0' to 150' blsd)
◑ Mixed
● Deep (Greater than 150')
⊗ Construction unknown

W63   Well number

Figure 2.   Continued.

E

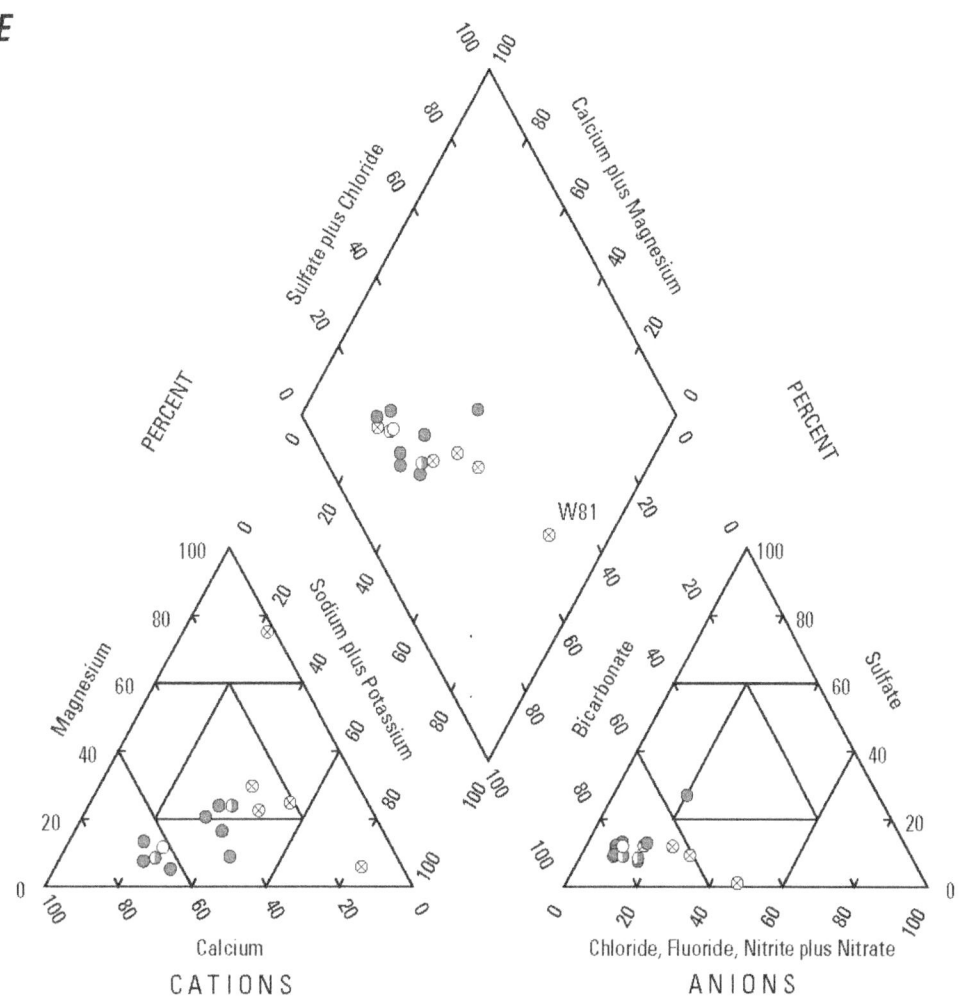

PERCENTAGE OF TOTAL MILLIEQUIVALENTS PER LITER

EXPLANATION

Wilson Grove storage unit

Well perforated interval

○ Shallow (0′ to 150′ blsd)

◑ Mixed

● Deep (Greater than 150′)

⊗ Construction unknown

W81 Well number

Figure 2.   Continued.

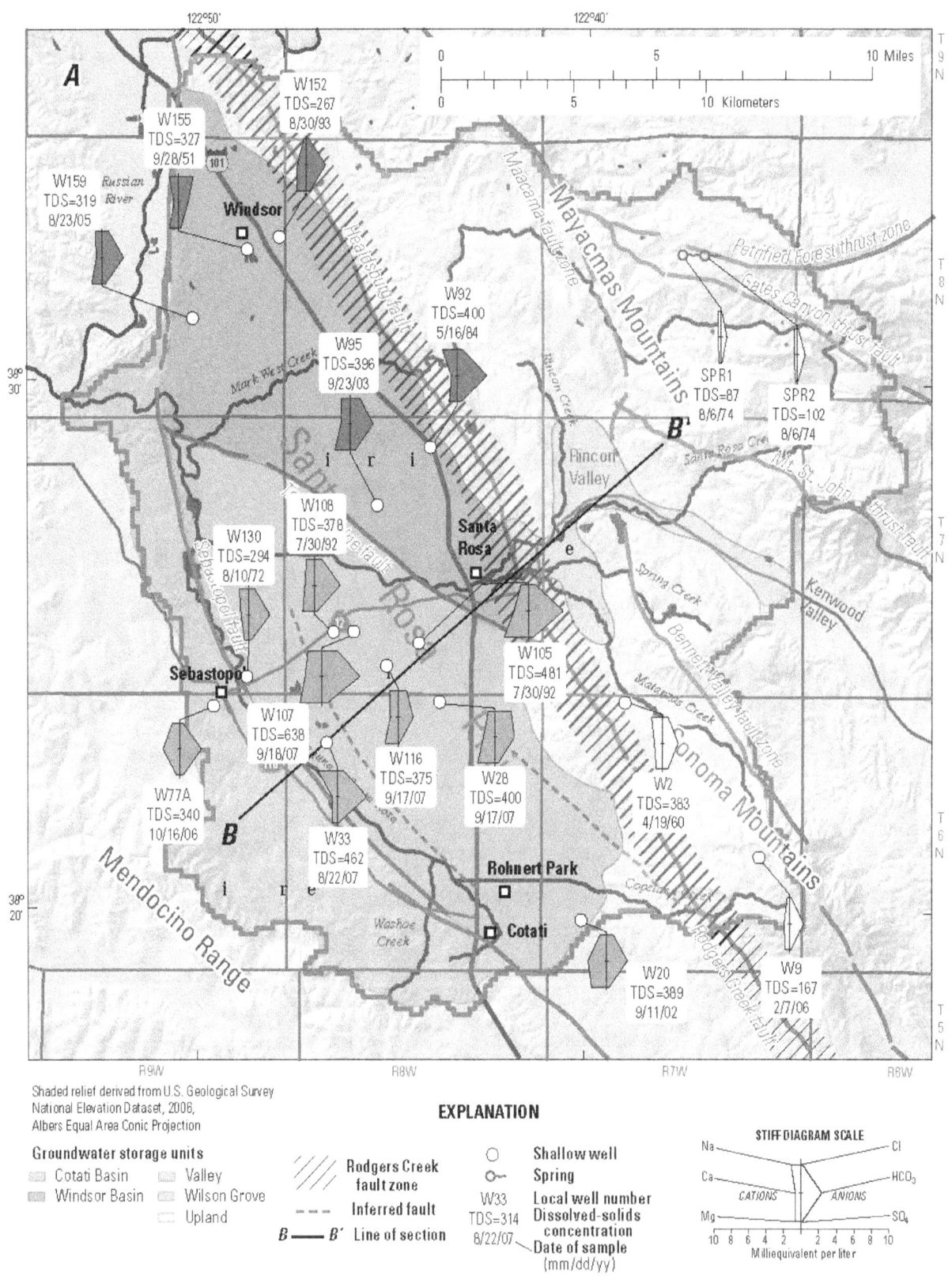

Figure 3.   Stiff diagrams showing water-type for selected in the Santa Rosa Plain watershed, Sonoma County, California: *A,* shallow wells; *B,* deep wells; and *C,* wells along section B–B'.

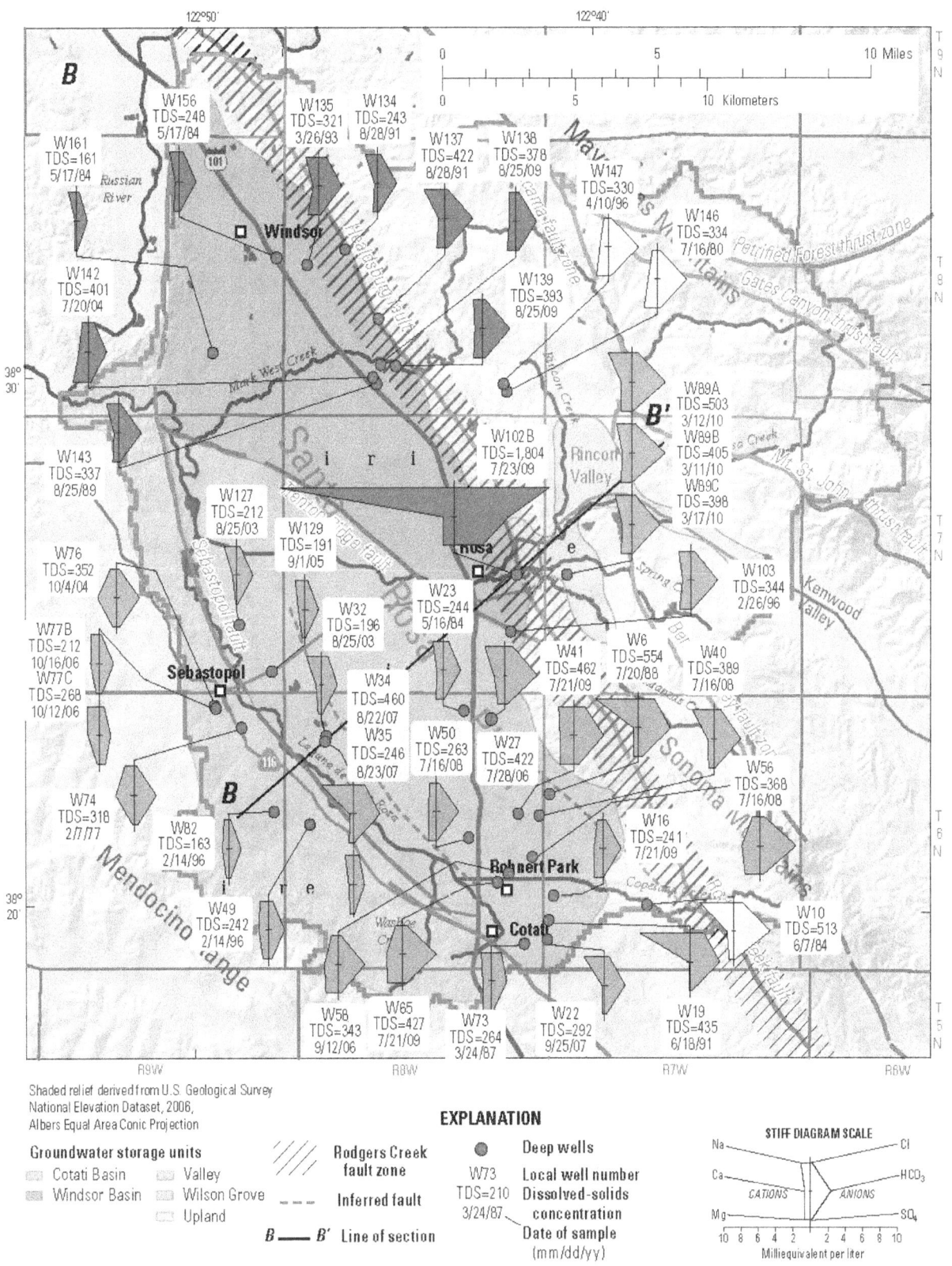

Shaded relief derived from U.S. Geological Survey
National Elevation Dataset, 2006,
Albers Equal Area Conic Projection

EXPLANATION

Groundwater storage units

- Cotati Basin
- Windsor Basin
- Valley
- Wilson Grove
- Upland

/// Rodgers Creek fault zone

--- Inferred fault

B —— B' Line of section

● Deep wells

W73 Local well number
TDS=210 Dissolved-solids concentration
3/24/87 Date of sample (mm/dd/yy)

STIFF DIAGRAM SCALE

Na ———— Cl
Ca ———— HCO₃
CATIONS   ANIONS
Mg ———— SO₄
10 8 6 4 2 2 4 6 8 10
Milliequivalent per liter

Figure 3. Continued.

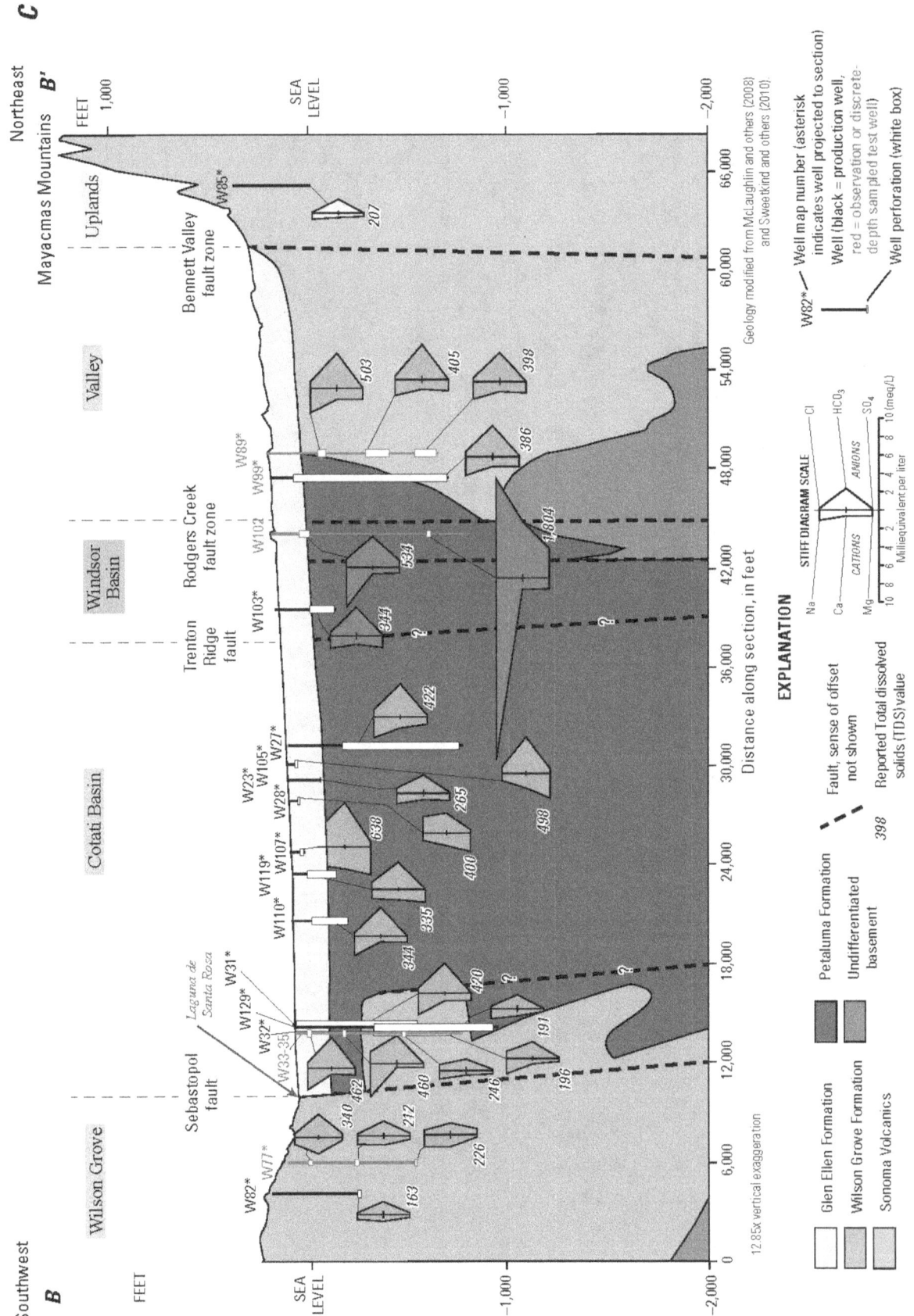

Figure 3. Continued.

Areal and Depth-Related Patterns

As described in *chapter B*, the study area is subdivided into five groundwater storage units, (1) the mountainous upland areas east of the city of Santa Rosa (UPL); (2) valleys located to the east of the SRP Valley (VAL), including Rincon Valley, Bennett Valley and the northern half of Kenwood Valley; (3) Windsor Basin (WB); (4) Cotati Basin (CB); and (4) the Wilson Grove area (WG; fig. 1). The WB and CB storage units compose the SRP. The major-ion composition and dissolved-solids concentration of the most recent sample compiled for each well (appendix A) were used to describe the chemical character in each groundwater storage unit.

Uplands (UPL) Storage Unit

Samples from springs SPR1 and SPR2 in the Mayacmas Mountains are a mixed cation-bicarbonate type water with dissolved-solids concentrations of about 100 mg/L or less—the lowest dissolved-solids concentrations in the SRPW (fig. 3A; appendix A). These samples represent recharge that has undergone minor water-rock interactions, has migrated through geologic materials that are resistant to weathering, or both. As groundwater migrates through the UPL and downgradient storage units, the dissolved-solids concentration of the groundwater increases as a result of water-rock interactions and anthropogenic inputs, such as irrigation return flows and septic-tank discharge. The median dissolved-solids concentration of samples from wells in the UPL storage unit was 330 mg/L (table 4).

All well samples in the UPL storage unit were a mixed cation-bicarbonate or calcium/magnesium-bicarbonate type water, except for the sample from well W10, which was a sodium-bicarbonate type water (fig. 2A). The sample from W10 had a dissolved-solids concentration of 513 mg/L— the highest dissolved-solids concentration in the storage unit (fig. 3B and appendix A). The sample also had a boron concentration of 1,800 µg/L, which was significantly higher than the median concentration of 100 µg/L for the storage unit (table 4). Cardwell (1958) hypothesized that high boron concentrations in the SRPW could be attributable to deeply circulating groundwater rising along a fault zone. Well W10 is within the Rodgers Creek fault zone (fig. 1); therefore, Cardwell's hypothesis could explain the high boron concentration in the sample from well W10.

Valley (VAL) Storage Unit

All well samples from the VAL storage unit were a mixed cation-bicarbonate type water that had a higher percentage of sodium than samples from the UPL storage unit, with the exception of the samples from W84 and W97 (figs. 2A and 2B). The sample from well W84 was a sodium/calcium-bicarbonate/sulfate type water, and the sample from well W97 was a mixed cation-chloride/sulfate type water (fig. 2B). None of the samples were from wells perforated solely in the shallow aquifer, which could account for the higher percentage of sodium in the well samples. In an aquifer system, it is common for the dominant cation to shift from calcium to sodium as depth increases as a result of cation exchange, in which calcium in solution is exchanged with sodium on clay surfaces in the aquifer sediments (Hem, 1992). The median dissolved-solids concentration of samples from wells in the VAL storage unit was 392 mg/L, which was about 20 percent higher than the median of samples from wells in the UPL storage unit (table 4). This higher dissolved-solids in the VAL storage unit compared to the UPL storage unit could result from a greater number of sediment and water interactions in the alluvial and volcanic deposits of the VAL compared to the weathering-resistant hard rocks of the UPL.

Windsor Basin (WB) Storage Unit

Most of the well samples from the WB storage unit were mixed cation-bicarbonate and sodium-bicarbonate type waters (fig. 2C). A comparison of major-ion composition and depth revealed a subtle shift in major cations, from greater proportions of calcium and magnesium in shallow groundwater to greater proportions of sodium in deeper groundwater (fig. 2C). This finding is consistent with ion exchange between clays and groundwater as distance and depth from recharge sources increases. The median dissolved-solids concentration of samples from wells in the WB storage unit was 321 mg/L, which was slightly lower than the median of samples from wells in the UPL and VAL storage units (table 4).

Several samples near Windsor (W154, W155, and W158) were a mixed cation-chloride type water (fig. 2C). The sample from W158 had a nitrate concentration, reported as nitrogen (NO_3-N), of about 10 mg/L (appendix A). The greater proportions of chloride and high nitrate concentration were consistent with the effects of recharge of septic-tank effluent or agricultural return flow.

Sample W102B was a sodium-mixed anion type water, with chloride as the most abundant anion (fig. 2C). This sample was collected from 786–806 ft bls in the open borehole of well W102 when it was drilled (table 3). The sample contained the highest dissolved-solids (1,804 mg/L) and boron (2,100 µg/L) concentrations of all the samples in this report (appendix A). Well W102 is within the Rodgers Creek fault zone (fig. 3B). The high dissolved-solids and boron concentrations are probably attributable to deeply circulating groundwater rising along the fault zone.

Cotati Basin (CB) Storage Unit

Most well samples in the CB storage unit were a mixed cation-bicarbonate water type or a sodium-bicarbonate water type (fig. 2D). As in the WB storage unit, there was a subtle shift in major cations from greater proportions of calcium and magnesium in shallow groundwater to greater proportions of sodium plus potassium in deeper groundwater. The median dissolved-solids concentration of samples from wells in the CB storage unit was 362 mg/L, which was close to the median dissolved-solids concentration of samples from wells in the UPL and VAL storage units (table 4).

Table 4.   Summary of groundwater quality for most recent complete sample from selected wells, classified by storage unit and depth category, Santa Rosa Plain Watershed, Sonoma County, California, 1947–2010.

[See figure 1 for locations of wells. **Abbreviations**: MCL, maximum contaminant level; mg/L, milligram per liter; NL, notification level; SMCL, secondary maximum contaminant level; µg/L, microgram per liter; µS/cm, microsiemen per centimeter; —, no data]

Number of wells in shallow aquifer	Number of wells in deep aquifer	Total number of wells	Parameter units MCL, SMCL, or NL	Constituent							
				Specific conductance (µS/cm) [1]900	Chloride (mg/L) [1]250	Total dissolved solids computed (mg/L) [1]500	Nitrate as N (mg/L) [2]10	Arsenic (µg/L) [3]10	Boron (µg/L) [4]1,000	Iron (µg/L) [5]300	Manganese (µg/L) [5]50
				Uplands groundwater storage unit							
2	—	—	Number of samples	2	2	2	1	0	1	1	1
2	—	—	Maximum	421	29.0	383	0.1	—	100	5,000	480
2	—	—	Minimum	230	4.0	167	0.1	—	100	5,000	480
2	—	—	Median	326	16.5	275	0.1	—	100	5,000	480
—	4	—	Number of samples	4	4	4	2	2	2	3	2
—	4	—	Maximum	616	42.0	513	0.4	10.0	1,800	195	110
—	4	—	Minimum	388	3.0	330	0.3	3.4	400	40	30
—	4	—	Median	471	10.2	378	0.3	6.7	1,100	80	70
—	—	9	Number of samples	9	9	9	5	3	5	6	5
—	—	9	Maximum	616	42.0	513	0.4	10.0	1,800	5,000	480
—	—	9	Minimum	211	3.0	167	0.1	1.4	28	40	5
—	—	9	Median	403	6.0	330	0.2	3.4	100	137	110.0
—	—	9	Wells above regulatory level	0	0	1	0	0	1	2	3
				Valley groundwater storage unit							
0	—	—	Number of samples	0	0	0	0	0	0	0	0
0	—	—	Maximum	—	—	—	—	—	—	—	—
0	—	—	Minimum	—	—	—	—	—	—	—	—
0	—	—	Median	—	—	—	—	—	—	—	—
—	3	—	Number of samples	3	3	3	3	1	3	3	3
—	3	—	Maximum	516	30.6	503	<0.5	2.3	440	540	170
—	3	—	Minimum	417	17.8	398	<0.5	2.3	420	110	52
—	3	—	Median	475	17.9	405	<0.5	2.3	430	120	82
—	—	8	Number of samples	8	8	8	3	2	7	6	5
—	—	8	Maximum	516	30.6	503	7.6	2.3	600	4,600	170
—	—	8	Minimum	117	7.1	75	0.2	1.1	50	110	52
—	—	8	Median	446	17.9	392	0.6	1.7	430	215	82
—	—	8	Wells above regulatory level	0	0	3	0	0	0	2	5
				Windsor groundwater storage unit							
7	—	—	Number of samples	7	7	7	5	1	6	2	2
7	—	—	Maximum	590	98.0	534	4.8	1.9	620	420	1,000
7	—	—	Minimum	339	8.0	267	0.2	1.9	<10	42	150
7	—	—	Median	477	39.0	396	0.9	1.9	100	231	575
—	9	—	Number of samples	6	9	9	2	5	8	5	7
—	9	—	Maximum	2,700	480.0	1,804	0.3	20.0	2,100	1,200	1,545
—	9	—	Minimum	206	10.5	161	0.1	6.0	100	50	230
—	9	—	Median	331	13.0	337	0.2	9.0	184	280	726
—	—	29	Number of samples	26	29	29	15	7	24	15	16
—	—	29	Maximum	2,700	480.0	1,804	17.6	20.0	2,100	1,200	1,545
—	—	29	Minimum	206	7.0	161	0.02	1.9	40	10	20
—	—	29	Median	377	26.0	321	0.5	6.0	106	320	675
—	—	29	Wells above regulatory level	1	1	3	1	2	2	8	15

Table 4. Summary of groundwater quality for most recent complete sample from selected wells, classified by storage unit and depth category, Santa Rosa Plain Watershed, Sonoma County, California, 1947–2010.—Continued

[See figure 1 for locations of wells. **Abbreviations**: MCL, maximum contaminant level; mg/L, milligram per liter; NL, notification level; SMCL, secondary maximum contaminant level; µg/L, microgram per liter; µS/cm, microsiemen per centimeter; —, no data]

Number of wells in shallow aquifer	Number of wells in deep aquifer	Total number of wells	Parameter units MCL, SMCL, or NL	Constituent							
				Specific conductance (µS/cm) [1]900	Chloride (mg/L) [1]250	Total dissolved solids computed (mg/L) [1]500	Nitrate as N (mg/L) [2]10	Arsenic (µg/L) [3]10	Boron (µg/L) [4]1,000	Iron (µg/L) [5]300	Manganese (µg/L) [5]50
Cotati groundwater storage unit											
10	—	—	Number of samples	10	10	10	9	1	4	2	1
10	—	—	Maximum	852	89.0	638	16.4	1.8	146	186	306
10	—	—	Minimum	390	22.0	294	0.1	1.8	80	40	306
10	—	—	Median	567	35.0	388	4.4	1.8	100	113	306
—	40	—	Number of samples	40	40	40	21	24	2	11	16
—	40	—	Maximum	700	81.0	554	7.1	15.0	26	1,900	460
—	40	—	Minimum	191	7.5	160	0.3	0.7	10	40	23
—	40	—	Median	365	17.0	273	1.0	4.3	18	310	69
—	—	87	Number of samples	86	87	87	57	40	18	21	27
—	—	87	Maximum	4,420	1,240.0	2,755	16.4	17.0	1,500	7,200	1,400
—	—	87	Minimum	191	7.5	160	0.1	0.7	10	40	23
—	—	87	Median	456	26.0	362	2.0	4.0	100	300	86
—	—	87	Wells above regulatory level	2	2	12	1	3	1	10	19
Wilson Grove groundwater storage unit											
1	—	—	Number of samples	1	1	1	1	1	1	1	1
1	—	—	Maximum	385	16.1	340	0.5	4.5	15	25	9
1	—	—	Minimum	385	16.1	340	0.5	4.5	15	25	9
1	—	—	Median	385	16.1	340	0.5	4.5	15	25	9
—	8	—	Number of samples	8	8	8	4	4	3	3	5
—	8	—	Maximum	372	22.1	352	1.3	13.0	16	3,354	120
—	8	—	Minimum	175	7.2	129	0.2	1.0	15	39	1
—	8	—	Median	352	12.6	238	0.9	9.3	16	280	13
—	—	15	Number of samples	15	15	15	11	6	7	6	7
—	—	15	Maximum	385	22.1	352	1.3	13.0	100	3,354	120
—	—	15	Minimum	91	7.2	105	0.1	1.0	10	10	1
—	—	15	Median	336	13.2	233	0.5	6.0	16	34	13
—	—	15	Wells above regulatory level	0	0	0	0	2	0	1	2
Total number of wells											
20	64	148	Total wells above regulatory level	3	3	19	2	7	4	23	44
20	64	148	Total number of samples	144	148	148	91	58	61	54	60

[1]State of California recommended SMCL.

[2]State MCL.

[3]Federal MCL.

[4]State of California notification level.

[5]State SMCL.

Samples from well W47 and W63 in the CB storage unit were a mixed cation-chloride type water (fig. 2D). These samples had the highest percentage of chloride (greater than 70 percent) in the CB storage unit, and contained dissolved-solids concentrations in excess of 1,550 mg/L (appendix A). These wells fall along the trace of an unnamed fault, east of the Sebastopol fault (figs. 1B and 3C). Unfortunately, boron concentrations were not analyzed for these samples to help determine if deeply circulating groundwater rising along the fault was a possible source of water to these samples.

Stiff diagrams of deep wells showed variability in water type and dissolved-solids concentrations in well samples from closely spaced wells in the southern part of the CB storage unit (fig. 3B). Multiple faults are mapped or inferred in the southern part of the CB storage unit (Langenheim and others, 2010), which could help explain the observed variability in water quality. Depth-dependent hydrologic, chemical, and isotopic data are needed to better understand this variability.

Wilson Grove (WG) Storage Unit

Most well samples in the WG storage unit were calcium-bicarbonate type or mixed cation-bicarbonate type waters with dissolved-solids concentrations less than 300 mg/L (fig. 2E, table 4). Unlike the WB and CB storage units, there was no shift in major cations from greater proportions of calcium and magnesium in shallow groundwater to greater proportions of sodium plus potassium in deeper groundwater. The sample from well W81 was a sodium-bicarbonate/chloride type water, which was significantly different than other samples in this storage unit (fig. 2E). Wells in the WG storage unit are perforated in the Wilson Grove Formation, which is composed of fine to medium-grained, moderately to well-sorted, tan to gray, uncemented to weakly cemented marine sandstone (Sweetkind and others, 2010). Unlike the Glen Ellen and Petaluma Formations, the Wilson Grove Formation has only minor clay deposits, which provide limited opportunities for the loss of calcium from solution by means of ion exchange with sodium attached to clays.

Changes in Chemical Character along Geologic Section B–B'

Stiff diagrams and dissolved-solids concentrations of well samples along geologic section B–B', modified from Sweetkind and others (2010, see fig. A1-4) and *chapter B* (see figs. 2 and 25), were used to illustrate changes in water quality as groundwater flows through the aquifer system. A total of 21 wells, including depth-dependent samples from wells W77, W89, and W102, were projected onto geologic section *B–B'* to compare water quality to geologic units and structure (fig. 3C).

Geologic Units along Section B–B'

Geologic section B-B' extends from the Mayacmas Mountains in the UPL storage unit on the east to the highlands in the WG storage unit on the west (fig. 3C). The UPL storage unit consists primarily of undifferentiated basement rocks and Sonoma Volcanics. In general, the basement rocks have low permeability and are not considered a major water-bearing unit, except in fractures and weathered zones. The Sonoma Volcanics are a heterogeneous assemblage of continental volcanic and volcaniclastic rocks, including basalt, andesite, and rhyolite lavas interbedded with air-fall and ash-flow tuffs, debris-flow deposits, and lacustrine deposits (Sweetkind and others, 2010).

The VAL storage unit consists primarily of the Glen Ellen Formation and the Sonoma Volcanics. For purpose of water-quality characterization in this chapter, all late Pliocene and younger nonmarine deposits are considered part of the Glen Ellen Formation, including the surficial Quaternary deposits. The Glen Ellen Formation consists of heterogeneous mixtures of tuffaceous clay, mud, bouldery to pebbly gravel, and sand and silt deposits with interbedded conglomerates (Sweetkind and others, 2010). Recent investigations indicated its thickness is about 100 to 150 ft throughout most of the SRPW (Sweetkind and others, 2010; *chapter B*).

The WB and CB storage units consist of 100–150 ft thick alluvial deposits of the Glen Ellen Formation underlain, for the most part, by the Petaluma Formation and Sonoma Volcanics and by the Wilson Grove Formation along their western edge (fig. 3C). The Sonoma Volcanics, however, were not identified along the geologic section in the upper 2,000 ft of the aquifer system, and water-quality data from wells were only available from about the upper 1,200 ft. The Pliocene and Miocene Petaluma Formation is dominated by deposits of moderately to weakly consolidated silty to clayey mudstone with local beds and lenses of poorly sorted sandstone (Sweetkind and others, 2010). The Petaluma Formation generally is finer grained then the overlying Glen Ellen Formation. These heterogeneous clay-rich deposits interfinger with the Sonoma Volcanics to the east and the Wilson Grove Formation to the west.

The WG storage unit consists almost entirely of the Wilson Grove Formation. The Pliocene and Late Miocene Wilson Grove Formation consists of consolidated, to weakly consolidated deposits of massive or thick-bedded, gray to buff, fine-grained to very fine grained, fossiliferous sand or sandstone (Sweetkind and others, 2010). Compared to the Petaluma Formation, the Wilson Grove Formation is coarser-grained and more permeable.

Geologic Structure along Section B–B'

Geologic section B-B' crosses multiple faults, including, from east to west, the Bennett Valley fault zone, Rodgers Creek fault zone, Trenton Ridge fault, and the Sebastopol fault. The Rodgers Creek and Bennett Valley fault zones are right-lateral faults and are branches of the San Andreas transform system in northern California (Langenheim and others, 2010). The Rodgers Creek fault zone forms the eastern boundary of the WB and CB storage units where they meet the UPL and VAL storage units (fig. 1). The Trenton Ridge fault is a northwest-southeast trending thrust fault that dips to the northeast and forms the boundary between the WB and CB storage units (fig. 1). The Sebastopol fault forms the boundary between the WG and CB storage units and the western boundary of the WB storage unit (fig. 1). An unnamed fault, parallel and east of the Sebastopol fault, appears to truncate the eastern extent of the Wilson Grove Formation along geologic section B–B' (fig. 3C).

Groundwater Flow along Section B–B'

Groundwater-level contour maps presented in *chapter B* (see fig. 29A) indicate that prior to significant groundwater development in the SRP, groundwater along the cross section flowed from the Mayacmas and Sonoma Mountains, in the UPL storage unit, westward toward the Laguna de Santa Rosa, on the western edge of the SRP, and eastward from the highlands in the WG storage unit toward the Laguna de Santa Rosa. In the UPL and VAL storage units, water-level contour lines bend in the upstream direction where they cross the confluence of the Santa Rosa, Spring, and Matanzas creeks east of the Rodgers Creek fault, indicating that streams are receiving water from the groundwater system (gaining streams). In contrast, immediately west of the Rodgers Creek fault zone in the WB and CB storage units, the water-level contour lines bend in the downstream direction where they cross the Santa Rosa Creek, indicating the stream is losing water to (recharging) the groundwater system (see chapter B, fig. 29A). About 5 mi west of the Rodgers Creek fault, water-level contour lines again bend in the downstream direction where they cross the Santa Rosa Creek, indicating that the stream is a gaining stream in the western part of the SRP (see *chapter B*, fig. 29A). Limited depth-dependent water-level data were available to determine the vertical flow gradient between geologic units along the cross section.

General Chemical Composition

The variation in major-ion composition of groundwater along geologic section B-B' is represented by Stiff diagrams and dissolved-solids concentrations for representative samples (fig. 3C). The groundwater was a mixed cation-bicarbonate type water, and most samples had relatively low dissolved-solids concentrations (about 230 mg/L or less) on both ends of the geologic section. As groundwater flows from the Mayacmas Mountains to the west, its dissolved-solids concentration increased with increasing distance from the mountains. Depth-dependent samples from the Sonoma Volcanics in well W89 had dissolved-solids concentrations of about 400 to 500 mg/L, which was about twice that in the sample from well W85 near the Mayacmas Mountains.

West of the VAL storage unit, in the Rodgers Creek fault zone, groundwater sampled from 786 to 806 ft bls in the open borehole of well W102 (sample W102B) when it was being drilled had a dissolved-solids concentration of about 1,800 mg/L and was a sodium-chloride type water. As stated previously, the high dissolved-solids concentration in well W102B was attributed to deeply circulating groundwater rising along the fault zone. The completed well W102, perforated from 130 to 180 ft bls in the Glen Ellen and Petaluma

Formations, yielded a sodium-bicarbonate type water with a dissolved-solids concentration of 534 mg/L. The relatively high concentrations of sulfate, boron, and dissolved solids in shallow water from well W102 (sample W102A) compared to the sample from well W85 at the east end of the section indicated possible upward flow along the fault zone through the deep aquifer and subsequent mixing with fresher groundwater in the shallow aquifer (appendix A). Additional depth-dependent samples are needed upgradient and downgradient of the Rodgers Creek fault zone to confirm this observation.

Comparison of data from the Rodgers Creek fault zone to available data from downgradient (west) of the fault zone indicated that the high dissolved-solids concentration of water in the fault zone did not significantly affect the water quality of wells in the WB and CB storage units. With the exception of the sample from well W107, all samples west of the Rodgers Creek fault zone had lower dissolved-solids concentrations than both the shallow and deep samples from well W102. The sample from well W103, the nearest downgradient well to well W102, was a mixed cation-bicarbonate type water with a dissolved-solids concentration of 344 mg/L. The difference in water type and its significantly lower dissolved-solids concentration indicated that groundwater underflow across the Rodgers Creek fault zone is minimal along geologic section B–B'. The Rodgers Creek fault zone likely restricts the lateral movement of water from the UPL and VAL storage units to the WB and CB storage units, forcing groundwater to discharge toward the stream channels east of the fault zone, as indicated by the groundwater-level contour maps presented in *chapter B* (see fig. 29A). The mixed cation-bicarbonate type water with low dissolved-solids concentration sampled from well W103, downgradient of the Rodgers Creek fault zone, is probably the result of recharge from Santa Rosa Creek west (downgradient) of the fault zone.

Stiff diagrams for samples from wells perforated within the Wilson Grove Formation, along the western end of section B-B', display a discernible difference in major-ion composition compared to samples from wells perforated in the Glen Ellen and Petaluma Formations to the east (fig. 3C). Calcium was the predominant cation of mixed cation-bicarbonate type water in samples from wells perforated in the Wilson Grove Formation. This water type is likely a reflection of recharge from comparatively low-elevation hills to the west through the marine sandstone of the Wilson Grove Formation. Calcium carbonate is commonly present as cementing material for fossil-rich sedimentary deposits and is a likely source of calcium in solution. Moreover, the relative lack of clay minerals in the Wilson Grove Formation provides little opportunity for the replacement of calcium in solution with sodium by means of ion exchange with clays.

The dissolved-solids concentrations in samples from wells perforated almost exclusively in the Wilson Grove Formation along the cross section (wells W32, W35, W76, W77, W82, and W129) were less than about 350 mg/l, and had a mean of about 200 mg/L (fig. 3C). The samples from wells with perforations in the Glen Ellen and Petaluma Formations immediately east of the WG storage unit

(W31, W33, and W34) were sodium-bicarbonate type water, and had dissolved-solids concentrations in excess of 420 mg/L (fig. 3C). The difference in water type and dissolved-solids concentrations indicated limited groundwater interaction between the Wilson Grove Formation and the neighboring Glen Ellen and Petaluma Formations. In addition, the low permeability clay layers in the Petaluma Formation limit the vertical groundwater movement between the Wilson Grove Formation and overlying Petaluma Formation.

Samples from shallow wells perforated or completed in the Glen Ellen Formation in the CB storage unit were primarily a mixed cation-bicarbonate type water in which magnesium was the most abundant cation, with the exception of the sample from well W33, which was a sodium-bicarbonate type water. Samples from wells perforated or completed in the Petaluma Formation, or in both the Glen Ellen and Petaluma Formations in the CB storage unit, were a mixed cation-bicarbonate type water in which sodium was the most abundant cation or were a sodium-bicarbonate water. In general, the dissolved-solids concentrations of samples from the shallow wells were higher than dissolved-solids concentrations of samples from the deep wells, which could be the result of anthropogenic factors, such as irrigation return flows. The samples from wells perforated in the Petaluma Formation, or in both the Glen Ellen and Petaluma Formations in the CB storage unit, had higher dissolved-solids concentrations and a greater proportion of sodium than samples from wells perforated in the Wilson Grove Formation on the western end of the CB storage unit (fig. 3C). The difference in dissolved-solids concentration and water type indicated that the unnamed fault, east of the Sebastopol fault, is at least a partial barrier to groundwater flow.

Temporal Trends

Temporal trends in water quality in the study area were evaluated by using analyses of specific conductance and chloride concentrations from 33 wells presented in appendix A that had a minimum of 8 records spanning 20 or more years and included at least one analysis between 2000 and 2010. Specific conductance is a general indicator of the water quality of a sample because it is directly related to the dissolved-solids concentration of the sample (Hem, 1992). Chloride was used in this study as a single-constituent indicator of general water quality because it is nonreactive geochemically and highly soluble. As a result of these properties, variations in chloride concentration are generally attributed to mixing with water from different sources (Hem, 1992). Because chloride is both a conservative ion and a major component of specific conductance, the two are usually strongly correlated.

The specific conductance of water from almost 75 percent of the 33 wells (appendix A) assessed for this study increased over time, with about half increasing by more than 10 percent of the initial value during the period of record (24–60 years for individual wells). Chloride behaved similarly, with concentrations increasing in about 67 percent of the wells evaluated,

and just over half increasing by more than 10 percent. Specific conductance and chloride concentration decreased in approximately 15 and 30 percent of evaluated wells, respectively, by more than 10 percent.

The largest increases in specific conductance, chloride concentration, or both were in wells W28, W41, W60, W65, and W73, located in the vicinity of the cities of Rohnert Park and Cotati (fig. 4; appendix A). The specific conductance of water from these wells increased more than 50 percent: from 310–380 µS/cm, in the early 1980s, to 520–610 µS/cm by 2009 (fig. 4). Chloride concentrations in water from these same wells also increased by more than 50 percent, with the exception of well W73, where the increase was less than 10 percent (fig. 4). Possible sources of increased specific conductance in the Rohnert Park/Cotati area include, groundwater underflow of high dissolved-solids concentration water present along the Rodgers Creek fault zone (for example, W102B), irrigation return flow, and septic-tank effluent or leaking sewer pipes. Groundwater pumping and fluctuating water levels could contribute to this possible change in the source of water to wells; municipal pumpage in Rohnert Park alone, more than doubled from about 2,500 acre-ft in 1976 to about 5,500 acre-ft in 1995, before being reduced to nearly zero by 2006 (see *chapter B*, fig. 37). As a consequence, water levels generally decreased between the mid-1970s and mid-1990s and then remained unchanged for a few years before increasing in the early 2000s (see *chapter B*, fig. 36). However, comparison of specific conductance and chloride versus groundwater withdrawals and water levels indicated no better than a weak relationship. Because of the presence of numerous wells in this part of the study area, it is unclear whether observed water-quality changes were in response to local conditions at these particular wells or were in response to more regional effects. Depth-dependent hydrologic, chemical, and isotopic data are needed to better understand the source of the increased specific conductance and chloride values.

The greatest decreases in specific conductance and chloride concentration were in samples from well W107 (fig. 4). Well W107 historically yielded some of the highest values in the SRP for specific conductance and chloride concentration, with maximum values in excess of 1,200 µS/cm and 150 mg/L, respectively (fig. 4; appendix A). The well is relatively shallow (perforated from 47 to 67 ft bls) and is in an area that has been classified as residential since 1974. Anthropogenic inputs could account for the relatively high specific conductance and chloride concentration values; however, because land use has not changed substantially in the vicinity of this well, it is unclear why water quality improved recently.

Water-Quality Constituents of Potential Concern

Specific conductance, chloride, dissolved solids, nitrate, arsenic, boron, iron, and manganese are water-quality constituents of potential concern in the SRPW because concentrations of these constituents in samples from some wells exceed state

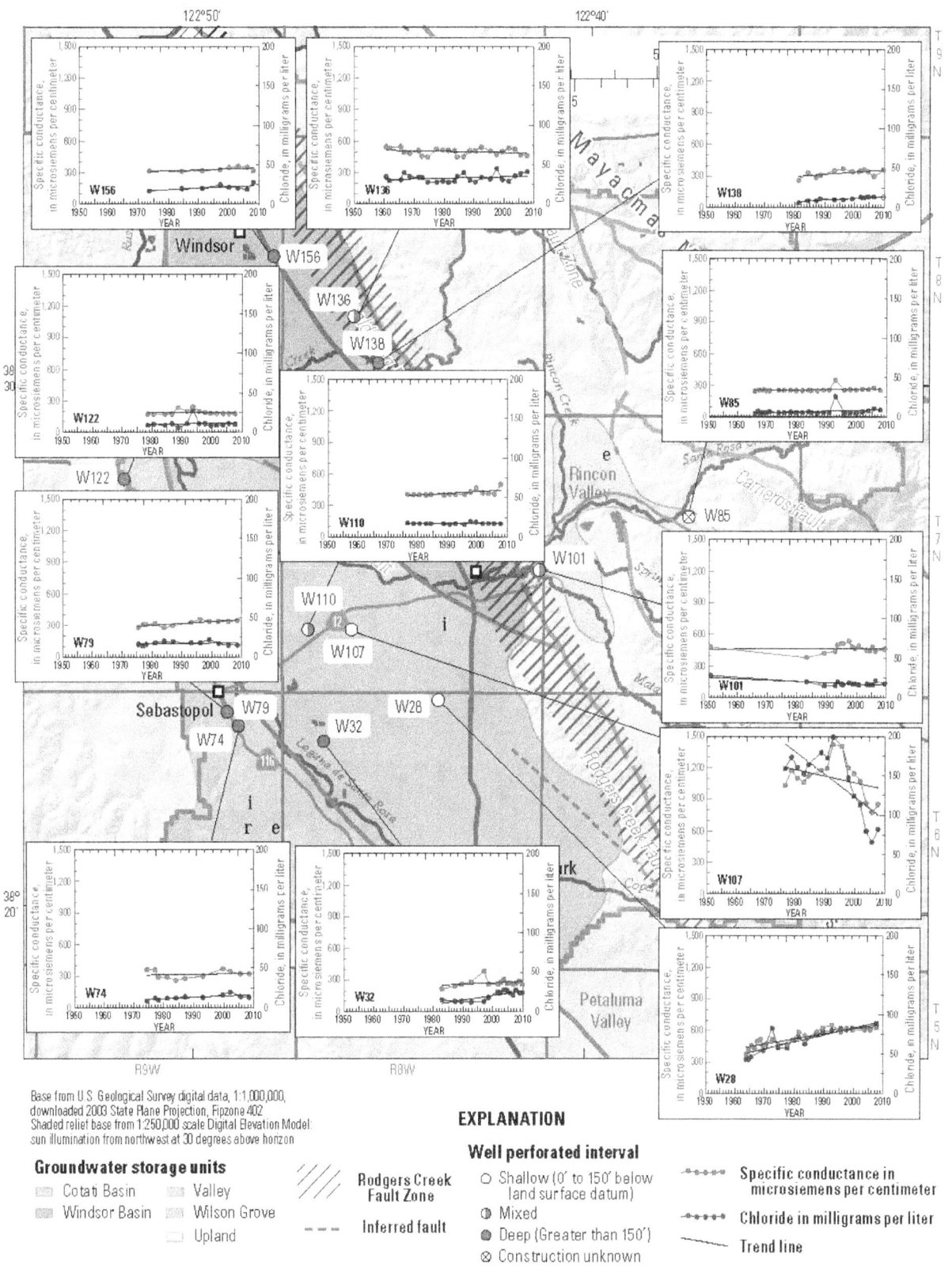

Figure 4. Time-series plots of specific conductance and chloride for selected wells in the Santa Rosa Plain watershed, Sonoma County, California, 1950–2009.

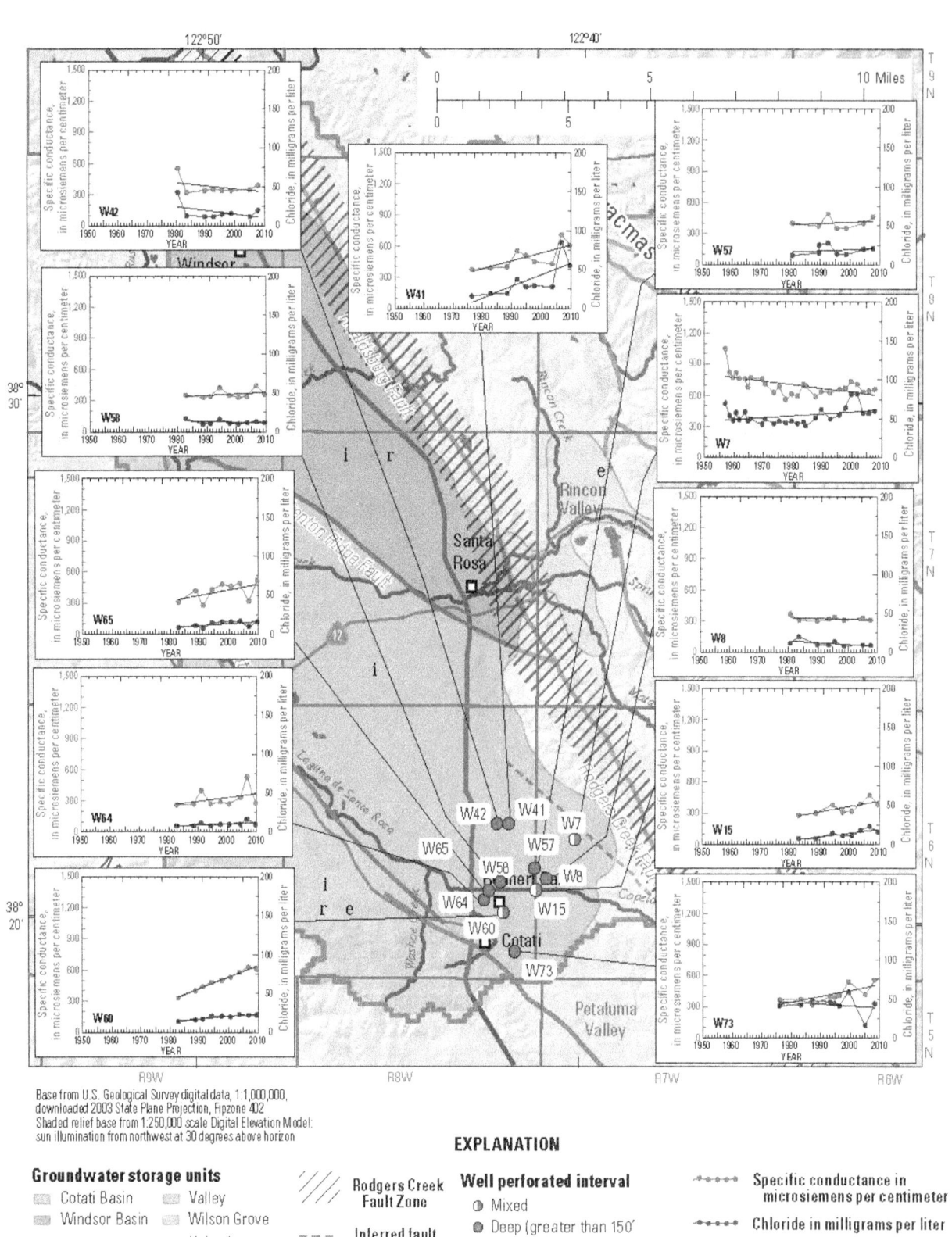

Base from U.S. Geological Survey digital data, 1:1,000,000,
downloaded 2003 State Plane Projection, Fipzone 402
Shaded relief base from 1:250,000 scale Digital Elevation Model:
sun illumination from northwest at 30 degrees above horizon

EXPLANATION

Groundwater storage units

- Cotati Basin
- Windsor Basin
- Valley
- Wilson Grove
- Upland

- Rodgers Creek Fault Zone
- Inferred fault

Well perforated interval
- Mixed
- Deep (greater than 150' below land surface datum)

- Specific conductance in microsiemens per centimeter
- Chloride in milligrams per liter
- Trend line

Figure 4. Continued.

or federal recommended or mandatory regulatory standards for drinking water (table 4). This section describes the sources, properties, and concentrations of these constituents in the SRPW. The range, median, and number of wells above regulatory standards for each constituent were summarized by storage unit and depth (table 4). Only the most recent complete analysis from each well (appendix A) was used to compile this summary.

Specific Conductance

The presence of charged ionic species in solution makes the solution conductive. As ion concentrations increase, conductance increases; thus, specific conductance at 25 used in this study provides an indication of total-ion or dissolved-solids concentration. Specific conductance has a recommended secondary maximum contaminant level for California (SMCL-CA) of 900 µS/cm (U.S. Environmental Protection Agency, 2009; California Department Public Health, 2011).

In the SRPW study area, only 3 of the 144 samples from wells with specific conductance measurements (about 2 percent) had values greater than or equal to the SMCL of 900 µS/cm (table 4). The highest specific conductance reported in table 4 is 4,420 µS/cm, which was measured in a sample from well W47 in the CB storage unit (appendix A).

Chloride

Chloride occurs naturally in groundwater from the weathering and leaching of sedimentary rocks and soils, and the dissolution of salt deposits, such as the evaporate deposit halite or sodium chloride (NaCl; Hem, 1992; Mullaney and others, 2009). An additional source is saline "connate" groundwater. The term connate implies that the solute source is fossil seawater trapped in the geologic formations when they were deposited (Hem, 1992). Anthropogenic sources of chloride include manufacturing, power generation, leachate from landfills, water softener backwash, and wastewater (Hem, 1992). Chloride has a recommended SMCL-CA of 250 mg/L (U.S. Environmental Protection Agency, 2009; California Department Public Health, 2011). The SMCL for chloride is based on aesthetic properties, including a salty taste and a common association with sodium, which is of concern to individuals susceptible to hypertension. Although not toxic to humans, chloride presents a common toxicity problem for plants (Ayers and Westcot, 1994). Because chloride is not absorbed or held back by soils, it is very mobile in groundwater and soil water; therefore, it is taken up by the crop, moves through the phloem, and accumulates in the leaves. Chloride tolerance varies broadly across different crops. Some of the more sensitive crops include avocado, strawberries, berries, and certain types of grapes and citrus (Ayers and Westcot, 1994).

Only 3 of the 148 samples from wells in the SRPW study area with chloride analyses (about 2 percent) had concentrations greater than or equal to the SMCL of 250 mg/L (table 4). The highest chloride concentration reported in table 4 is 1,240 mg/L, which was measured in a sample from well W47 in the CB storage unit (appendix A).

Dissolved Solids

Total dissolved solids (or dissolved solids) are a measure of all dissolved chemicals in water (Hem, 1992; California State Water Resources Control Board, 2010). Dissolved solids can be measured or calculated in several ways; most commonly, a summation of dissolved ions is used, if a complete analysis is available, or the residue that remains after evaporation at a specific temperature is weighed (residue-on-evaporation, or ROE, at 180 in this study). The computed dissolved solids can differ from the ROE value by 10–20 mg/L in either the positive or negative direction when the solids concentration is 100–500 mg/L (Hem, 1992); thus, a direct comparison of dissolved-solids concentrations from different labs is difficult, unless the method used by the analyzing laboratory is known. For this report, the total dissolved-solids concentration was calculated as the summation of the reported major cations (calcium, magnesium, sodium, and potassium) and anions (chloride, sulfate, bicarbonate, fluoride, silica, and nitrate, if available). Total dissolved solids has a recommended SMCL-CA of 500 mg/L (U.S. Environmental Protection Agency, 2009; California Department Public Health, 2011).

Of the 148 samples from wells in the SRPW study area with dissolved-solids analyses, 19 (about 13 percent) had concentrations greater than or equal to the SMCL of 500 mg/L (table 4). The highest dissolved-solids concentration reported in table 4 is 2,755 mg/L, which was measured in a sample from well W47 in the CB storage unit (appendix A). In the WB and CB storage units, the median dissolved-solids concentration was higher in samples from shallow wells than in samples from deep wells (table 4), which could reflect anthropogenic sources of dissolved-solids concentrations, such as irrigation-return flows, to the shallow aquifer.

Nitrate

Nitrate (NO_3) is one of the most frequently identified contaminants in groundwater and is attributable to both anthropogenic and natural sources (Freeze and Cherry, 1979). Natural sources of nitrate include the atmosphere and decomposition (oxidation, or mineralization) of organic material (Hem, 1992). Anthropogenic sources include fertilizers, septic-tank effluent, leaking sewers, and atmospheric deposition of nitrogen emissions (Hem, 1992). Nitrate concentrations in drinking water in excess of the maximum contaminant level (MCL) of 10 mg/L as nitrogen (N), equivalent to 45 mg/L as NO_3, are considered hazardous and can result in methemoglobinema (blue-baby syndrome) in small children (Hem, 1992).

Only 2 of the 91 samples from wells in the SRPW study area with nitrate analyses (about 2 percent) had NO_3-N concentrations greater than or equal to the MCL of 10 mg/L (table 4). NO_3-N concentrations in the UPL storage unit, where anthropogenic contributions are expected to be low, were all less than or equal to 0.4 mg/L (table 4). On the basis of the NO_3-N concentrations in the UPL storage unit, concentrations greater than 1 mg/L in the SRPW likely reflect anthropogenic contributions. Based on this assumption, anthropogenic nitrate was present in some shallow wells of the WB and CB storage units that had median concentrations of 0.9 and 4.4 mg/L, respectively (table 4). As expected, NO_3-N concentrations were lower in the deep wells of these storage units (median concentrations of 0.2 and 1.0, respectively; table 4). Concentrations are typically lower at depth because of greater distance from the anthropogenic source at land surface and increased opportunity over time and distance for nitrate to be diluted by mixing with low-nitrate groundwater and to be denitrified under anoxic conditions in the aquifer system.

Arsenic

Arsenic is a tasteless, odorless, semi-metallic element which can be present in surface and groundwater from natural and anthropogenic sources. Arsenic in the natural environment is most commonly associated with sulfide and with ferromanganese-oxide minerals, particularly in areas characterized by geothermal water or high evaporation rates (Welch and others, 2000). Anthropogenic sources of arsenic include wood preservatives, pesticides, and used as a semi-conductor in the manufacture of microelectronics (Agency for Toxic Substances and Disease Registry, 2009). Industrial and agricultural uses of arsenic have been reduced in the United States over the last several decades because of the increasing recognition of its threat to human health (Agency for Toxic Substances and Disease Registry, 2009). The MCL for arsenic was lowered from 50 µg/L to 10 µg/L in 2006. Long-term exposure to arsenic in water can cause a variety of dermal, cardiovascular, respiratory, and neurological ailments, as well as an increased risk of some cancers (U.S. Environmental Protection Agency, 2009; California Department of Public Health, 2011).

Of the 58 samples from wells in the SRPW study area with arsenic analyses, 7 (about 12 percent) had arsenic concentrations greater than or equal to the MCL of 10 µg/L (table 4). The highest arsenic concentration reported in table 4 is 20 µg/L, which was measured in a sample from well W156 in the WB storage unit (appendix A). About 30 percent of the wells sampled for arsenic in the WB and WG storage units had arsenic concentrations in excess of the MCL.

Boron

Boron is a naturally occurring metalloid in many minerals. Natural sources of boron include the mineral tourmaline, which is present in igneous rocks, and evaporite minerals, such as borax, colemanite, and kernite (Hem, 1992; Reimann

and Caritat, 1998). The most prevalent sources of boron in water are from the leaching of rocks and soils that contain borate or borosilicate minerals, wastewater with cleaning agents containing boron, and fertilizers and pesticides (California State Water Resources Control Board, 2010). High boron concentrations (in excess of 48,000 mg/L) have been associated with thermal springs (Hem, 1992). Although boron is a trace micronutrient necessary for metabolism of important substances such as calcium and magnesium, high concentrations of boric acid, the most common form in drinking water, can be toxic to humans. Symptoms of boric acid ingestion include gastrointestinal tract distress, vomiting, abdominal pain, diarrhea, and nausea (State Water Resources Control Board, 2010). Once boron compounds dissolve, they behave like a salt (dissolved anion) and are difficult to remove by treatment. Boron has a health-based notification level (NL-CA) of 1,000 µg/L. NL-CAs are established by the CDPH for some constituents in drinking water that lack MCLs. If a constituent is detected at concentrations greater than its NL-CA, California State law requires timely notification of local governing bodies and recommends consumer notification. Boron is essential to plant nutrition; however, concentrations in a small excess of the required amount are toxic to some plants (Hem, 1992). For example, the optimal boron concentration range is 300–500 µg/L for citrus and grapes, but concentrations slightly higher are detrimental to these crops (Kabay and others, 2006).

Of the 61 samples from wells in the SRPW study area with boron analyses, 4 (about 7 percent) had boron concentrations greater than or equal to the NL-CA of 1,000 µg/L (table 4). The highest boron concentration reported in table 4 is 2,100 µg/L, which was measured in the deep sample from well W102 in the WB storage unit (sample W102B; appendix A). The median boron concentration was about 100 µg/L in samples from wells with boron analyses in the WB and CB storage units, areas where most of the irrigated agriculture exists.

Iron

Iron is derived from natural weathering of many rocks and minerals whose iron content is relatively high, including pyroxenes, amphiboles, biotite, magnetite, and olivine (Hem, 1992). When exposed to anoxic or suboxic conditions and low to near neutral pH, iron is readily dissolved in the form of ferrous (Fe^{2+}) iron. Under oxidizing conditions (dissolved oxygen greater than 1–2 mg/L), and at all but a very low pH, iron exists as ferric (Fe^{3+}) iron, which is much less soluble (Hem, 1992). The SMCL of iron is 300 µg/L (U.S. Environmental Protection Agency, 2009; California Department of Public Health, 2011). Excessive amounts of iron can cause aesthetic effects, such as an objectionable metallic taste and turning water a reddish or orange color that can stain plumbing fixtures and other surfaces, but normally, it does not pose a health risk. Excessive iron can also have economic implications, including corrosion and encrustation of steel well casing, perforations, and related infrastructure.

Of the 54 samples from wells in the SRPW study area with iron analyses, 23 (about 43 percent) had iron concentrations greater than or equal to the SMCL of 300 µg/L (table 4). The highest iron concentration reported in table 4 is 7,200 µg/L, which was measured in a sample from well W119 in the CB storage unit (appendix A). About 50 percent of the samples from wells with iron analyses in the WB and CB storage units were at or above the SMCL.

Manganese

Manganese is derived from natural weathering of many rocks and minerals, including basalt, many olivines, pyroxene, and amphibole (Hem, 1992). Similar to iron, manganese is readily dissolved under reducing conditions (dissolved oxygen is absent) and low to near-neutral pH. In the presence of oxygen and carbonate or silicate at high pH, manganese will precipitate out of solution and form black-colored deposits that, in drinking water, are unpleasant in appearance and taste (Hem, 1992). At concentrations above the SMCL of 50 µg/L (U.S. Environmental Protection Agency, 2009; California Department of Public Health, 2011), manganese can have a metallic taste and cause staining of plumbing fixtures and other surfaces, but normally, it does not pose a health risk.

Of the 60 samples from wells in the SRPW study area with manganese analyses, 44 (about 73 percent) had manganese concentrations greater than or equal to the SMCL of 50 µg/L (table 4). The highest manganese concentration reported in table 4 is 1,545 µg/L, which was measured in a sample from well W142 in the WB storage unit (appendix A). Of the samples from wells with manganese analyses in the WB and CB storage units, about 94 and 70 percent, respectively, were at or above the SMCL.

Source and Age of Groundwater

Stable and radioactive isotopes, including oxygen-18 (^{18}O), deuterium (^2H), tritium (^3H), carbon-14 (^{14}C), and carbon-13 (^{13}C), were used to determine the sources and ages of water in the SRPW. Analyses from samples collected for this study, and selected data from the NSF GAMA study (Kulongoski and others, 2010, appendix B9, tables 6 and 7), were used to gain insight into recharge processes and the evolution of water quality in the study area.

Oxygen-18 and Deuterium

Two sets of samples were collected from 15 surface-water sites during the fall of 2008 and spring of 2009 for analysis of oxygen-18 and deuterium (fig. 1A, table 5). Additional samples were collected from one surface-water site (SW21) during 2003–07 independent of this study (fig. 1A, table 5). Groundwater samples were collected from 32 wells, including discrete-depth zones in 3 wells (W77, W89, and W102)

during 2004–10 (fig. 5A, table 5). Twenty-five of the groundwater samples were collected as part of the NSF GAMA study (Kulongoski and others, 2006).

Background

Oxygen-18 (^{18}O) and deuterium (^2H or D) are stable (nonradioactive), naturally occurring isotopes of oxygen and hydrogen. The abundance of heavier oxygen-18 and deuterium relative to isotopically lighter oxygen-16 (^{16}O) and hydrogen-1 or protium (^1H) can be used to infer the source and the evaporative history of water. Oxygen-18 and deuterium data are expressed in delta notation (δ) as per mil (parts per thousand, ‰) differences in the ratios of $^{18}O/^{16}O$ and $^2H/^1H$ in samples relative to a standard known as Vienna Standard Mean Ocean Water (VSMOW; Gat and Gonfiantini, 1981). By convention, the value of VSMOW is 0 per mil. Oxygen-18 ($\delta^{18}O$) and deuterium (δ^2H or δD) ratios are useful in a wide variety of hydrologic studies (Gat and Gonfiantini, 1981). Analytical precision is generally within about 0.2 and 2 per mil for $\delta^{18}O$ and δD, respectively (Coplen, 1994).

Because the source of much of the world's precipitation is derived from the evaporation of seawater, the $\delta^{18}O$ and δD compositions of precipitation near coasts throughout the world cluster along a line known as the global meteoric water line (GMWL; Craig, 1961), such that the following is found:

$$\delta D = 8\delta^{18}O + 10 \qquad (1)$$

Plotting the isotopic composition of precipitation shows a general trend from heavier (having more oxygen-18 and deuterium and, therefore, less negative values) to lighter (having less oxygen-18 and deuterium and, therefore, more negative values) from the equator to the poles (Gat and Gonfiantini, 1981). Storms that originate over cold waters in the Gulf of Alaska have a lighter isotopic composition than storms that originate over warm tropical waters in the vicinity of Hawaii or the Gulf of Mexico. Local meteoric water lines (LMWL) can be slightly offset from the GMWL by a different intercept term in equation 1 (deuterium excess), but are parallel to the GMWL. In the case of this study, there was no well-constrained LMWL, so the GMWL was used as a surrogate for comparison with local samples.

Differences also result from fractionation (partitioning of molecules of different isotopic composition of the same element) as moist air masses move inland. As storms move inland from coastal areas, the concentration of heavier isotopes relative to lighter isotopes decreases because heavier isotopes are preferentially concentrated in the liquid phase, and lighter isotopes are preferentially concentrated in the vapor phase during repeated cycles of evaporation and condensation. In addition, precipitation that condenses at higher altitudes and at cooler temperatures tends to be isotopically lighter than precipitation that condenses at lower altitudes and warmer temperatures (Muir and Coplen, 1981).

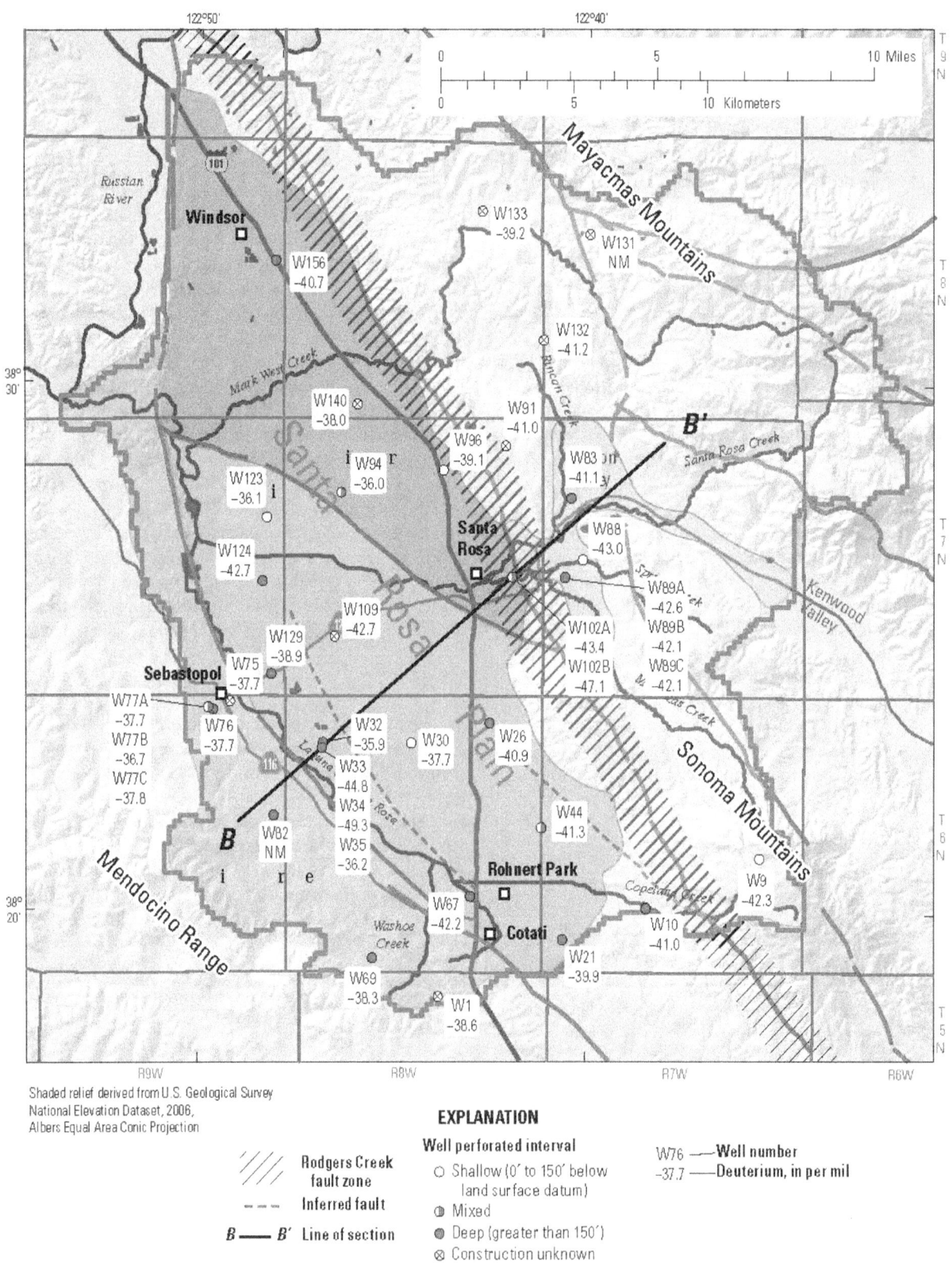

Figure 5. Delta deuterium and oxygen-18 data from selected wells in the Santa Rosa Plain watershed, Sonoma County, California, 2003–10: *A*, areally and *B*, along section B-B′.

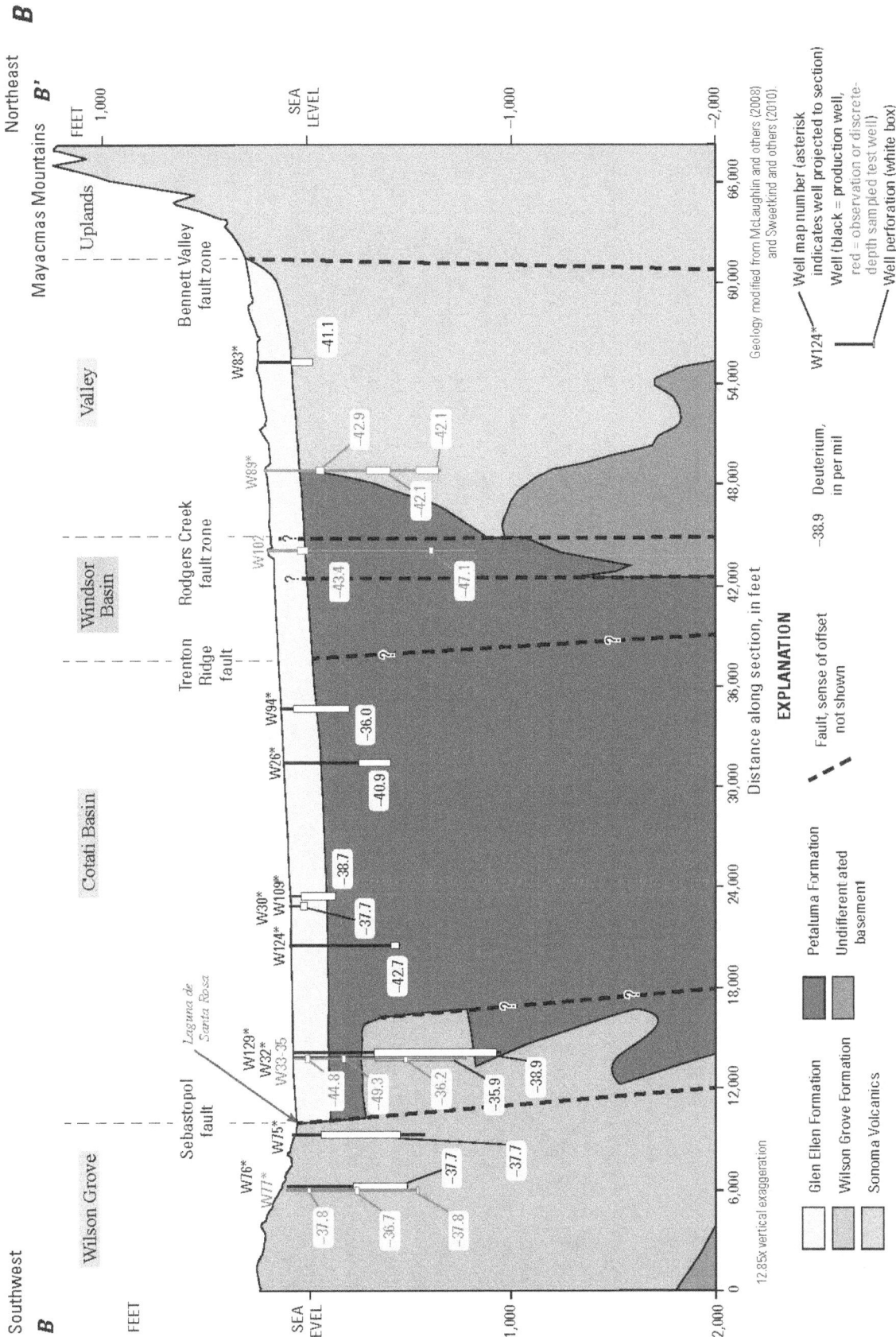

Figure 5. Continued.

Table 5. Summary of isotopic data in samples from selected surface-water sites and wells, Santa Rosa Plain watershed and neighboring area, Sonoma County, California, 2003–2010.

[See figure 1 for locations of surface-water sites and figure 7 for locations of selected wells. Sample depths listed for multi-zone sampled wells and piezometers. δE is delta notation, the ratio of the heavier isotope (i) to the more common lighter isotope of an element (E), relative to a standard reference material, expressed as per mil (parts per thousand). **Abbreviations**: Ave, avenue; Cr, circle; Crk; creek; Dr., drive; mm/dd/yyyy, month/day/year; Mtn., mountain; N/A, not applicable; no., number; nr, near; pCi/L, picocuries per liter; Rd., road; Sprs, springs; St., street; TU, tritium units; Unk, unknown; USGS, U.S. Geological Survey; –, no data; @, at; <, less than]

Map no.	Groundwater storage unit	State well number or surface-water site name	USGS identification number	Depth category	Sampled interval (feet)	Sample date (mm/dd/yyyy)	Delta deuterium (per mil)	Delta oxygen-18 (per mil)	Tritium (pCi/L)	Tritium[2] (TU)	Carbon-14 (percent modern)	Uncorrected carbon-14 age (years before present)	Delta carbon-13 (per mil)
Surface-water sites													
SW1	Uplands	Crane Creek nr Rohnert Park	382051122383801	–	–	10/24/2008	-35.5	-5.33	–	–	–	–	–
SW1	Uplands	Crane Creek nr Rohnert Park	382051122383801	–	–	04/02/2009	-39.2	-6.10	–	–	–	–	–
SW2	Cotati	Laguna de Santa Rosa @ Stony Point	11465680	–	–	10/23/2008	-25.1	-3.30	–	–	–	–	–
SW2	Cotati	Laguna de Santa Rosa @ Stony Point	11465680	–	–	04/01/2009	-33.1	-4.92	–	–	–	–	–
SW3	Uplands	Santa Rosa Crk @ Wildwood Mtn. Rd.	382801122371801	–	–	10/24/2008	-36.8	-5.83	–	–	–	–	–
SW3	Uplands	Santa Rosa Crk @ Wildwood Mtn. Rd.	382801122371801	–	–	04/02/2009	-41.8	-6.76	–	–	–	–	–
SW5	Uplands	Spring Lake @ pier	382710122391401	–	–	10/24/2008	-13.1	0.41	–	–	–	–	–
SW5	Uplands	Spring Lake @ pier	382710122391401	–	–	04/02/2009	-36.2	-4.54	–	–	–	–	–
SW6	Valley	Brush Cr @ Hwy 12	11466065	–	–	10/24/2008	-23.7	-3.44	–	–	–	–	–
SW6	Valley	Brush Cr @ Hwy 12	11466065	–	–	04/02/2009	-37.4	-5.80	–	–	–	–	–
SW7	Valley	Matanzas Crk @ Bethards Dr.	382543122394601	–	–	10/24/2008	-38.2	-5.89	–	–	–	–	–
SW7	Valley	Matanzas Crk @ Bethards Dr.	382543122394601	–	–	04/02/2009	-40.0	-6.06	–	–	–	–	–
SW8	Windsor	Santa Rosa Crk @ Pierson St.	11466200	–	–	10/24/2008	-37.9	-5.45	–	–	–	–	–
SW8	Windsor	Santa Rosa Crk @ Pierson St.	11466200	–	–	04/02/2009	-40.1	-6.37	–	–	–	–	–
SW9	Windsor	Matanzas Cr @ Santa Rosa	11466170	–	–	10/24/2008	-38.0	-5.54	–	–	–	–	–
SW9	Windsor	Matanzas Cr @ Santa Rosa	11466170	–	–	04/02/2009	-39.0	-5.98	–	–	–	–	–
SW10	Cotati	Santa Rosa Crk @ Willowside Ave	11466320	–	–	10/23/2008	-31.2	-4.53	–	–	–	–	–
SW10	Cotati	Santa Rosa Crk @ Willowside Ave	11466320	–	–	04/01/2009	-38.9	-6.09	–	–	–	–	–
SW13	Wilson Grove	Laguna de Santa Rosa nr Sebastopol	11465750	–	–	10/23/2008	-23.8	-3.07	–	–	–	–	–
SW13	Wilson Grove	Laguna de Santa Rosa nr Sebastopol	11465750	–	–	04/01/2009	-33.8	-5.32	–	–	–	–	–
SW14	N/A	Jonive Cr nr Sebastopol	382417122512901	–	–	10/23/2008	-31.4	-5.05	–	–	–	–	–
SW14	N/A	Jonive Cr nr Sebastopol	382417122512901	–	–	04/01/2009	-34.6	-5.60	–	–	–	–	–
SW15	Uplands	Mark West Cr @ Mark West Sprs	383258122431201	–	–	10/24/2008	-37.3	-5.79	–	–	–	–	–
SW15	Uplands	Mark West Cr @ Mark West Sprs	383258122431201	–	–	04/01/2009	-40.6	-6.56	–	–	–	–	–

Table 5. Summary of isotopic data in samples from selected surface-water sites and wells, Santa Rosa Plain watershed and neighboring area, Sonoma County, California, 2003–10.—Continued

[See figure 1 for locations of surface-water sites and figure 7 for locations of selected wells. Sample depths listed for multi-zone sampled wells and piezometers. δE is delta notation, the ratio of the heavier isotope (i) to the more common lighter isotope of an element (E), relative to a standard reference material, expressed as per mil (parts per thousand). **Abbreviations**: Ave, avenue; Cr, circle; Crk, creek; Dr., drive; mm/dd/yyyy, month/day/year; Mtn., mountain; N/A, not applicable; no., number; nr, near; pCi/L, picocuries per liter; Rd., road; Sprs, springs; St., street; TU, tritium units; Unk, unknown; USGS, U.S. Geological Survey; –, no data; @, at; <, less than]

Map no.	Groundwater storage unit	State well number or surface-water site name	USGS identification number	Depth category	Sampled interval (feet)	Sample date (mm/dd/yyyy)	Delta deuterium (per mil)	Delta oxygen-18 (per mil)	Tritium (pCi/L)	Tritium2 (TU)	Carbon-14 (percent modern)	Uncorrected carbon-14 age (years before present)	Delta carbon-13 (per mil)
						Surface-water sites—Continued							
SW17	Windsor	Mark West Cr nr Windsor	11465500	–	–	10/24/2008	-29.4	-4.19	–	–	–	–	–
SW17	Windsor	Mark West Cr nr Windsor	11465500	–	–	04/01/2009	-40.6	-6.48	–	–	–	–	–
SW18	N/A	Russian River @ River Front Park	383101122514901	–	–	10/23/2008	-43.7	-6.40	–	–	–	–	–
SW18	N/A	Russian River @ River Front Park	383101122514901	–	–	04/01/2009	-44.5	-6.68	–	–	–	–	–
SW21	Wilson Grove	Mark West Cr nr Mirabel Heights	11466800	–	–	09/07/2003	-34.5	-5.07	–	–	–	–	–
SW21	Wilson Grove	Mark West Cr nr Mirabel Heights	11466800	–	–	06/02/2004	-31.7	-4.86	–	–	–	–	–
SW21	Wilson Grove	Mark West Cr nr Mirabel Heights	11466800	–	–	07/19/2004	-31.3	-4.45	–	–	–	–	–
SW21	Wilson Grove	Mark West Cr nr Mirabel Heights	11466800	–	–	09/15/2004	-34.0	-4.54	–	–	–	–	–
SW21	Wilson Grove	Mark West Cr nr Mirabel Heights	11466800	–	–	09/14/2005	-34.1	-4.80	–	–	–	–	–
SW21	Wilson Grove	Mark West Cr nr Mirabel Heights	11466800	–	–	06/14/2006	-32.4	-4.91	–	–	–	–	–
SW21	Wilson Grove	Mark West Cr nr Mirabel Heights	11466800	–	–	07/20/2006	-30.3	-4.14	–	–	–	–	–
SW21	Wilson Grove	Mark West Cr nr Mirabel Heights	11466800	–	–	08/23/2006	-33.2	-4.65	–	–	–	–	–
SW21	Wilson Grove	Mark West Cr nr Mirabel Heights	11466800	–	–	09/22/2006	-33.6	-4.80	–	–	–	–	–
SW21	Wilson Grove	Mark West Cr nr Mirabel Heights	11466800	–	–	06/06/2007	-32.7	-4.71	–	–	–	–	–
SW21	Wilson Grove	Mark West Cr nr Mirabel Heights	11466800	–	–	08/22/2007	-31.2	-4.12	–	–	–	–	–
SW21	Wilson Grove	Mark West Cr nr Mirabel Heights	11466800	–	–	10/23/2008	-23.1	-3.68	–	–	–	–	–
SW21	Wilson Grove	Mark West Cr nr Mirabel Heights	11466800	–	–	04/01/2009	-37.3	-5.68	–	–	–	–	–
SW21	Wilson Grove	Mark West Cr nr Mirabel Heights	11466800	–	–	06/16/2010	-36.3	-5.49	–	–	–	–	–
SW21	Wilson Grove	Mark West Cr nr Mirabel Heights	11466800	–	–	08/25/2010	-35.1	-4.82	–	–	–	–	–
						Wells							
W1	Wilson Grove	5N/8W-03G1	381827122434601	Unk	–	09/01/2004	-38.6	-6.03	<1	<0.3	–	–	–
W9	Uplands	6N/7W-23H1	382038122354001	Shallow	76-136	10/05/2004	-42.3	-6.79	7.7	2.4	–	–	–
W10	Cotati	6N/7W-28L1	382009122382601	Deep	441-862	08/31/2004	-41.0	-6.23	1.6	0.5	–	–	–
W21	Cotati	6N/7W-31C1	381934122403601	Deep	170-680	09/01/2004	-39.9	-6.01	1.6	0.5	–	–	–
W26	Cotati	6N/8W-02I2	382337122422902	Deep	369-530	09/20/2004	-40.9	-6.28	1.9	0.6	–	–	–
W30	Cotati	6N/8W-04R1	382315122443001	Shallow	55-85	08/31/2004	-37.7	-5.89	1.9	0.6	–	–	–
W32	Cotati	6N/8W-07A2	382307122463801	Deep	650-800	09/13/2004	-35.9	-5.68	ND	–	–	–	–
W33	Cotati	6N/8W-07A4	382308122463903	Shallow	60-80	08/22/2007	-44.8	-6.61	[2]<0.3	<0.1	16.7	14,000	-20.23
W34	Cotati	6N/8W-07A5	382308122463902	Deep	237-257	08/22/2007	-49.3	-7.33	[2]<0.3	<0.1	1.3	34,000	-21.10
W35	Cotati	6N/8W-07A6	382308122463901	Deep	550-570	08/23/2007	-36.2	-5.67	[2]<0.3	<0.1	58.5	4,000	-22.86
W44	Cotati	6N/8W-13R3	382141122410901	Mixed	130-450	09/01/2004	-41.3	-6.28	<1	<0.3	–	–	–
W67	Cotati	6N/8W-26C2	382021122425701	Deep	295-670	09/02/2004	-42.2	-6.29	1	0.3	–	–	–

Table 5. Summary of isotopic data in samples from selected surface-water sites and wells, Santa Rosa Plain watershed and neighboring area, Sonoma County, California, 2003–10.—Continued

[See figure 1 for locations of surface-water sites and figure 7 for locations of selected wells. Sample depths listed for multi-zone sampled wells and piezometers. δE is delta notation, the ratio of the heavier isotope (i) to the more common lighter isotope of an element (E), relative to a standard reference material, expressed as per mil (parts per thousand). **Abbreviations:** Ave, avenue; Cr, circle; Crk, creek; Dr., drive; mm/dd/yyyy, month/day/year; Mtn., mountain; N/A, not applicable; no., number; nr, near; pCi/L, picocuries per liter; Rd., road; Sprs, springs; St., street; TU, tritium units; Unk, unknown; USGS, U.S. Geological Survey; –, no data; @, at; <, less than]

Map no.	Groundwater storage unit	State well number or surface-water site name	USGS identification number	Depth category	Sampled interval (feet)	Sample date (mm/dd/yyyy)	Delta deuterium (per mil)	Delta oxygen-18 (per mil)	Tritium (pCi/L)	Tritium[1] (TU)	Carbon-14 (percent modern)	Uncorrected carbon-14 age (years before present)	Delta carbon-13 (per mil)
						Wells—Continued							
W69	Wilson Grove	6N/8W-33M1	381915122452701	Shallow	229–295	10/27/2004	-38.3	-5.92	ND	–	–	–	–
W75	Wilson Grove	6N/9W-02B1	382400122491201	Mixed	138–528	10/05/2004	-37.7	-6.14	2.2	0.7	45.2	6,000	-15.48
W76	Wilson Grove	6N/9W-02C1	382352122493301	Deep	332–600	10/04/2004	-37.7	-6.12	2.6	0.8	57.5	4,000	-17.99
W77A	Wilson Grove	6N/9W-02C2	382352122493801	Shallow	100–120	10/16/2006	-37.8	-6.06	2.1	0.6	57.8	4,000	-17.12
W77B	Wilson Grove	6N/9W-02C2	382352122493801	Deep	340–360	10/16/2006	-36.7	-6.04	[2]<0.3	<0.1	38.5	7,000	-18.82
W77C	Wilson Grove	6N/9W-02C2	382352122493801	Deep	640–650	10/12/2006	-37.8	-6.07	1.7	0.5	59.2	4,000	-16.91
W82	Wilson Grove	6N/9W-13J2	382153122480301	Deep	432–452	09/22/2004	-41.1	–	<1	<0.3	–	–	–
W83	Valley	7N/7W-07K1	382754122402501	Deep	160–265	09/23/2004	-43.0	-6.49	<1	<0.3	–	–	–
W88	Valley	7N/7W-18R2	382647122400702	Shallow	50–100	09/15/2004	-42.6	-6.57	ND	–	–	–	–
W89A	Valley	7N/7W-19G2	382623122403301	Deep	250–290	03/12/2010	-42.1	-6.49	3.0	0.9	34.4	8,000	-14.77
W89B	Valley	7N/7W-19G2	382623122403301	Deep	500–620	03/17/2010	-42.1	-6.46	0.7	0.2	27.1	10,000	-15.98
W89C	Valley	7N/7W-19G2	382623122403301	Deep	750–860	03/11/2010	-41.0	-6.54	[2]<0.3	<0.1	26.9	10,000	-16.33
W91	Uplands	7N/8W-01M1	382849122420501	Unk	–	09/15/2004	-36.0	-6.46	<1	<0.3	68.7	3,000	-19.88
W94	Windsor	7N/8W-08L1	382759122462001	Mixed	65–341	08/31/2004	-39.1	-5.65	2.2	0.7	–	–	–
W96	Windsor	7N/8W-10A1	382825122434101	Shallow	65–85	09/22/2004	-43.4	-6.16	<1	<0.3	–	–	–
W102A	Windsor	7N/8W-24F1	382622122414901	Mixed	130–180	08/24/2009	-47.1	-6.57	[2]<0.3	<0.1	26.2	10,000	-17.13
W102B	Windsor	7N/8W-24F1	382622122414901	Deep	786–806	07/23/2009	-38.7	-7.03	[2]<0.3	<0.1	3.4	27,000	-22.28
W109	Cotati	7N/8W-19P1	382510122462701	Mixed	61–231	09/30/2004	-36.1	-5.93	1.6	0.5	84.7	1,000	-18.12
W123	Cotati	7N/9W-12K1	382751122482501	Shallow	40–60	09/14/2004	-42.7	-5.82	5.1	1.6	–	–	–
W124	Cotati	7N/9W-24G1	382619122482101	Deep	507–547	09/14/2004	-38.9	-6.53	<1	<0.3	–	–	–
W129	Cotati	7N/9W-36K2	382429122480501	Deep	410–1,020	09/14/2004	–	-6.25	<1	<0.3	24.0	11,000	-19.53
W131	Uplands	8N/7W-07R1	383257122400501	Unk	–	09/15/2004	–	–	1.6	0.5	–	–	–
W132	Uplands	8N/7W-30E1	383053122410801	Unk	–	09/29/2004	-41.2	-6.59	<1	<0.3	–	–	–
W133	Uplands	8N/8W-11G1	383319122424301	Unk	–	09/14/2004	-39.2	-6.30	<1	<0.3	–	–	–
W140	Windsor	8N/8W-32K1	382937122460601	Shallow	70–110	09/16/2004	-38.0	-6.01	6.7	2.1	–	–	–
W156	Windsor	8N/9W-13J2	383220122480501	Deep	260–400	09/02/2004	-40.7	-6.37	<1	<0.3	–	–	–

[1] Calculated from tritium concentration in pCi/L using conversion factor of 3.2 pCi/L.

[2] Below long term minimum detection level (LT-MDL) of 0.3 pCi/L.

Isotopic Composition of Samples

The δ^{18}O and δD values ranged from +0.41 to −7.33 and −13.1 to −49.3 per mil, respectively, for surface- and groundwater samples in the SRPW (fig. 6A, table 5) and plotted on either side and along the GMWL (fig. 6A). Water that has undergone evaporation plots substantially to the right of the GMWL because as water evaporates, the isotopic values of the residual water become heavier, which forms an evaporative trend line with a slope of 3 to 6, in contrast to the slope of 8 for the GMWL (fig. 6A). The effects of evaporation are evident in samples from surface-water sites (fig. 6A). The surface-water samples that plot furthest to the right of the GMWL were collected during the warmer, drier months— note, particularly, the samples collected in fall 2008, when antecedent effects of seasonal evaporation would be the greatest, from Spring Lake (SW5) and the Laguna de Santa Rosa (SW2 and SW13) (fig. 6A). Surface-water samples collected during cooler and wetter months, including samples collected in spring 2009, were generally isotopically lighter (more negative) and clustered closer to the GMWL, indicating little evaporation (fig. 6B). The isotopically lightest samples from the SRPW were collected in spring 2009 from SW3 on Santa Rosa Creek and SW15 on Mark West Creek (fig. 6B, table 5). Both sites receive runoff from the Mayacmas Mountains, where maximum elevations in the SRWP exceed 2,000 ft above sea level. As stated previously, precipitation that condenses at higher altitudes and cooler temperatures tends to be isotopically lighter than precipitation that condenses at lower altitudes and warmer temperatures. Therefore, precipitation and runoff from the Mayacmas Mountains would be expected to be isotopically lighter than precipitation that falls directly on the lower elevations of the SRP and the lower Mendocino Range to the west. Fall and spring δ^{18}O and δD compositions of water from the Russian River, as represented by SW18 (fig. 6B), however, were comparable, which is a reflection of the river's comparatively constant flows from a large drainage area, of which the SRPW is only a portion, coupled with continual dry-season releases from a large upstream reservoir.

Values of δ^{18}O and δD ranged from −5.65 to −7.33 per mil and from −35.9 to −49.3 per mil, respectively, for groundwater samples collected in the study area (fig. 6C, table 5). In general, the isotopic values of the well samples were similar to the range in isotopic values of the spring surface-water samples. With the exception of well samples from the UPL and WG storage units, the isotopic values generally plotted slightly below the GMWL (fig. 6D). This could indicate that these samples have experienced evaporation or have mixed with evaporated surface waters, or it could reflect a LMWL that plots parallel to, but slightly to the right of, the GMWL. The position of LMWLs relative to the GWML in California can differ slightly as a function of elevation, latitude, and other factors (Kendall and Coplen, 2001; Bowen and Revenaugh, 2003). Groundwater samples, however, generally plotted closer to the GMWL compared to fall surface-water samples, indicating that the groundwater undergoes less evaporation seasonally than surface-water (figs. 6B and 6D).

Isotopic values exhibited no direct relationship to the depth of the well's perforated interval. Samples from deep wells had both the heaviest and lightest measured values (fig. 6C).

In general, the isotopic values for water samples from the UPL and VAL storage units grouped together and, with the exceptions of samples from well W34 and W102B, fell within the lighter range of isotopic values for all well samples in the SRPW study area. The isotopic values of well samples in the UPL and VAL storage units were similar to the spring 2009 isotopic values from surface-water sites SW3 and SW15 (figs. 6B and 6D, table 5), indicating that precipitation and runoff from the Mayacmas Mountains is the probable source of recharge to these storage units. The isotopic values for samples from the WG storage unit also grouped together, but fell within the heavier range of isotopic values for wells in the study area (fig. 6D). The difference in isotopic values between the WG storage unit and the UPL and VAL storage units indicates that these storage units have different sources of recharge. Infiltration of precipitation and runoff from the Wilson Grove highlands, which are at a lower altitude than the Mayacmas Mountains and, therefore, likely have precipitation with heavier isotopic values, is the probable source of recharge to wells in the WG storage unit.

The isotopic values for well samples from the WB and CB storage units extended over the entire range of isotopic values for all wells in the study area, indicating a mixture of water sources (fig. 6D). Most of the samples from wells in the WB and CB storage units were heavier (less negative) than well samples in the UPL and VAL storage units, indicating that groundwater underflow from the latter cannot be the sole source of recharge to the downgradient WB and CB storage units. The heavier isotopic values, which showed only slight deviation from the GMWL, indicate that at least some of the recharge to the WB and CB storage units originates as precipitation falling directly on the lower elevations of the SRP.

The isotopic values in well samples W34 and W102B were significantly lighter (more negative) than the other surface-water or well samples, indicating that recharge to both wells originated at higher altitudes or under cooler conditions than modern precipitation (fig. 6A, table 5). Uncorrected carbon-14 dates for these well samples, presented in the "Groundwater Age" section of this chapter, indicate these wells were recharged near the end of the last North American glaciation, when it likely was colder, wetter, or both, which would cause isotope ratios to be lighter (more negative) than modern precipitation.

Changes in Isotopic Composition along Geologic Section B–B′

δD values of well samples were plotted along geologic section B–B′ (modified from Sweetkind and others, 2010, fig. A1–4, and *chapter B*, figs. 2 and 25) to illustrate changes in the isotopic composition as groundwater flows through the aquifer system (figs. 5A and 5B). A total of 16 wells, including

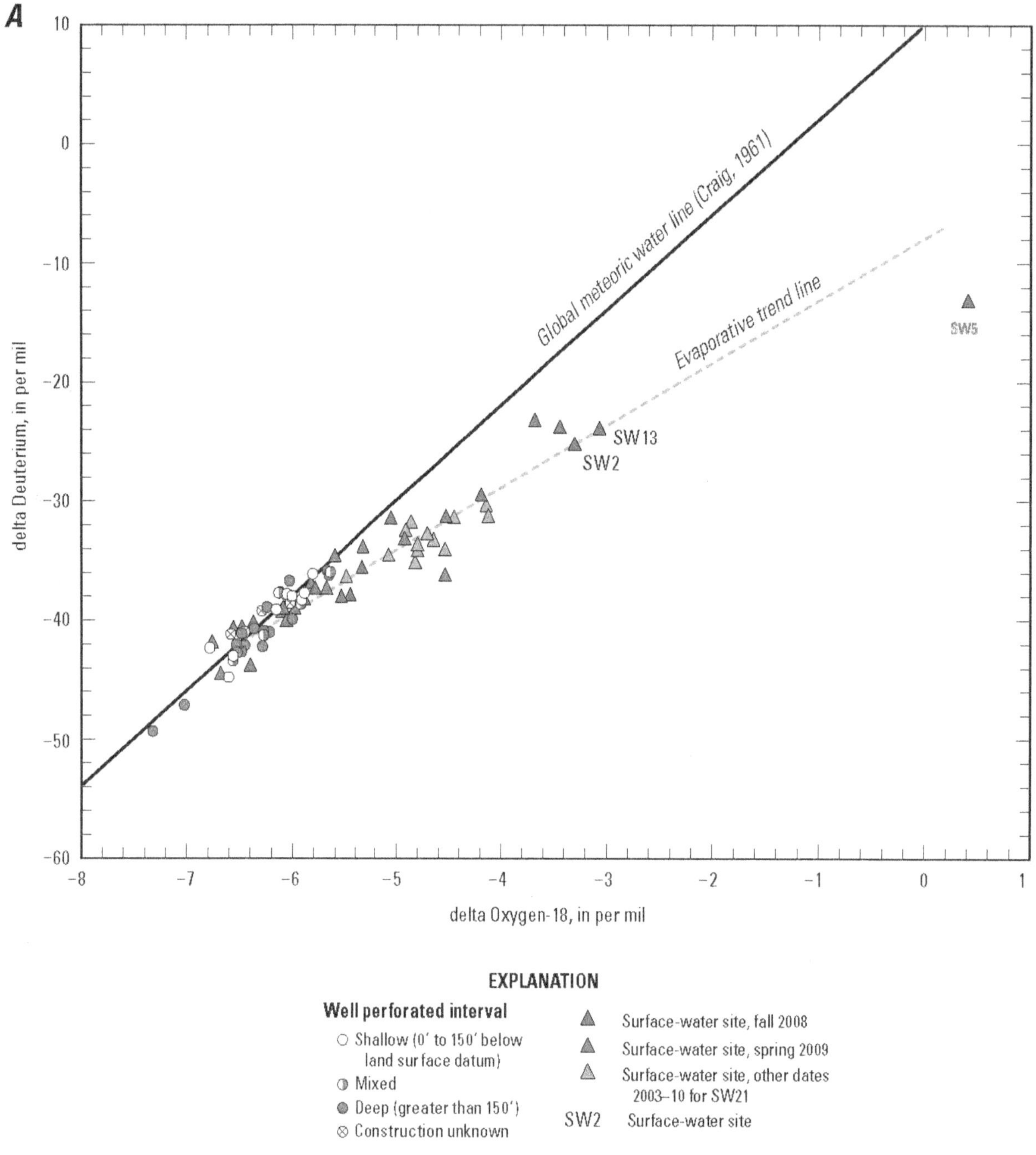

Figure 6.   Relation between delta deuterium and delta oxygen-18 for water samples from in the Santa Rosa Plain watershed, Sonoma County, California, 2003–10: *A*, all water samples; *B*, selected surface-water samples; *C*, selected groundwater samples by perforated interval; and *D*, selected groundwater samples by groundwater storage unit.

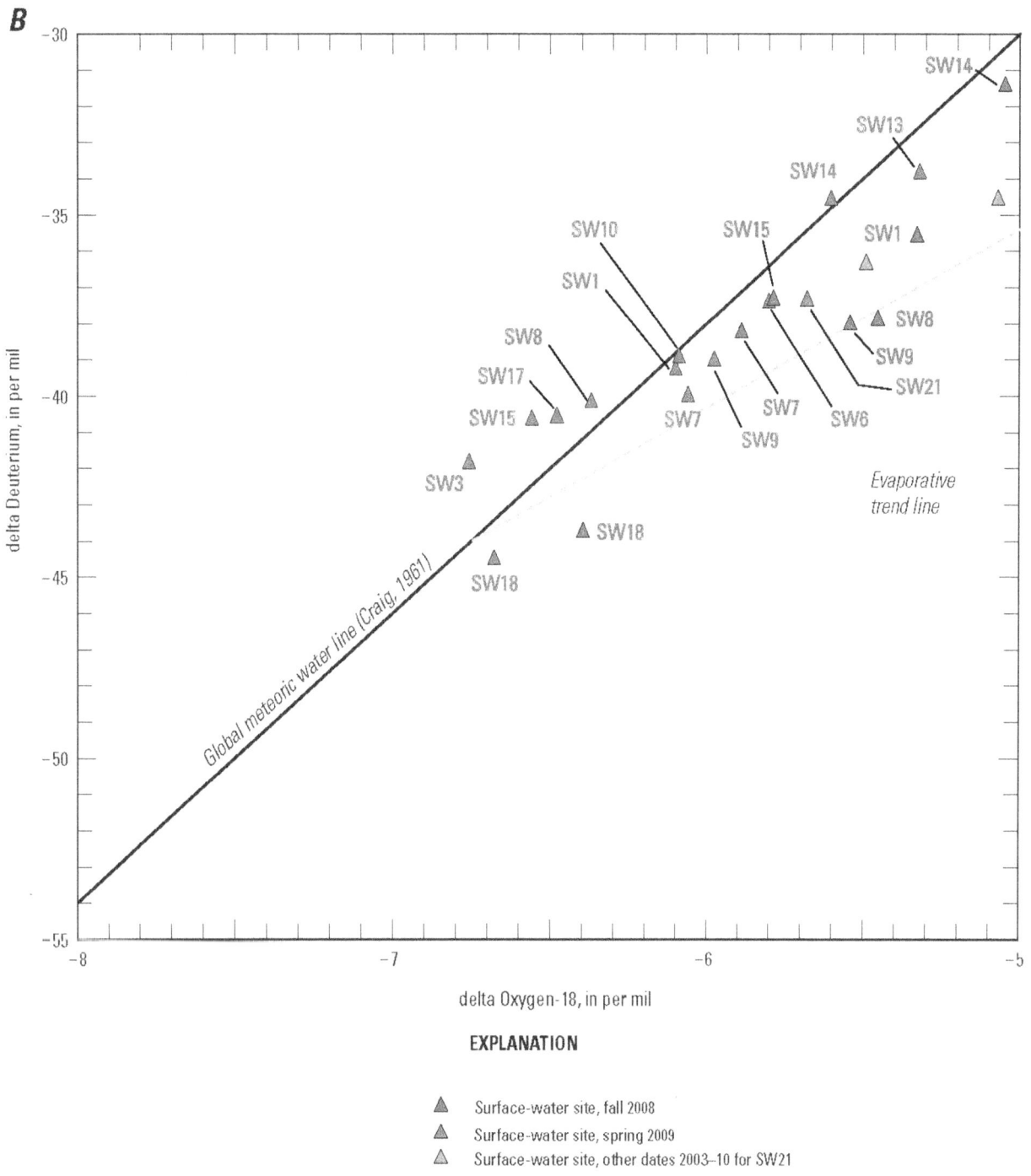

Figure 6. Continued.

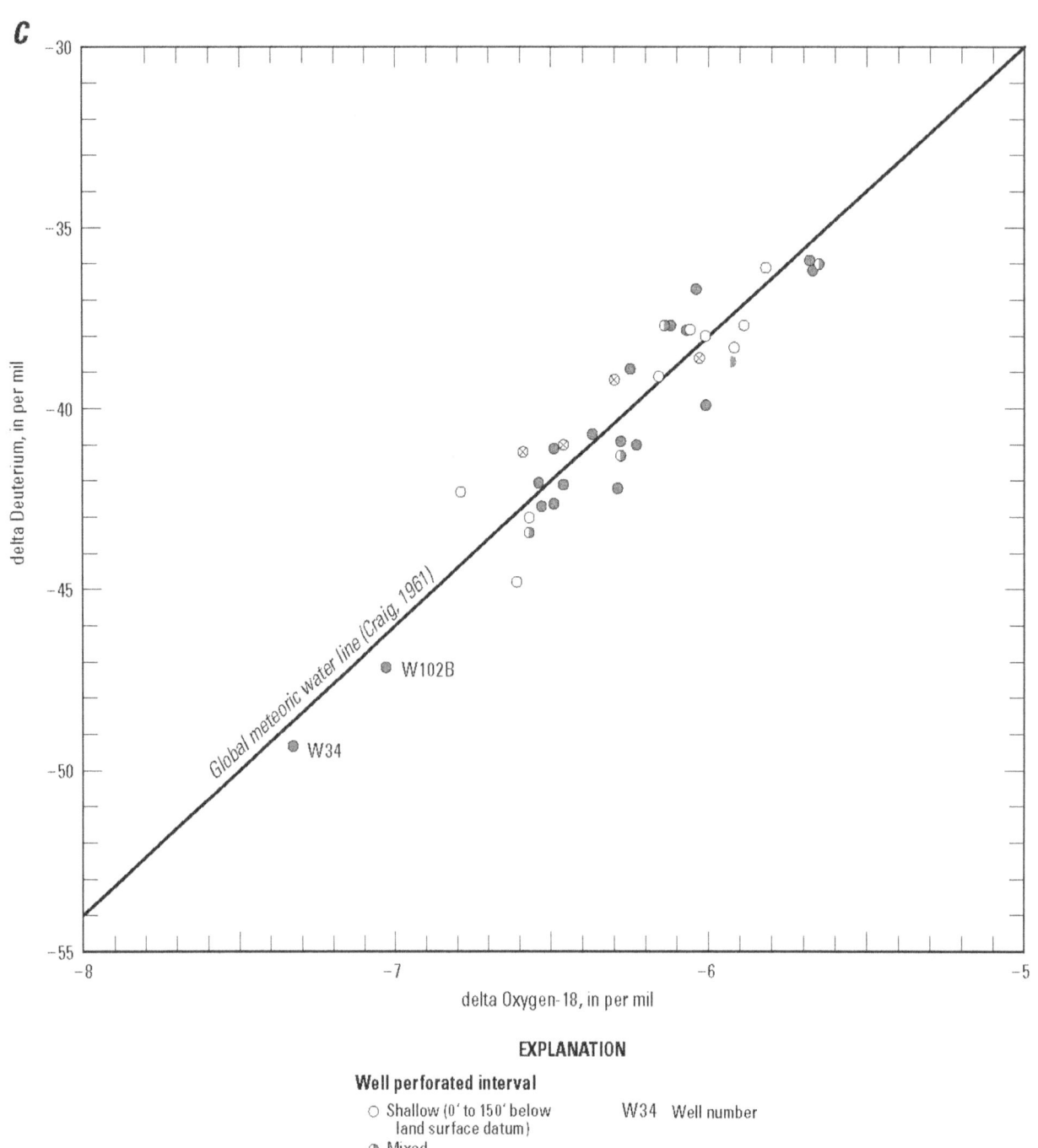

Figure 6.   Continued.

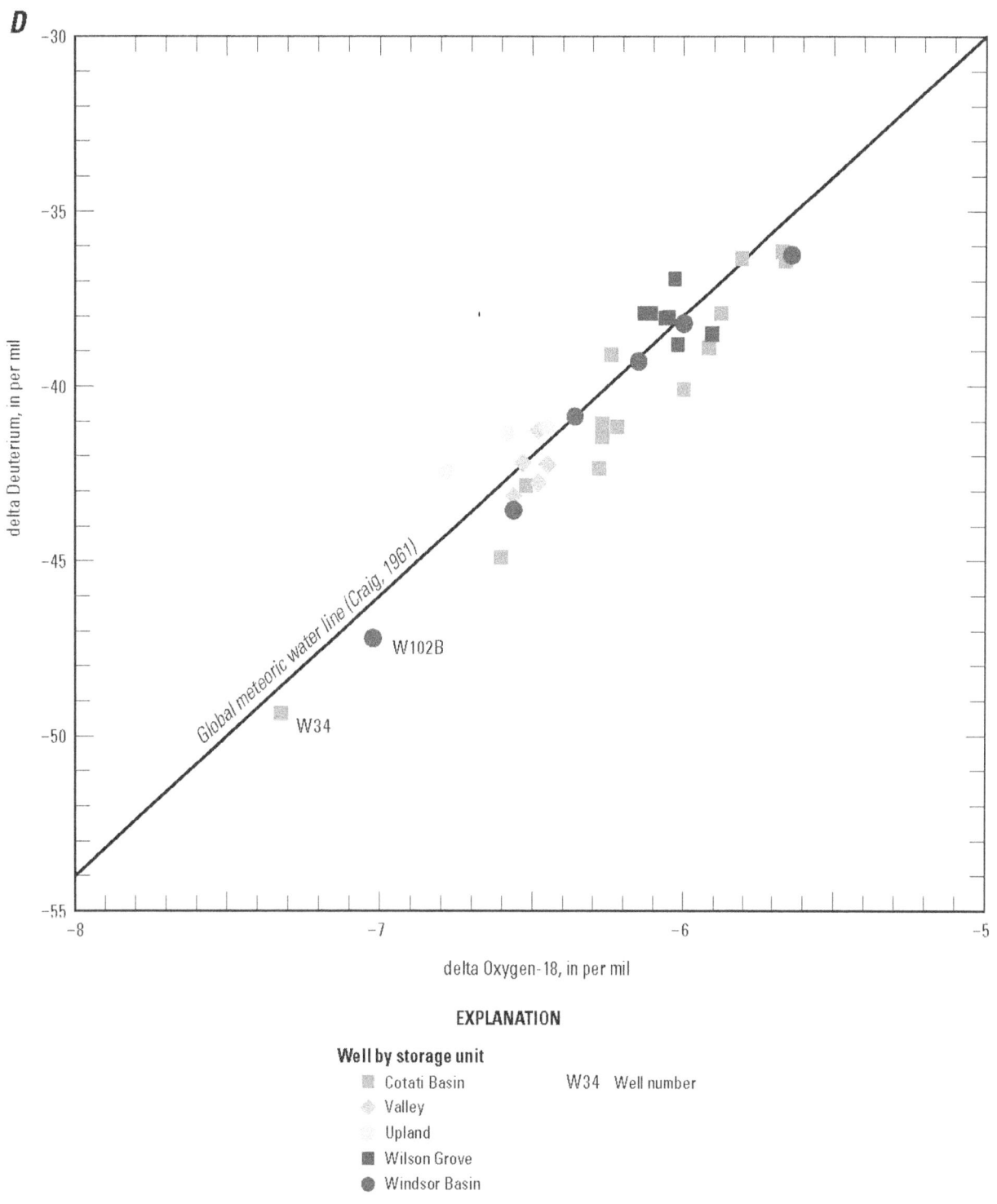

EXPLANATION

Well by storage unit

▨ Cotati Basin	W34	Well number
◈ Valley		
▨ Upland		
■ Wilson Grove		
● Windsor Basin		

Figure 6. Continued.

depth-dependent samples from wells W77, W89, and W102, were projected onto geologic section B–B' for the comparison of δD values with geologic units and structure (fig. 5B).

Isotopic values from wells on both the east and west ends of the geologic section were relatively similar at all depths within each respective area (fig. 5B): δD values east of the Rodgers Creek fault ranged from −41.1 to −42.9 per mil, with a mean composition of −42.0 per mil, whereas δD values west of the Sebastopol fault ranged from −36.7 to −37.8 per mil, with a mean composition of −37.5 per mil. The relatively uniform isotopic composition of samples within each area indicates water recharged under similar conditions with substantial mixing in the subsurface and contrasts with the greater variability in areas between the faults. The mean δD value of well samples east of the Rodgers Creek fault zone was about 4.5 per mil lighter (more negative) than for well samples west of the Sebastopol fault. This difference can be explained by the higher altitude of the Maycamas and Sonoma Mountains on the east side of the SRPW, relative to the Mendocino Range on the west side. The altitude effect, which was inferred from measurements taken on the western flank of the Sierra Nevada Mountains, is −0.56 per mil/100 ft for δD (Ingraham and Taylor, 1991; Rose and others, 1996). The Mayacmas and Sonoma Mountains are at least 1,000 ft higher than the Mendocino Range, which would equate to a −5.6 per mil difference in δD.

Greater variation in isotopic composition was found along the central part of geologic section B-B', between the Rodgers Creek and Sebastopol faults, than between the mountains and the faults (fig. 5B, table 5). δD values from of the central part of the geologic section span the range of isotopic composition for all wells sampled in the study area, from −35.9 to −49.3 per mil. The deep sample from well W102 (W102B) had a δD of −47.1 per mil, the second lightest sample analyzed for this study. Well W102 lies within the Rodgers Creek fault zone (fig. 5A). As described in the "Chemical Character of Groundwater" section of this chapter, the sample from well W102B had high dissolved-solids and boron concentrations, which were attributed to deeply circulating groundwater rising along the fault zone. The shallow sample from well W102 (W102A) had a δD value similar to samples from wells upgradient (east) of the Rodgers Creek fault.

δD values from well samples along the geologic section between the Trenton Ridge fault and the unnamed fault east of the Sebastopol fault ranged from −36.0 to −42.7 per mil, with a mean composition of −39.2 per mil. These values were heavier than the δD values in both the shallow and deep samples from well W102 and the mean isotopic composition for well samples east of the Rodgers Creek fault (−42.0 per mil). Samples from wells perforated solely in the Glen Ellen or in the Glen Ellen and Petaluma formations between the faults have δD values slightly heavier (less negative) than samples perforated solely in the Petaluma Formation. This shift toward more negative δD values in the Petaluma Formation could result from recharge originating at a higher altitude, during a cooler climate, or both.

Samples from wells perforated in the Wilson Grove Formation, east of the Sebastopol fault (W32, W35, and W129) and west of the unnamed fault, had similar δD values as samples from wells in the WG storage unit west of the fault (fig. 5B). This similarity in isotope values indicates that these wells are receiving water from the Wilson Grove storage unit to the west, not groundwater underflow from the Petaluma Formation to the east. These data indicate that the Sebastopol fault is not a complete barrier to groundwater underflow in the Wilson Grove Formation. Samples from wells W33 in the Glen Ellen Formation and W34 in the Petaluma Formation, however, which also are located between these two faults, have significantly lighter δD values (−44.8 and −49.4 per mil, respectively) than the well samples from the Wilson Grove Formation (−35.9 to −38.9 per mil) (fig. 5B). The isotopically light water in the samples from wells W33 and W34 indicates that this water probably was derived from precipitation that fell at higher altitudes in the eastern part of the SRPW, was recharged during a cooler and wetter climate, or both. As stated in the "Major-Ion Composition" section of this chapter, the water type of samples from wells W33 and W34 was similar to the water type of samples from deep wells in the CB east of the unnamed fault (fig. 3C). The similarity of isotopic signature and water type for wells W33 and W34 and the deep wells in the CB east of the unnamed fault is indicative of a similar source for these samples.

Tritium and Carbon-14

Tritium (^{3}H) was analyzed in samples collected from 30 wells, including discrete-depth zones in 3 of the wells (W77, W89, and W102), for a total of 35 samples (fig. 7A, table 5). Carbon-14 (^{14}C) was analyzed in samples from 11 wells, including discrete-depth zones in the same 3 wells (W77, W89, and W102), for a total of 16 samples (fig. 7A, table 5). Most of the samples were collected and analyzed for the NSF GAMA study. Both ^{3}H and ^{14}C were used in this study to qualitatively and semi-quantitatively estimate the amount of time since the water was last in contact with the atmosphere (time since recharge).

Background

Tritium

Tritium is both a naturally occurring and an anthropogenically generated, short-lived, radioactive isotope of hydrogen, whose presence can be used to identify relatively young (post-1952) water (Clark and Fritz, 1997). It has a half-life of 12.32 years (Lucas and Unterweger, 2000). Because of its short half-life, it is useful for distinguishing between water that has been in the hydrologic cycle more or less than about 60 years. Tritium activity is measured in disintegrations per unit of time and, in this study, is reported in both tritium units (TU) and picocuries per liter (pCi/L); a tritium unit is one ^{3}H atom in 10^{18} atoms of hydrogen and equals

3.2 pCi/L. The interaction of cosmic radiation with the upper atmosphere results in the creation of a global steady-state inventory of 3.5–4.5 kgs of ^3H in the atmosphere, from which small amounts reach the Earth's surface in precipitation (Lal and Peters, 1967; O'Brien and others, 1991). Approximately 800 kg of ^3H were released as a result of atmospheric testing of thermonuclear weapons during 1952–62 (Michel, 1976). This produced a spike in ^3H concentrations in precipitation and in groundwater recharged during that time that was much higher than the natural level. Much smaller amounts of thermonuclear ^3H were released until 1980, after which all atmospheric testing ceased. By 1990, this anthropogenic source of ^3H had been largely washed out of the atmosphere, and concentrations in precipitation had decreased close to natural, pre-bomb levels (Clark and Fritz, 1997). Minor amounts of anthropogenic ^3H continue to be released to the atmosphere from nuclear power plants and related facilities that process nuclear material.

Because ^3H is part of the water molecule, and its concentration is unaffected by reactions other than radioactive decay, it is an excellent tracer of modern water. Tritium data can be used in conjunction with data for its daughter product helium-3 (^3He) to determine an absolute age. However, this method might not yield reliable tritium-based ages where crystalline rock is present because water from this source typically yields concentrations of terrigenic helium that greatly exceed concentrations derived from thermonuclear ^3H (Solomon and Cook, 2000). In the absence of ^3He concentrations, ^3H data alone were used in this study to interpret qualitative and "mixed" ages of water.

Carbon-14

Carbon-14 is the naturally occurring, long-lived (half-life of 5,730 years), radioactive isotope of carbon that can sometimes be used to determine the age of groundwater far beyond the range for ^3H. Carbon-14 data are expressed as percent modern carbon (pmc) by comparing ^{14}C activities to the specific activity of National Bureau of Standards oxalic acid; 12.88 disintegrations per minute per gram of carbon in the year 1950 equals 100 percent modern carbon (Stuiver and Polach, 1977). Carbon-14 is created in the upper atmosphere by the bombardment of nitrogen atoms by cosmic radiation. Thermonuclear testing and nuclear power plants have also contributed ^{14}C to the atmosphere, and the burning of fossil fuels during the industrial age has "diluted" the level of ^{14}C (Clark and Fritz, 1997). Atmospheric ^{14}C is present as carbon dioxide (CO_2), which can then be incorporated into various hydrospheric (oceans, lakes, and groundwater) and biospheric (plants and animals) reservoirs. Whether through infiltration of water or the decay and release of biomass into the soil zone, once these intermediate sources of carbon are isolated from the atmosphere, the ^{14}C content in the dissolved carbon steadily decreases.

In reality, ^{14}C that has been isolated from the atmosphere is seldom only affected by just radioactive decay. Chemical reactions along a groundwater flowpath can dilute ^{14}C by either the addition of dissolved inorganic carbon (DIC, which is carbonate plus bicarbonate) that lacks ^{14}C, or by the removal of DIC that contains ^{14}C (Clark and Fritz, 1997). ^{14}C concentrations can be decreased, for example, when carbon is added to groundwater by the dissolution of calcite or dolomite. Because these minerals were formed long ago, they are devoid of ^{14}C and are often said to contain "dead" carbon (Freeze and Cherry, 1979). The addition of DIC from these sources dilutes the original ^{14}C content to give the appearance of older water, as does the production of DIC from oxidation of organic matter that is devoid of ^{14}C.

Ratios of the stable isotopes carbon-13 to the far more abundant carbon-12 (^{12}C) were used in this study as indications of biogeochemical and carbon-exchange processes that can affect estimates of ^{14}C ages. Because carbonate minerals and DIC exchange carbon isotopes (equilibration), albeit slowly, groundwater can acquire a less negative delta carbon-13 ($\delta^{13}C$) value as it moves along a flowpath from infiltration to discharge. Stable carbon isotopes can also be affected by decomposition (oxidation or mineralization) of organic matter buried in the aquifer because organic material has a more negative $\delta^{13}C$ composition than does inorganic carbon so that carbon isotopes would become lighter. Stable carbon isotopes were used to make qualitative inferences about the extent to which these processes have caused the "calculated" ^{14}C age to overestimate the actual time elapsed since recharge in this study.

To obtain a more accurate estimate of "true" age for groundwater (the time elapsed between recharge and discharge) based on ^{14}C, carbonate dissolution and exchange and organic mineralization need to be corrected for through the use of coupled groundwater-flow and geochemical-reaction models that incorporate $d^{13}C$ and related chemical data. As geochemical modeling was beyond the scope of this study, ^{14}C ages in this report are "uncorrected" ages.

Groundwater Age in the Santa Rosa Plain Watershed

Tritium in water from wells in the study area ranged from less than the detection limit of 0.1 (for samples collected during 2006–10) to 2.4 TU (table 5). For this study, groundwater that had ^3H values greater than 0.3 TU (detection limit of samples collected during 2004) was interpreted to be water recharged after 1952, or modern recharge. A total of 15 of the 35 well samples (about 43 percent) had ^3H concentrations in excess of 0.3 TU (table 5). As expected, modern recharge was more prevalent in the shallow well samples. Of the 7 samples from shallow wells, 5 (about 70 percent) had ^3H concentrations in excess of 0.3 TU, and 3 of the samples had ^3H concentrations of 1.6 TU or more; in contrast, only 7 of the 19 samples from deep wells (about 37 percent) had ^3H concentrations in excess of 0.3 TU, and all samples had ^3H concentrations of 0.9 TU or less (table 5). The vertical migration of

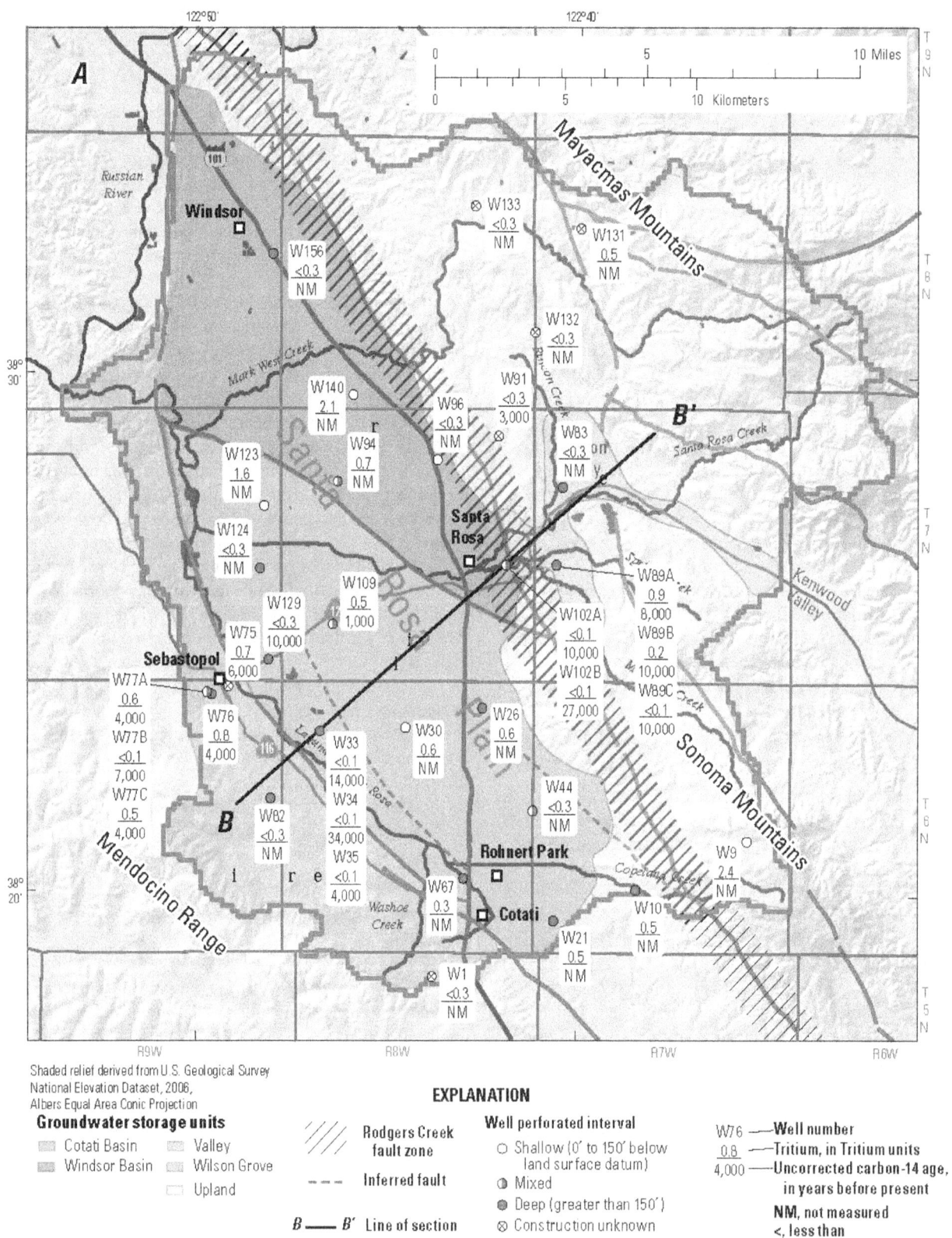

Figure 7.   Tritium and carbon-14 values from selected surface-water sites and wells in the Santa Rosa Plain watershed, Sonoma County, California, 2004–10: *A*, areally and *B*, along section B–B'.

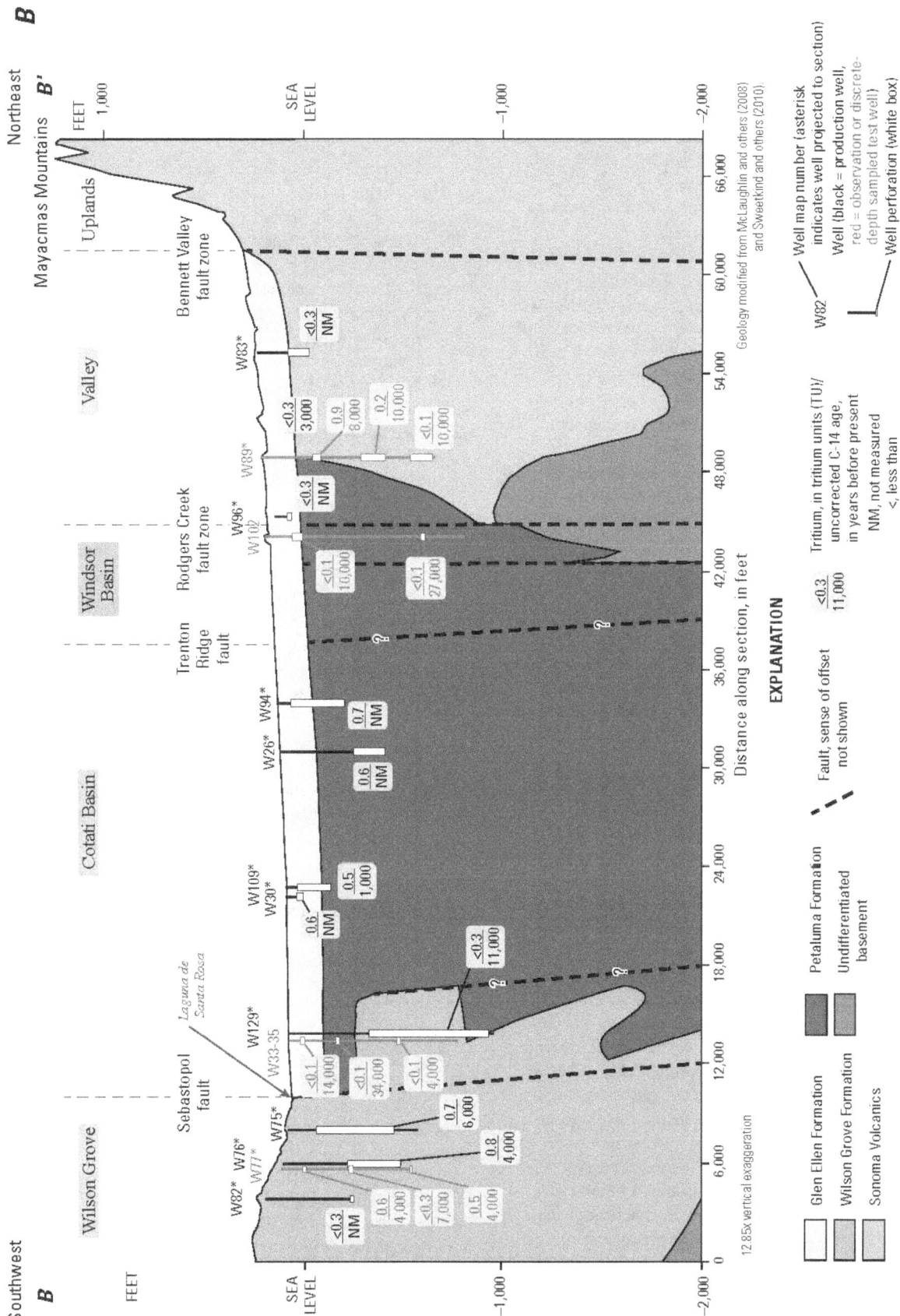

Figure 7. Continued.

recharge in the SRP probably is retarded by the presence of low permeability clay deposits in the Glen Ellen and Petaluma Formations. Only 3 of the 9 (about 33 percent) samples from the UPL and VAL storage units had ^{3}H concentrations in excess of 0.3 TU (fig. 7A, table 5). The absence of modern water in many of these samples probably reflects the low vertical permeability of the basement rocks and Sonoma Volcanics that compose a large part of the UPL and VAL storage units. Most of the precipitation on these storage units probably becomes runoff that contributes to streamflow and potential recharge in the downstream storage units.

Measured ^{14}C activities for the 16 water samples ranged from 1.3 to 84.7 pmc (table 5). These ^{14}C activities represent uncorrected ages of 1,000 to 34,000 years before present (fig. 7A, table 5). Of the 16 samples, 11 were from deep wells (entire perforated or open interval below 150 ft bls), of which all had uncorrected ^{14}C ages of 4,000 years or older; 5 of the deep well samples were 10,000 years or older (table 5). The relatively old age of the deep well samples supports the inference from the ^{3}H data that the vertical migration of modern recharge probably is retarded by low permeability clay deposits in the Glen Ellen and Petaluma Formations.

The two oldest uncorrected ^{14}C ages were from samples collected from the borehole at well W102 (786–806 ft bls) when the well was being constructed, (sample W102B; 27,000 years) and from well W34 (34,000 years). Well W102 is located along the eastern edge of the SRP within the Rodgers Creek fault zone (fig. 7A). As described in the "Chemical Character of Groundwater" section of this chapter, sample W102B had high dissolved-solids and boron concentrations, which were hypothesized to be from deeply circulating groundwater rising along the fault zone. The old age of the water supports this hypothesis. Well W34 is in the western part of the CB storage unit near the Laguna de Santa Rosa, which is an area of natural groundwater discharge. The well is perforated in the Petaluma Formation from 237 to 257 ft bls. Groundwater-level contour maps, presented in *chapter B* (see fig. 29A), indicate that groundwater moved from the UPL storage unit westward toward the Laguna de Santa Rosa and eastward from the highlands in the WG storage unit toward the Laguna de Santa Rosa, prior to significant groundwater development in the SRP. The sample from W34 more closely resembles that of groundwater in the eastern part of the SRP than groundwater in the WG storage unit (fig. 3C). The similarity of water type to well samples to the east and its old age indicate that well W34 receives water at the end of a long groundwater-flow path through the CB storage unit. Because radiocarbon dates presented in this report are "uncorrected" for dissolution and exchange with calcium carbonate and for oxidation of organic carbon in the sediment, they are to be considered as only a semi-quantitative indication of the groundwater age. Nevertheless, the implication that samples from W102 and W34 represent groundwater that was recharged during the Pleistocene epoch (more than about

10,000 years before present) is supported by isotopically light oxygen-18 and deuterium ratios (table 5). The oxygen-18 and deuterium ratios are consistent with recharge during a cooler and wetter time than present.

Groundwater Age Along Geologic Section B–B'

Tritium and uncorrected ^{14}C ages were plotted along geologic section B-B' to illustrate changes in groundwater age as groundwater flows through the aquifer system (figs. 7A and 7B). Unfortunately, none of the samples were collected from wells perforated solely in the Glen Ellen Formation in the VAL storage unit, on the eastern side of the geologic section, where modern recharge would be expected. The samples from the underlying Sonoma Volcanics had ^{3}H concentrations less than 0.3 TU, except for the shallow sample at well W89 (W89A), which had a ^{3}H concentration of 0.9 TU. The uncorrected ^{14}C age for this sample was 8,000 years before present, indicating that the sample from this well is a mixture of modern and old water. The deeper samples from this well (W89B and W89C) had uncorrected ^{14}C ages of 10,000 years before present.

The shallow and deep samples from well W102, in the Rodgers Creek fault zone, had ^{3}H concentrations less than 0.1 TU, indicating these samples contained water that was recharged before 1952. The uncorrected ^{14}C ages of the samples ranged from 10,000 years in the shallow sample to 27,000 years in the deep sample (fig. 7B). As indicated previously, high dissolved-solids and boron concentrations in the deeper sample were attributed to deeply circulating groundwater rising along the fault zone. The uncorrected ^{14}C age in the shallow sample was similar to the uncorrected ^{14}C age in the deeper samples from well W89 to the east. These data indicate upward movement of groundwater upgradient (east) of the Rodgers Creek fault zone, which is supported by groundwater-level contour maps presented in *chapter B* (see fig. 29A).

Available samples from all wells west of the Rodgers Creek fault zone and east of the unnamed fault east of the Sebastopol fault had ^{3}H concentrations in excess of 0.3 TU, indicating that these samples contained at least some water that was recharged after 1952. Most of the wells with ^{3}H data in this portion of the cross-section, however, are relatively shallow. Carbon-14 was analyzed only for the sample from well W109 in this group of wells, which yielded an uncorrected ^{14}C age of 1,000 years before present. Only one of these wells (well W26) was perforated solely in the Petaluma Formation, but, unfortunately, ^{14}C was not analyzed on the sample from this well. The ^{3}H data from these wells indicated modern recharge in the SRP downgradient of the Rodgers Creek fault zone.

Samples from wells perforated in the Wilson Grove Formation (W35, W75, W76, W77A–C, W82, and W129), on the western side of the geologic section, had ^{3}H concentrations ranging from less than 0.1 to 0.8 TU and uncorrected

[14]C ages ranging from 4,000 to 11,000 years before present (fig. 7B, table 5). The shallow (100–120 ft bls) and deep (640–650 ft bls) samples from well W77 and the samples from long-screened wells W75 (138–528 ft bls) and W76 (332–600 ft bls) were the only groundwater samples from wells in the Wilson Grove Formation that had ^{3}H concentrations in excess of 0.3 TU (fig. 7B, table 5). The samples from these wells were a mixture of modern and old water. One possible explanation for this mixture is movement of modern recharge from the shallow aquifer system to the deeper aquifer system through the long-screened wells when the wells are not being pumped. Groundwater samples from wells in the Wilson Grove Formation probably were younger than the reported uncorrected ^{14}C age because of groundwater-aquifer interactions. As stated previously, the Wilson Grove Formation consists of fossiliferous sand or sandstone (Sweetkind and others, 2010). Carbon-14 present in the groundwater in the Wilson Grove Formation can interact with "dead" carbon (devoid of ^{14}C) present in the fossils and calcium carbonate cement in the sandstone, which would reduce the measured ^{14}C activity in the groundwater and result in an apparent older ^{14}C age.

Wells W35 and W129 are located between the Sebastopol fault and the unnamed fault to the east (fig. 7B). Both wells are totally (well W35) or partially (well W129) perforated in the Wilson Grove Formation (fig. 7B). Major ion data (fig. 3C) indicate that these wells predominately receive groundwater from the WG storage unit to the west. The sample from well W35 had a similar uncorrected ^{14}C age as samples from the WG storage unit (4,000 years before present); however, the uncorrected ^{14}C age for the sample from well W129 was significantly older (11,000 years before present) (fig. 7B). The older age of the sample from well W129 could be the result of its greater depth and the fact that it is partially perforated in the Petaluma Formation (fig. 7B).

Wells W33 and W34 are also located between the Sebastopol fault and the unnamed fault to the east, but are not perforated in the Wilson Grove Formation (fig. 7B). Well W33 is perforated in the Glen Ellen Formation from 60 to 80 ft bls, and well W34 is perforated in the Petaluma Formation from 237 to 257 ft bls. In contrast to the samples from wells W35 and W129, the major-ion data for samples from wells W33 and W34 indicated that these wells predominately received groundwater from the CB storage unit to the east (fig 3C). The uncorrected ^{14}C ages for wells W33 and W34 were 14,000 and 34,000 years before present, respectively (fig. 7B, table 5). The similarity of water type to samples from wells to the east and the very old age indicate that wells W33 and W34 receive water from the end of long groundwater flowpaths that traverse the CB storage unit. The age of water could be younger in well W33 than in well W34 because it passes through a shorter groundwater flowpath.

Summary and Conclusions

Groundwater quality was characterized for the SRPW by using analyses from previous investigations for selected physical properties and inorganic constituents compiled from databases maintained by the CDPH, the CDWR, and public-supply purveyors from 1947 to 2010. These data were used to characterize areal, vertical, and temporal variations in groundwater quality and to identify constituents of potential concern. Stable and radioactive isotopes measured in groundwater samples collected in 2004 as part of the GAMA Program or measured in surface- and groundwater samples collected from 2006 to 2010, as part of this study or concurrently by private consultants, were used to help identify the recharge source and age of groundwater in the study area.

The major ion composition and dissolved-solids concentration of well samples were used to characterize the water quality of the Uplands (UPL), Valley (VAL), Windsor Basin (WB), Cotati Basin (CB), and Wilson Grove (WG) storage units of the SRPW. Samples from springs in the Mayacmas Mountains were a mixed cation-bicarbonate type water with dissolved-solids concentrations less than about 100 mg/L— the lowest dissolved-solids concentrations in the SRPW. As groundwater migrates through the UPL and downgradient storage units, the dissolved-solids concentration of the groundwater increased as a result of water-rock interactions and anthropogenic inputs, such as irrigation return flows and septic-tank discharge. The median dissolved-solids concentrations of well samples in the UPL and VAL storage units were 330 and 392 mg/L, respectively.

Most of the well samples from the WB and CB storage units are mixed cation-bicarbonate and sodium-bicarbonate type waters. A comparison of major-ion composition and depth shows a subtle shift in major cations from greater proportions of calcium and magnesium in shallow groundwater, to greater proportions of sodium plus potassium in deeper groundwater. This finding is consistent with increasing mineralization and ion exchange between clays and groundwater, with increasing distance and depth from recharge sources.

Most well samples in the WG storage unit were calcium-bicarbonate type or mixed cation-bicarbonate type waters with dissolved-solids concentrations less than 300 mg/L. Unlike the WB and CB storage units, there is no shift in major cations from greater proportions of calcium and magnesium in shallow groundwater, to greater proportions of sodium plus potassium in deeper groundwater. Wells in the WG storage unit are perforated in the Wilson Grove Formation, which is composed of marine sandstone. Unlike the Glen Ellen and Petaluma Formations, the Wilson Grove Formation has only minor clay deposits, so there is less potential for the loss of calcium from solution by means of ion exchange with sodium in association with clay minerals.

Water from almost 75 percent of the 33 wells with records spanning 20 or more years had increased specific conductance over time, and about half had increases of more than 10 percent during the period of record. Chloride behaved similarly; concentrations increased in about 67 percent of the wells evaluated, and just over half increased by more than 10 percent. In approximately 15 and 30 percent of evaluated wells, specific conductance and chloride concentration, respectively, decreased by more than 10 percent in approximately. The greatest increases in specific conductance, chloride concentration, or both were in wells located in the vicinity of the cities of Rohnert Park and Cotati. Possible causes of increased specific conductance in the Rohnert Park/Cotati area include groundwater underflow of high dissolved-solids concentration water present along the Rodgers Creek fault zone, irrigation return flow, and septic-tank effluent or leaking sewer pipes. Depth-dependent hydrologic, chemical, and isotopic data are needed to better understand the cause of the increased specific conductance and chloride values.

Specific conductance, chloride, dissolved solids, nitrate, arsenic, boron, iron, and manganese are water-quality constituents of potential concern in the SRPW because concentrations of these constituents in samples from some wells exceed state or federal recommended or mandatory regulatory standards for drinking water. About 43 percent of the samples analyzed for iron had concentrations greater than or equal to the SMCL of 300 µg/L, and about 73 percent of the samples analyzed for manganese, had concentrations greater than or equal to the SMCL of 50 µg/L. About 12 percent of the samples analyzed for arsenic had concentrations greater than or equal to the MCL of 10 µg/L, with about 30 percent of the wells sampled in the WB and WG storage units exceeding the MCL. About 12 percent of the samples analyzed for dissolved solids had concentrations greater than or equal to the SMCL of 500 mg/L. Boron concentrations were greater than or equal to regulatory standards in 7 percent of the samples with analyses. Specific conductance, chloride, and nitrate values were greater or equal to regulatory standards in only about 2 percent of the samples with analyses.

Samples were collected from 15 surface-water sites during the fall of 2008 and spring of 2009, and from 32 wells during 2004–10, for the analysis of the stable isotopes of oxygen and hydrogen ($\delta^{18}O$ and δD). In general, the isotopic values of the well samples were similar to the range of isotopic values measured in the spring surface-water samples. With the exception of well samples from the UPL and WG storage units, the isotopic values of groundwater samples generally plotted slightly below the GMWL, indicating that the samples could have been subject to some evaporation, been mixed with evaporated surface water, or been derived from recharge source areas with somewhat different meteoric water lines because of differing altitudes. The isotopic values displayed no relationship to the depth of a well's perforated interval—samples from deep wells had both the heaviest and lightest isotopic values analyzed. In general, the isotopic values of samples from the UPL and VAL storage units grouped together and were in the

lighter range of all measured isotopic values. The isotopic values for well samples from the WG storage unit also grouped together, but fell within the heavier range of all isotopic values from wells in the SRPW. This difference can be explained by the higher altitude of the Maycamas and Sonoma Mountains on the east side of the SRPW relative to the Mendocino Range on the west side. Precipitation and subsequent recharge at the higher altitudes in the Maycamas and Sonoma Mountains would be isotopically lighter than precipitation and subsequent recharge at the lower altitudes in the Mendocino Range.

The isotopic values from wells in the WB and CB storage units spanned the entire range of isotopic values in the SRWP, indicating a mixture of water sources. Most of the isotopic values for samples from wells in the WB and CB storage units were heavier (less negative) than for well samples in the UPL and VAL storage units, indicating that groundwater underflow from the UPL and VAL storage units cannot be the sole source of recharge to the WB and CB storage units. The heavier isotopic values, which only deviated slightly from the GMWL, indicated that at least some of the recharge to the WB and CB storage units originates as precipitation directly falling on the lower elevations of the SRP. The isotopic values of samples collected from a borehole in the Petaluma Formation within the Rodgers Creek fault zone, on the east side of the SRP, and from a well perforated in the Petaluma Formation near the Laguna de Santa Rosa, on the west side of the SRP, were significantly lighter (more negative) than the other surface-water or well samples. The more negative values are consistent with recharge at higher altitudes or recharge under cooler conditions than have prevailed in the post-glacial period.

Tritium (^3H) concentrations were analyzed in 35 samples collected from 30 wells and ranged from less than 0.3 (detection limit of samples collected in 2004) to 2.4 TU. About 43 percent of the samples had detectable ^3H concentrations, indicating that these samples contained some modern water (water recharged since 1952). As expected, modern recharge was more prevalent in the shallow well samples. Five of the seven samples from shallow wells (about 70 percent) had ^3H concentrations greater than 0.3 TU, and three of the samples had ^3H concentrations of 1.6 TU or more. In contrast, only 7 of the 19 samples from deep wells (about 37 percent) had ^3H concentrations greater than 0.3 TU, and all samples had ^3H concentrations of 0.9 TU or less. The vertical migration of recharge in the SRP is probably retarded by the presence of low permeability clay deposits in the Glen Ellen and Petaluma Formations. Only three of the nine (about 33 percent) samples from the UPL and VAL storage units had detectable ^3H concentrations. The absence of modern water in many of these samples probably reflects the low vertical permeability of the basement rocks and Sonoma Volcanics that compose the UPL and VAL storage units. Most of the precipitation on these storage units probably becomes runoff that contributes to streamflow and potential recharge in the downstream storage units.

Measured ^{14}C activities for the 16 well samples analyzed ranged from 1.3–84.7 pmc. These ^{14}C activities represent uncorrected ages of 1,000–34,000 years before present. The deep well samples all had uncorrected ^{14}C ages of 4,000 years or older, and five of the deep well samples were 10,000 years or older. The relatively old age of the deep well samples supports the inference made from the ^{3}H data that the vertical migration of modern recharge probably is retarded by low permeability clay deposits in the Glen Ellen and Petaluma formations.

The two oldest uncorrected carbon-14 ages (27,000 and 34,000 years before present) were from samples collected from a borehole in the Petaluma Formation within the Rodgers Creek fault zone on the east side of the SRP (well W102) and from a well perforated in the Petaluma Formation near the Laguna de Santa Rosa on the west side of the SRP (well W34). The sample from well W102 in the Rodgers Creek fault zone had high dissolved-solids and boron concentrations, which were hypothesized to be from deeply circulating groundwater rising along the fault zone. The very old age of the water supports this hypothesized source of water to this well. The sample from well W34, near the Laguna de Santa Rosa, was similar in water type to wells in the CB storage unit to the east. The similarity of water type to well samples to the east and the very old age of water indicate that well W34 receives water from the end of a long groundwater-flowpath passing through the CB storage unit. Because radiocarbon dates presented in this report were "uncorrected" for dissolution and exchange with calcium carbonate and for oxidation of organic carbon in the sediment, they are to be considered as only a semi-quantitative indication of the groundwater age. Nevertheless, the deduction that these samples represent groundwater recharged during the Pleistocene epoch (more than about 10,000 years before present) is supported by isotopically light oxygen-18 and deuterium ratios, which indicate recharge during a cooler and wetter period than the present, such as the Pleistocene.

References Cited

Agency for Toxic Substances and Disease Registry, 2009, Case Studies in Environmental Medicine: Arsenic Toxicity, accessed May 1, 2011, available online at *http://www.atsdr.cdc.gov/csem/arsenic/docs/arsenic.pdf*

American Public Health Association, 2005, Standard methods for the examination of water and wastewater, (21st ed.): Washington, D.C., American Water Works Association and Water Pollution Control Federation, 1,368 p.

Ayers, R.S., and Westcot, D.W., 1994, Water quality for agriculture, Food and Agriculture Organization of the United Nations: Irrigation and Drainage Paper 29 Rev 1; accessed on January 9, 2013 at URL: *http://www.fao.org/docrep/003/T0234E/T0234E05.htm*

Belitz, Kenneth, Dubrovsky, N.M., Burow, K.R., Jurgens, B.C., and Johnson, T., 2003, Framework for a groundwater quality monitoring and assessment program for California: U.S. Geological Survey Water Resources Investigations Report 03–4166, 28 p.

Bowen, G. J., and J. Revenaugh, 2003, Interpolating the isotopic composition of modern meteoric precipitation: Water Resources Research, v. 39, no. 10, 1299, doi:10.1029/2003WR002086.

California Department of Public Health, 2011, Title 22 of the California Code of Regulations, accessed March 1, 2011, at URL *http://www.cdph.ca.gov/certlic/drinkingwater/pages/Chemicalcontaminants.aspx*

California State Water Resources Control Board, 2010, Groundwater Information Sheet—Salinity; accessed on January 9, 2013, at URL: *http://www.swrcb.ca.gov/gama/docs/coc_salinity.pdf.*

Cardwell, G.T., 1958, Geology and Ground Water in Santa Rosa and Petaluma Valley Areas, Sonoma County, California: U.S. Geological Survey Water-Supply Paper 1427, 273 p.

Clark, I.D, and Fritz, Peter, 1997, Environmental isotopes in hydrogeology: New York, Lewis Publishers, 328 p.

Coplen, T.B., 1994, Reporting of stable hydrogen, carbon, and oxygen abundances: Pure and Applied Chemistry, v. 66, p. 273–276.

Coplen, T.B., Wildman, J.D., and Chen, J., 1991, Improvements in the gaseous hydrogen-water equilibration technique for hydrogen isotope ratio analysis: Analytical Chemistry, v. 63, p. 910–912.

Craig, Harmon, 1961, Isotopic variations in meteoric waters: Science, v. 133, p. 1702–1703.

Epstein, S., and Mayeda, T., 1953, Variation of O-18 content of water from natural sources: GeoChimica et Cosmochimica Acta, v. 4, p. 213–224.

Fifield, L.K., 1999, Accelerator mass spectrometry and its applications: Reports on Progress in Physics, v. 62, no. 8, p. 1223–1274.

Fishman, M.J., 1993, Methods of analysis by the U.S. Geological Survey National Water Quality Laboratory—Methods for the determination of inorganic and organic constituents in water and fluvial sediments: U.S. Geological Survey Open-File Report 93–125, 217 p.

Fishman, M.J., and Friedman, L.C., eds., 1989, Methods for determination of inorganic substances in water and fluvial sediments: U.S. Geological Techniques of Water-Resources Investigations, book 5, chap. A1, 545 p.

Ford, R.S., 1975, Evaluation of ground water resources: Sonoma County, volume 1: geologic and hydrologic data: California Department of Water Resources Bulletin 118–4, 177 p., 1 plate.

Fox, K.F., Jr., 1983, Tectonic setting of late Miocene, Pliocene, and Pleistocene rocks in part of the Coast Ranges north of San Francisco, California: U.S. Geological Survey Professional Paper 1239, 33 p.

Freeze, R.A., and Cherry, J.A., 1979, Groundwater: Englewood Cliffs, New Jersey, Prentice-Hall, 604 p.

Garbarino, J.R., 1999, Methods of analysis by the U.S. Geological Survey National Water Quality Laboratory—Determination of dissolved arsenic, boron, lithium, selenium, strontium, thallium, and vanadium using inductively coupled plasma-mass spectrometry: U.S. Geological Survey Open-File Report 99–093, 31 p.

Garbarino, J.R., Kanagy, L.K., and Cree, M.E., 2006, Determination of elements in natural-water, biota, sediment, and soil samples using collision/reaction cell inductively coupled plasma–mass spectrometry: U.S. Geological Survey Techniques and Methods, book 5, sec. B, chap. 1, 88 p.

Gat, J.R., and Gonfiantini, R., 1981, Stable isotope hydrology, deuterium and oxygen-18 in the water cycle: International Atomic Energy Agency, Technical Reports Series No. 210, 339 p.

Hem, J.D., 1992, Study and interpretation of the chemical characteristics of natural water (3d ed.): U.S. Geological Survey Water-Supply Paper 2254, 263 p., 3 plates.

Herbst, C.M., Jacinto, D.M., and McGuire, R.A., 1982, Evaluation of ground water resources, Sonoma County, v. 2, Santa Rosa Plain: California Department of Water Resources Bulletin 118–4, 107 p., 1 plate.

Herd, D.G., and Helley, E.J., 1976, Faults with Quaternary displacement northwestern San Francisco Bay region, California: U.S. Geological Survey Miscellaneous Field Studies Map MF-818, 1 sheet, scale 1:125,000.

Ingraham, N.L., and Taylor, B.E., 1991, Light stable isotope systematic of large-scale hydrologic regimes in California and Nevada: Water Resources Research, v. 27, p. 77–90.

Kabay, N., Yilmaz, I., Bryjak, M. and Yuksel, M., 2006, Removal of boron from aqueous solutions by a hybrid ion exchange-membrane process: Desalination, v. 198, p. 158–165.

Kendall, C., and Coplen, T.B., 2001, Distribution of oxygen-18 and deuterium in river waters across the United States: Hydrological Processes, v. 15, p. 1363–1393.

Kulongoski, J.T., Belitz, Kenneth, and Dawson, B. J., 2006, Ground-water quality data in the North San Francisco Bay hydrologic provinces, California, 2004, Results from the California Ground-Water Ambient Monitoring and Assessment (GAMA) Program: U.S. Geological Survey Data Series Report 167, 100 p.

Kulongoski, J.T., Belitz, Kenneth, Landon, M.K., and Farrar, C.D., 2010, Status and understanding of groundwater quality in the North San Francisco Bay groundwater basins, 2004, California GAMA Priority Basin Project: U.S. Geological Survey Scientific Investigations Report 2010–5089, 88 p.

Lal, D., and Peters, B., 1967, Cosmic ray produced radioactivity on the earth: Encyclopedia of Physics, New York, Springer-Verlag, v. 46, p. 407–434.

Langenheim, V.E., Graymer, R. W., Jachens, R.C., McLaughlin, R.J., Wagner, D.L., and Sweetkind, D.S., 2010, Geophysical framework of the Northern San Francisco Bay region, California: Geosphere, v. 6, no. 5, p. 594–620.

Lucas, L.L., and Unterweger, M.P., 2000, Comprehensive review and critical evaluation of the half-life of tritium: Journal of Research of the National Institutes of Standards and Technology, v. 105, no. 4, p. 541–549.

Michel, R.L., 1976, Tritium inventories of the world oceans and their implications: Nature, v. 263, p.103–106.

Muir, K.S., and Coplen T.B., 1981, Tracing ground-water movement by using the stable isotopes of oxygen and hydrogen, upper Penitencia Creek alluvial fan, Santa Clara Valley, California: U.S. Geological Survey Water-Supply Paper 2075, 18 p.

Mullaney, J.R., Lorenz, D.L., Arntson, A.D., 2009, Chloride in groundwater and surface water in areas underlain by the glacial aquifer system, northern United States: U.S. Geological Survey Scientific Investigations Report 2009–5086, 41 p.

O'Brien, K., Lerner, A.D., Shea, M.A., Smart, D.F., 1991, The production of cosmogenic isotopes in the Earth's atmosphere and their inventories, *in* Sonett, C.P., Giampapa, M.S., Matthews, M.S., eds., The Sun in Time, University of Arizona Press, p. 317–342.

Ostlund, H.G., and Dorsey, H.G., 1975, Rapid electrolytic enrichment of hydrogen gas proportional counting of tritium: International Conference on Low Radioactivity Measurement and Applications, High Tatras, Czechoslovakia, October 1975, Proceedings, 6 p.

Patton, C.J. and Truitt, E.P., 2000, Methods of analysis by the U.S. Geological Survey National Water Quality Laboratory—Determination of ammonium plus organic nitrogen by a Kjeldahl digestion method and an automated photometric finish that includes digest cleanup by gas diffusion: U.S. Geological Survey Open-File Report 00–170, 31 p.

Piper, A.M., 1944, A graphic procedure in the geochemical interpretation of water-analyses: American Geophysical Union Transactions, v. 25, p. 914–923.

Reimann, C., and de Caritat, P., 1998, Chemical elements in the environment. Factsheets for the Geochemist and Environmental Scientist: Berlin, Springer-Verlag, 398 p.

Rose, T.P., Davisson, M.L., and Criss, R.E., 1996, Isotope hydrology of voluminous cold springs in fractured rock from an active volcanic region, northeastern California: Journal of Hydrology, v. 179, p. 207–236.

Solomon, D.K., and Cook, P.G., 2000, ^{3}H and ^{3}He (Chapter 13), p. 397–424, *in* Cook, P.G., and Herczeg, A.L., eds., Environmental tracers in subsurface hydrology: Boston, Kluwer Academic Publishers, 529 p.

Stiff, H.A., Jr., 1951, The interpretation of chemical analysis by means of patterns: Journal of Petroleum Technology, v. 3, no. 10, p. 15–17.

Struzeski, T.M., DeGiacomo, W.J., and Zayhowski, E.J., 1996, Methods of analysis by the U.S. Geological Survey National Water Quality Laboratory-Determination of dissolved aluminum and boron in water by inductively coupled plasma-atomic emission spectrometry: U.S. Geological Survey Open-File Report 96–149, 17 p.

Stuiver, Minze, and Polach, H.A., 1977, Discussion: Reporting of 14C Data: Radiocarbon, v. 19, no. 3, p. 355-363.

Sweetkind, D.S., Taylor, E.M., McCable, C.A., Langenheim, V.E., and McLaughlin, R.J., 2010, Three-dimensional geologic modeling of the Santa Rosa Plain, California: Geosphere, v. 6, no. 3, p. 237–274.

University of Miami Tritium Laboratory, 2010, Tritium Procedures and Standards, accessed December 13, 2010, data available online at URL *http://www.rsmas.miami.edu/groups/tritium/analytical-services/procedures-and-standards/tritium/*

U.S. Environmental Protection Agency, 1993, Methods for the determination of inorganic substances in environmental samples: Cincinnati, Ohio, Environmental Monitoring and Support Laboratory, EPA/600/R-93/100.

U.S. Environmental Protection Agency, 1994, Methods for the determination of metals in environmental samples, supplement 1: Cincinnati, Ohio, Environmental Monitoring and Support Laboratory, EPA/600/R-94/111.

U.S. Environmental Protection Agency, 2009, National primary drinking water regulations, EPA 816-F-09–004, 6 p., accessed November 1, 2010, at URL *http://water.epa.gov/drink/contaminants*

U.S. Geological Survey, 2010, National field manual for the collection of water-quality data: U.S. Geological Survey Techniques of Water-Resources Investigations, book 9, chaps. A1-A9, available *online at http://pubs.water.usgs.gov/twri9A.*

Welch, A.H., Westjohn, D.B., Helsel, D.R., and Wanty, R.B., 2000, Arsenic in Ground Water of the United States: Occurrence and Geochemistry: Ground Water, v. 38, no. 4, p. 589–604.

Chapter D. Conceptual Model of Santa Rosa Plain Watershed Hydrologic System

By Tracy Nishikawa, Joseph A. Hevesi, Donald S. Sweetkind, and Peter Martin

Introduction—Purpose of Conceptualization

A conceptual model of the hydrologic system for the Santa Rosa Plain watershed (SRPW) will be used (1) to better understand the movement and storage of water in the SRPW; (2) to aid in the interpretation of stream-gage records, well hydrographs, and water-quality data; and (3) to provide a framework for the development of a numerical-flow model needed to evaluate the response of the SRPW to current and historic hydrologic conditions and water uses and to predict the response of the hydrologic system to potential future conditions. The conceptual model of the SRPW is based on known and estimated physical and hydrologic characteristics of the surface-water and groundwater systems and how these characteristics influence the flow and storage of water in the SRPW. Following Markstrom and others (2008), the hydrologic system of the SRPW is conceptualized as having three regions: (1) region 1, which consists of the plant canopy, the land surface, and the soil zone; (2) region 2, which consists of streams, lakes, and wetlands; and (3) region 3, which is the subsurface zone that consists of an unsaturated zone and an underlying saturated zone (fig. 1). Water is stored in each region, and the regions are linked by flow processes. The flow of water into and out of each region, as well as the flow of water within each region, is a function of mechanisms specific to each region.

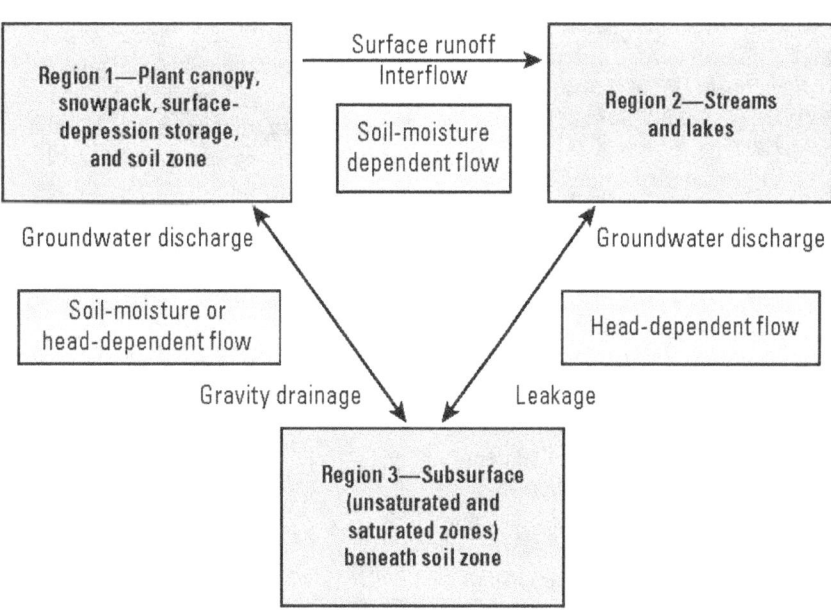

Figure 1. Schematic diagram of the general conceptual model of the Santa Rosa Plain watershed hydrologic system (modified from Markstrom and others, 2008).

Boundaries and Boundary Conditions

The SRPW covers about 260 square miles (mi²) and includes a combination of previously defined surface-water drainage divides and groundwater subbasin boundaries (fig. 2). Most of the SRPW area is contained within the surface-water drainage basin defined by the Mark West Creek drainage (MWCD), the largest tributary, by area, to the Russian River. The SRPW also includes several smaller, peripheral surface-water drainages to the northwest and south of the MWCD, where the study area was extended to better match the boundary of the Santa Rosa Plain groundwater subbasin. The surface watershed area overlies all of the Santa Rosa Plain and Rincon Valley groundwater subbasins, as well as the northern half of the Kenwood Valley basin, the eastern part of the Wilson Grove Formation Highlands groundwater basin, the southern part of the Healdsburg area groundwater subbasin, and the southern part of the Alexander Valley groundwater subbasin.

The SRPW boundary includes mostly naturally-defined topographic drainage divides, such that surface-water inflows from outside the boundary are not expected. Surface-water flows in the SRPW can exit as runoff, evapotranspiration (ET), or consumptive use. Most runoff from the SRPW flows to the Russian River drainage, and the majority of the runoff discharges from Mark West Creek (see *chapter B*, fig. 7); however, a few small areas along the southern boundary drain southward into San Pablo Bay.

As described in *chapter B* of this report, the groundwater basin underlying the SRPW is subdivided into five groundwater storage units: (1) the mountainous uplands areas (UPL) north, east, and southeast of the city of Santa Rosa; (2) valleys located to the east of the SRP Valley (VAL), including Rincon Valley, Bennett Valley and northern half of Kenwood Valley; (3) Windsor Basin (WB); (4) Cotati Basin (CB); and (4) the Wilson Grove area (WG; fig. 2). About 75 percent of the length of the defined SRPW boundary is interpreted to be a no-flow boundary, where low-permeability basement rock, a groundwater divide, or both prevent hydraulic communication between the SRPW groundwater basin and neighboring groundwater subbasins. The remaining SRPW boundary includes four segments where there is groundwater communication with adjacent groundwater subbasins or basins. These four boundary segments include (1) the VAL storage unit and southern half of Kenwood Valley basin in the east, (2) the CB storage unit and Petaluma Valley basin in the south, (3) the WG storage unit and Wilson Grove Formation Highlands basin in the southwest, and (4) the WB storage unit and the Healdsburg area subbasin in the northwest (fig. 2).

The lower boundary of the SRPW groundwater system is the contact with the low-permeability basement rock. It is assumed that the basement rock contributes negligible water to the groundwater system and does not store appreciable volumes of water. The theoretical upper boundary of the groundwater system is land surface, but varies spatially and temporally with water level. Inflows from the upper boundary come from groundwater recharge. Outflows across the upper boundary are ET and groundwater discharge to surface-water bodies.

Region 1: Plant Canopy, Land Surface, and Soil Zone

Region 1 of the SRPW—the plant canopy, the land surface, and the soil zone—provides the primary link between climatic factors and the SRPW hydrologic system. The plant canopy includes natural vegetation, crops (for example, row crops, pastures, vineyards, and orchards), and landscaped urbanized areas (for example, residential areas, parks, and golf courses). The three types of land surfaces included in the SRPW conceptual model are (1) areas covered by soil; (2) naturally exposed bedrock areas free of soil cover; and (3) areas covered by anthropogenic features such as buildings, roads, and parking lots. The soil zone represents the upper unsaturated zone and is conceptually defined as the layer extending from the ground surface to the base of the root zone (Markstrom and others, 2008). The soil zone stores and transmits water between the atmosphere and the underlying unsaturated and saturated zones.

Inflows to Region 1

In the SRPW, the primary sources of water to region 1 include (1) precipitation (primarily in the form of rainfall), (2) irrigation, (3) groundwater discharge, and (4) surface water. Inflows from precipitation and irrigation to the soil zone only occur in pervious areas and are limited by the infiltration capacity of the soil (or for some locations exposed bedrock). There is no natural surface-water inflow (including both runoff and interflow) into the SRPW from neighboring areas; however, on a local scale within the SRPW, surface-water inflows can be lateral redistribution of overland flow or interflow.

Precipitation

Precipitation is primarily rainfall in the SRPW (Cardwell, 1958) and is the main source of water to region 1. According to the 30-year (1971–2000) normal precipitation estimated by PRISM (Daly and others, 2004), the mean annual precipitation for the SRPW is approximately 40 inches (in.), or 560,000 acre-feet per year (acre-ft/yr), distributed over the watershed. Precipitation can be snow on rare occasions, especially at the higher altitudes (Cardwell, 1958); however, snow accumulations are negligible in the SRPW.

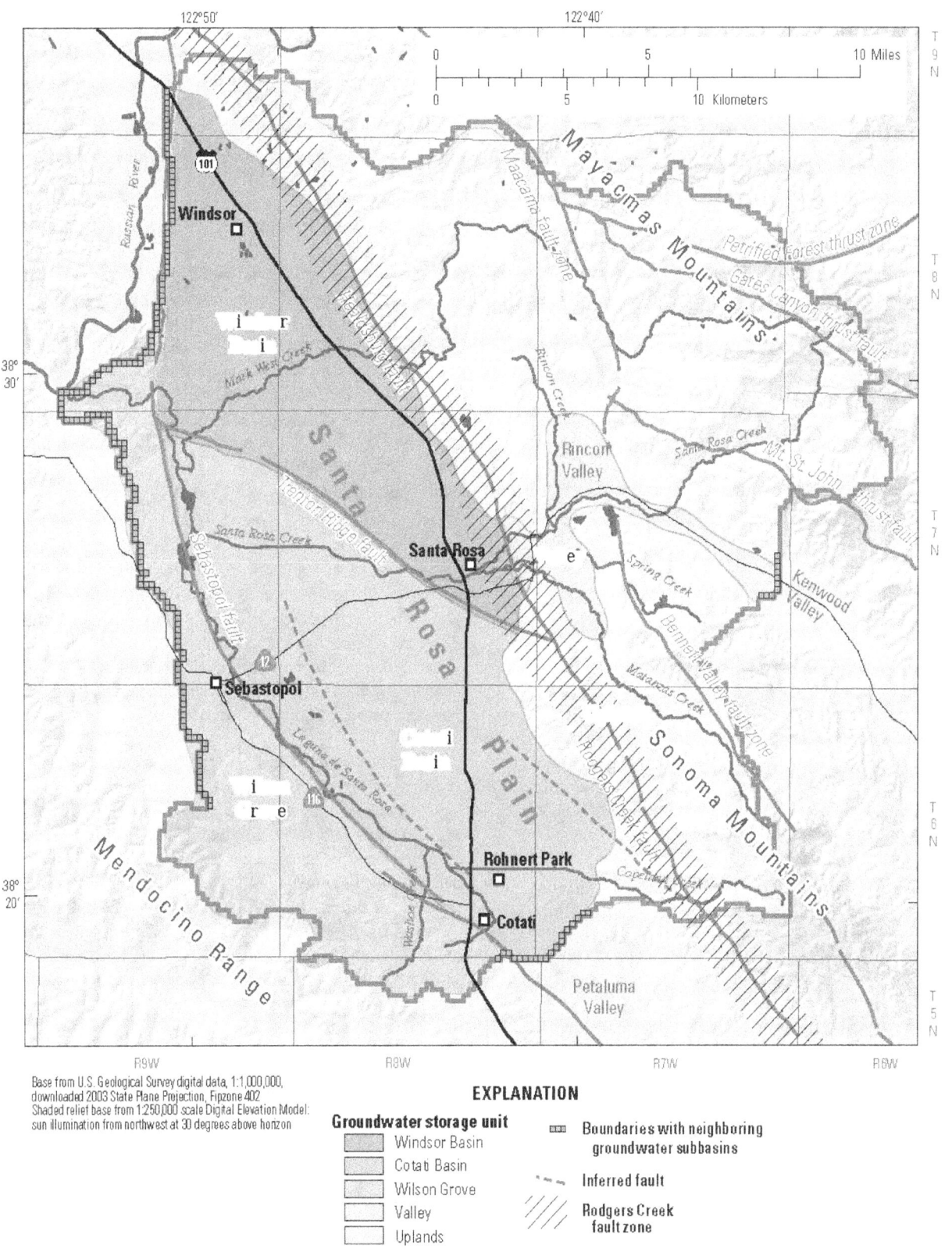

Figure 2. Locations where groundwater in the Santa Rosa Plain watershed could be in hydraulic communication with neighboring groundwater subbasins, Santa Rosa Plain watershed, California.

Irrigation Return Flows

On a local scale, irrigation return flows are a significant inflow to region 1. The source for most of the irrigation is groundwater pumped from within the SRPW; this represents a transfer of water from the saturated zone of region 3 to the soil zone. Agricultural pumpage is unreported; however, Hevesi and others (2011) estimated agricultural pumpage in the SRPW to range from 8,900 acre-foot (acre-ft) in 1974 to 46,600 acre-ft in 2008 by using a calibrated watershed model of the study area.

Other sources of irrigation water, including reclaimed municipal wastewater and imported water from the Russian River, are primary sources of irrigation water for some locations. The use of imported Russian River water for irrigation is less than about 1,000 acre-ft/yr and does not represent a significant inflow to the SRPW (Donald Seymour, Sonoma County Water Agency, written commun., 2010).

Inflow from irrigation is minimal to nonexistent during the wet, winter months for most years (except during droughts) and is highest during the late spring growing season and the warm, dry summer months. In the SRPW conceptual model, it is assumed that about 80 percent of water applied as irrigation is used by plants to satisfy crop demand, about 15 percent becomes irrigation-return flow to regions 2 and 3, and about 5 percent becomes surface-water runoff.

Groundwater Discharge

Groundwater in the SRPW often is discharged directly to a stream channel, lake, or spring in region 2. However, if the water table rises at an intra-channel location, groundwater can flow directly into the soil zone in region 1. Groundwater flow to the soil zone increases the water content of the soil, which in turn provides more water available for ET. Cardwell (1958) estimated that groundwater discharge lost to ET in the Laguna de Santa Rosa ranged from 4,000 to 6,000 acre-ft/yr; this value is probably lower than ET for the entire SRPW. If the soil zone becomes fully saturated, groundwater can discharge to surface-water bodies and contribute to runoff, or can simply discharge to land surface. The runoff also can re-infiltrate the soil zone downslope of the discharge zone.

Temporally, groundwater discharge can be an important inflow component to region 1 during wetter than normal periods. Spatially, groundwater discharge can be important in the low-lying areas of the SRPW, such as the Laguna de Santa Rosa, and in the mountains where springs can discharge.

Outflows from Region 1

The primary outflows from region 1 in the SRPW conceptual model are ET, surface-water runoff, and infiltration to the unsaturated zone.

Evapotranspiration

ET is the combined water loss to the atmosphere from evaporation of water and soil water and plant transpiration. ET is a function of potential evapotranspiration (PET), water availability, soil texture, vegetation type, vegetation density, and root depth. PET is the rate of ET that is possible given an unlimited water supply (California Irrigation Management Information System, 2004) under specific climate conditions (Maidment, 1993). If water supply is limited, ET will be less than PET.

The average PET rate is 46–48 inches per year (in./yr), or about 660,000 acre-ft/year, distributed over the SRPW (California Irrigation Management Information System, 2004). Woolfenden and others (2011) estimated the total actual ET from the soil, unsaturated, and saturated zones (regions 1 and 3) to be about 265,700 acre-ft/yr by using a preliminary watershed model. ET also occurs along many other stream channels in the SRPW where the water table is shallow enough for transpiration by phreatophytes.

Surface-Water Runoff

Precipitation can fall on pervious or impervious surfaces in the SRPW. Precipitation on pervious soils either infiltrates, ponds on the soil surface (eventually infiltrating or running off), runs off directly, or evaporates. Precipitation on impervious surfaces either ponds, runs off, or evaporates. For the purposes of the SRPW conceptual model, surface-water runoff includes overland flow and shallow subsurface flow (interflow).

The magnitude of surface-water runoff relative to rainfall is a primary driver for flow processes within region 1 and between region 1 and regions 2 and 3. Where runoff is high, less water is available for infiltration, ET, and, ultimately, recharge. The spatial distribution of runoff in the SRPW depends on soil thickness and type, the permeability of the soil and underlying bedrock, land use and topography. For example, steep hillsides, thin soil and vegetation cover, and impervious surfaces in urban areas are characteristics that cause more rapid runoff. In the upper drainages in rugged areas of the mountains surrounding the SRPW, the soils are thinner, the bedrock is relatively impermeable, and the slopes are steep, making these areas more conducive to runoff. As defined by the U.S. Department of Agriculture (2007), runoff potential is low along the western boundary, high to moderately high in the southern part of the Santa Rosa Plain and throughout the upland areas of the Sonoma and Mayacmas Mountains, and moderately low and moderate to high in the vicinity of Santa Rosa and throughout the northern part of the SRPW (see *chapter A*, fig. 10).

Infiltration and Recharge

For this study, infiltration is defined as the rate of water entering the soil zone, and recharge is defined as the rate of water entering the saturated zone. The average annual rate of infiltration is unknown; however, Cardwell (1958) reported a field experiment in the SRPW where about 27 percent of precipitation infiltrated. Assuming the average annual precipitation is 40 in., the potential average annual infiltration rate is about 153,000 acre-ft/yr over the entire SRPW. Note that Cardwell (1958) assumed that the infiltration equaled the recharge rate and did not account for any losses such as ET.

Storage in Region 1

Water in region 1 is stored on and above the land surface on a seasonal basis (months) as retention storage and below the land surface as soil moisture in the soil zone. Retention storage includes three components: (1) interception storage, (2) pervious-area depression storage, and (3) impervious-area retention storage (Linsley and others, 1982). Interception storage is the water held by the surfaces of vegetation; this could be significant in the forested areas of the SRPW mountains. Pervious-area depression storage is water ponded within depressions on the pervious land surface; this could be significant over short periods in flat-lying areas of the SRP. Impervious-area retention storage includes precipitation ponded on very low permeability surfaces, such as rooftops, parking lots, and exposed bedrock.

Soil-zone storage includes two components: field-capacity storage and fully saturated storage. Field-capacity storage is the volume of water retained by a soil after gravity drainage (Hillel, 1982); that is, air, water, and soil co-exist in a given volume of soil zone. Fully saturated storage is the volume of infiltrated water stored in all available pore spaces; that is, only water and soil co-exist in a given volume of soil zone. Field-capacity and fully saturated storage are dependent on soil texture and thickness (Hillel, 1982); therefore, they are expected to be greater in areas with higher clay content and thicker soils—for example, the southern part of the SRPW (see *chapter A*, figs. 7 and 8).

Region 2: Surface-Water Zone

In the SRPW conceptual model, inflows to region 2, the surface-water zone, include (1) overland flow and interflow from region 1, (2) groundwater discharge, (3) precipitation, and (4) reclaimed municipal wastewater. Outflows include (1) surface-water discharge, (2) streambed losses (seepage), (3) evaporation, and (4) diversions (for example, irrigation).

In the SRPW conceptual model, the location of the hydrologic features of region 2 is defined by a combination of the 1/24,000 scale high-resolution National Hydrographic Dataset (NHD; Simley and Carswell, 2009) and an updated hydrography map provided by Sonoma County Water Agency (Christopher Delaney, Sonoma County Water Agency, written commun., 2009). The NHD dataset includes mapped stream channels and water bodies (lakes and ponds). The refined hydrography map is very similar to the NHD data, but provides an updated version of the hydrography (for example, updated engineered channels). As described in *chapter B* of this report, the majority of water bodies in the SRPW are small (less than 10 acres), and therefore, it is assumed that water bodies are not significant water sources, sinks, or storage components.

Inflows to Region 2

The primary inflow to region 2 of the SRPW conceptual model is surface-water runoff from region 1. The surface-water runoff includes overland flow and shallow subsurface interflow. A secondary source of inflow to region 2 is groundwater discharge; this source could be the only inflow to region 2 during the dry season. Groundwater discharge to streams between July and September, as represented by average streamflow during these months, can range from 30 to 3,100 acre-ft/yr (see *chapter B*, table 3). Note that these values do not address the possibility of urban runoff or other sources adding to streamflow.

The groundwater-level contours indicate that groundwater could be discharging to the streams in the lower reaches of Mark West and Santa Rosa Creeks and the upper reaches of Santa Rosa Creek near the Rodgers Creek fault zone (see *chapter B*, fig. 29). Precipitation and reclaimed municipal wastewater also are inflows to region 2; however, these sources are small compared to the other sources in the SRPW conceptual model.

Outflows from Region 2

The primary outflow from region 2 of the SRPW conceptual model is stream discharge from the SRPW (either to the Russian River watershed to the north or the Petaluma River watershed to the south). Secondary water outflows from region 2 are seepage into the underlying unsaturated zone, evaporation, and local diversions of streamflow for irrigation.

Streamflow data from a 5-year record on lower Mark West Creek, about 2 mi upstream of the confluence with the Russian River, provide some indication of the surface-water outflows from region 2. During the 5 year period of record (water-years 2006–2010), the average flow rate was 288 cubic feet per second (ft³/s), or about 210,000 acre-ft/yr. The area upstream of this gage includes about 90 percent of the total area of the SRPW and, therefore, provides a good estimate of the total surface-water outflow from region 2.

Region 3: Subsurface Zone

Region 3, the subsurface zone of the SRPW conceptual model, includes the unsaturated and saturated zones. The areally extensive unsaturated zone extends from the bottom of the soil zone in region 1, or the bottom of the streambed or lakebed in region 2, to the top of the saturated zone (the water table) and is composed of unconsolidated materials (gravel, sand, silt, and clay), consolidated rock, water, and air. The saturated zone extends from the water table to the top of the low-permeability basement and is composed of unconsolidated materials, consolidated rock, and water. The thicknesses of the unsaturated zone and saturated zone vary with of the altitude of the water table. During relatively wet periods at some locations, the unsaturated zone can be absent if the water table intersects regions 1 or 2, thus creating a groundwater-discharge zone.

Unsaturated Zone

Percolation through the bottom of the soil zone in region 1 and seepage through streambeds and lakebeds in region 2 become infiltration to the top of the unsaturated zone and are a function of the physical characteristics of the unsaturated zone (vertical hydraulic conductivity, thickness, effective porosity, and vertical-saturation profile).

In the SRPW conceptual model, water flow in the unsaturated zone is assumed to be vertical; however, in reality, lateral flow can result from vertical heterogeneity. The long-term average rates of deep percolation, in response to infiltration from regions 1 and 2, generally are equal to the long-term recharge rate; however, the annual, monthly, and daily recharge rates can be much different than rates of deep percolation as a result of the time delay and dampening effect of the unsaturated zone on transient, unsaturated flow.

Inflows to the Unsaturated Zone

Inflow to the unsaturated zone is infiltration from the overlying soil zone of region 1, from overlying streambeds from region 2, and from septic-tank effluent. Areally, water infiltrates the soil zone in locations where precipitation is high, soils are more permeable, and ET is minimal. Locally, water infiltrates from runoff that collects in stream channels and percolates through permeable streambeds in losing reaches of streams. The infiltration rate, or vertical flux, from regions 1 and 2 is limited by the vertical hydraulic conductivity of the unsaturated zone. Under conditions when infiltration from the overlying zone is limited by a low hydraulic conductivity of the unsaturated zone, a fraction of the infiltration is rejected, and this water remains in region 1. At the water table, capillary rise into the unsaturated zone contributes minor inflows; this is assumed to be insignificant on the scale of the SRPW.

In the SRPW, infiltration from the soil zone is the greatest in the WG, WB, CB, and VAL storage units in areas where the Wilson Grove and Glen Ellen Formations are present at land surface (see *chapter B*, fig. 1). In addition, infiltration is higher where precipitation is higher (for example, the Wilson Grove highlands shown in *chapter A*, fig. 6). Although precipitation is high in the UPL storage unit (see *chapter A*, fig. 6), the infiltration is probably low because the Sonoma Volcanics and Franciscan Formation are at land surface throughout most of the storage unit (see *chapter B*, fig. 1). These geologic units generally have low hydraulic conductivities, and the land surface is relatively steep, which precludes high infiltration rates. Infiltration from stream channels can be greatest in the transition from the UPL storage unit to the SRP, where streamflow is concentrated in the channel, the subsurface tends to be more permeable (see *chapter A*, fig. 9), and the unsaturated zone is thick. As described in *chapter B* of this report, domestic pumpage is large in comparison to municipal pumpage; therefore, the potential recharge from septic-tank effluent also could be large. However, basin-wide nitrate concentrations in groundwater are relatively low, as reported in *chapter C* of this report, which indicates that septic-tank effluent is not a primary component of unsaturated-zone inflow.

Outflows from the Unsaturated Zone

Outflows from the unsaturated zone include recharge to the saturated zone and transpiration by plants. Recharge to the saturated zone is the primary outflow from the unsaturated zone and is dependent on the vertical permeability of the unsaturated and saturated interface. That is, if the interface is permeable, there is high recharge potential; however, if the interface is not permeable (permeable unsaturated zone overlying a low-permeability saturated zone), there is low recharge potential. Under conditions where the water table is rising, there is a decrease in the thickness of the unsaturated zone, and water stored in the unsaturated zone is added to the saturated zone, which contributes to recharge (this component of recharge is in addition to vertical percolation through the unsaturated zone).

In the SRPW, outflows from the unsaturated zone are highest in the WG storage unit where the permeable Wilson Grove Formation is present and WB and CB storage units, immediately downgradient (west) of the Rodgers Creek fault zone, where coarse-grained deposits are present along the channels (see *chapter B*, fig. 1). In contrast, outflows from the unsaturated zone are lower in the western parts of the WB and CB storage units because the unsaturated zone is dominated by fine-grained deposits of the Glen Ellen Formation (see *chapter B*, fig. 1).

Saturated Zone

Important hydrologic characteristics of the saturated zone of the SRPW groundwater-flow system include those that reflect the ability of the groundwater system to transmit, store, and release water; that allow for vertical passage of water between hydrogeologic units; and that control the flow of water across geologic or hydrologic boundaries. The movement of water through the saturated groundwater system is controlled by topography, aquifer and aquitard properties (including thickness and orientation of layers, structures, and fault planes), and the magnitude and distribution of recharge and discharge, including pumping. Aquifer and aquitard properties depend on the type of sediments and rocks composing the hydrogeologic system. Geologic structures, such as fault planes or zones, either can be flow barriers or flow conduits, depending on age, orientation, mineralization along the fault, and the juxtaposition of aquifers and aquitards across the fault.

Inflows to the saturated zone of region 3 include groundwater recharge from the unsaturated zone and underflow from adjacent groundwater basins. Outflows from the saturated zone include ET, discharge to streams and lakes, discharge to the soil zone, pumping, and underflow to adjacent basins.

Inflows to the Saturated Zone

Inflows to the saturated zone include groundwater recharge from the unsaturated zone, direct recharge from streams, and underflow from adjacent basins. Herbst and others (1982) estimated the average annual recharge to the Santa Rosa Plain groundwater basin between 1960 and 1975 was about 29,300 acre-ft. Assuming this estimate is correct, the average annual recharge for the SRPW would be greater than this value because the SRPW includes areas not included in the 1982 estimate: the northern half of Kenwood Valley, the area west of the Sebastopol Fault, and the mountains that border the Santa Rosa Plain. A preliminary coupled watershed-groundwater-flow model of the SRPW for water-years 1976–2009 was used to estimate an average groundwater underflow to the SRPW of about 70 acre-ft/yr (Woolfenden and others, 2011).

Outflows from the Saturated Zone

Discharge from the groundwater system can be (1) underflow to adjacent groundwater subbasins, (2) ET, (3) discharge to springs and streams in region 2, (4) pumping, or (5) discharge to the soil zone in region 1. ET was discussed in the section entitled "Outflows from Region 1." The average annual discharge from groundwater underflow estimated by using a preliminary coupled watershed-groundwater-flow model of the SRPW was about 1,200 acre-ft/yr during water-years 1976–2009 (Woolfenden and others, 2011).

Groundwater discharges directly to streams where the water table is at or above the bottom of the streambed and contributes to the baseflow component of total streamflow in region 2. Water-level contour maps indicate that Santa Rosa Creek consistently gains streamflow in the VAL storage unit. Groundwater discharged from springs is a source of baseflow for streams or is lost to ET. Springs are sensitive to changes in groundwater levels caused by natural variations in climate, by the development of groundwater resources, or by land-use changes (increases in impervious areas), and their flows often diminish or stop during dry periods or when nearby wells are pumped.

In the SRPW, pumping for public-supply, agricultural, and domestic uses is the largest component of groundwater discharge. The total annual pumpage from 70 public-supply wells from 1975–2009 ranged from about 2,000 acre-ft to about 8,300 acre-ft. Agricultural pumpage is unreported; however, Hevesi and others (2011) estimated agricultural pumpage for 1,072 agricultural wells by using a calibrated watershed model of the study area. Total estimated agricultural water demand ranged from 8,900 acre-ft in 1974 to 46,600 acre-ft in 2008. Domestic pumpage is unreported also; however, it was estimated for 1974–2010 by using population density and census tracts in rural areas with an assumed per capita consumptive-use factor of 0.19 acre-ft per person (California Department of Water Resources, 1994). Annual domestic pumpage estimates ranged from 11,077 to 23,202 acre-ft during 1974–2010, and pumpage increased as the rural population grew. The average total pumpage from all sources for 1975–2010 was about 48,000 acre-ft/yr. The maximum was about 53,400 acre-ft in 2008.

In the SRPW conceptual model, about 15 percent of the groundwater used directly for agricultural irrigation is assumed to be returned to the saturated zone as recharge, and about 5 percent is assumed to become runoff. In addition, an unknown, but likely small, percentage of domestic and municipal-well water used for landscape irrigation becomes recharge. A fraction of domestic-well water can contribute to recharge as leakage through underground septic systems. Municipal-well water can also indirectly contribute to recharge as reclaimed water used to supplement agricultural and landscape irrigation.

Pumping removes water from the saturated zone and can lower the water table, which in turn decreases baseflow in streams. The magnitude and spatial distribution of baseflow in the stream channels can decrease seasonally as pumping increases during the dry summer months to provide water for irrigation. In addition, baseflow can decrease with sustained or increased pumping near streams. It has not been determined whether groundwater discharge to stream channels and springs has decreased in the SRPW in response to pumping from the large number of wells that were drilled between the 1950s and 2010.

Summary of Conceptual Model

The geologic, hydrologic, and water-quality data presented in *chapters A–D* were used to develop the conceptual model for the SRPW (fig. 1).

Precipitation

Precipitation, primarily as rainfall, is the major source of inflow to the SRPW. The mean annual precipitation for the SRPW is approximately 40 in., or about 560,000 acre-ft/yr distributed over the 167,400 acre watershed. Precipitation is greatest (about 42–57 in./yr) in the Mayacmas and Sonoma Mountains in the UPL storage unit (see *chapter A*, fig. 6); however, because of the low permeability of the basement rocks and Sonoma Volcanics that compose these mountains, and the steep slope of the storage unit, most of the precipitation on this storage unit probably becomes runoff that contributes to streamflow and potential recharge in the downstream storage units. In addition, there is direct infiltration of precipitation on the SRP.

Streamflow

Mark West Creek, Santa Rosa Creek, and Laguna de Santa Rosa are the major streams that drain the SRPW. The main channel of Mark West Creek originates in the Mayacmas Mountains and is perennial throughout much of its length in the UPL storage unit, with summer flows maintained by numerous springs near the headwaters. Santa Rosa Creek also originates in the Mayacmas Mountains, in steep terrain, with mostly-natural vegetation. The upper part of Santa Rosa Creek and Matanzas Creek, one of its tributaries, are perennial streams in the UPL storage unit. As the streams flow through the VAL storage unit, they gain flow until just east of the Rodgers Creek fault zone, where groundwater-level contours indicate groundwater from the VAL storage unit discharges to the stream channel. Immediately west of the Rodgers Creek fault zone, groundwater-level contours indicate that the Santa Rosa Creek loses water to (recharges) the SRP groundwater basin. As the Santa Rosa Creek reaches the western end of the SRP, the stream begins to gain flow and is perennial. The Laguna de Santa Rosa, which originates in the southern part of the SRPW, is perennial in most sections.

Streamflow leaves the SRPW from Mark West Creek into the Russian River. There is a stream gage located near the outlet (MWCM); however, the record is short (water-years 2006–2010). Therefore, the MWCM record was extended (water years 1960–2010) by using the MOVE.1 technique and data from a neighboring stream gage on the Napa River (NAPN). The long-term estimated mean discharge for the extended 51-year time series was 265 ft^3/s, or about 192,000 acre-ft/yr.

Aquifer System

The Glen Ellen, Wilson Grove, and Petaluma Formations, and the Sonoma Volcanics have distinct aquifer properties and constitute the four principal water-bearing aquifer units in the study area. In general, the aquifer units transition from Sonoma Volcanics interbedded with the Petaluma Formation in the UPL storage unit east of the Rodgers Creek fault zone, to the Glen Ellen Formation overlying the Sonoma Volcanics in the VAL storage unit, to heterogeneous continental sediments of the Glen Ellen and Petaluma Formations beneath the SRP, to dominantly fine-grained marine sands of the Wilson Grove Formation on the west in the WG storage unit (fig. 3).

Earlier work reported the Glen Ellen Formation was as thick as 3,000 ft (Cardwell, 1958); however, recent work indicated that the formation is highly variable, but generally is a few hundred feet thick or less. In general, the fluvial sediments of the Glen Ellen Formation have low to moderate permeability and tend to be more permeable in the VAL storage unit and on the eastern end of the SRP than the remainder of the SRPW. Wells perforated in the Glen Ellen Formation that include some beds of moderately- to well-sorted, coarse-grained materials yield large amounts of water to wells.

The 2,700-ft-thick, sand-dominated Wilson Grove Formation is exposed in the low hills of the WG storage unit, west of the SRP, and it continues to the east for an uncertain distance beneath the SRP, where it is concealed by variable thicknesses of alluvial materials and the Glen Ellen Formation and interfingers with the Petaluma Formation (fig. 3). The predominance of relatively well-sorted marine sand and the low degree of cementation in the Wilson Grove Formation result in moderate permeability and moderate to high storativity.

The Petaluma Formation is dominated by fine-grained materials, either in thick beds or as interstitial material in poorly sorted silty and clayey sands or gravels. The formation is at least 3,000 ft thick in places within the study area, and has been divided into three distinct members (lower, middle, and upper) on the basis, in part, of the dominant grain-size and sorting. The lower member, which is up to 750 ft thick, is largely made up of dense beds of mudstone and has the lowest hydraulic conductivity. The formation coarsens in the middle and upper members, in which beds of poorly-sorted sands and gravels increase the hydraulic conductivity. The upper member is about 500 ft thick and unconformably overlies the middle member in the CB storage unit.

The Sonoma Volcanics include a thick accumulation of andesitic and basaltic tuffs containing interbedded lavas and volcaniclastic rocks and are an important aquifer in the UPL and VAL storage units (fig. 3). Lithologies with the greatest permeability include rubble zones between lava flows, beds of scoria and coarse tephra, air-fall tuffs, and some coarse-grained facies of volcaniclastic units.

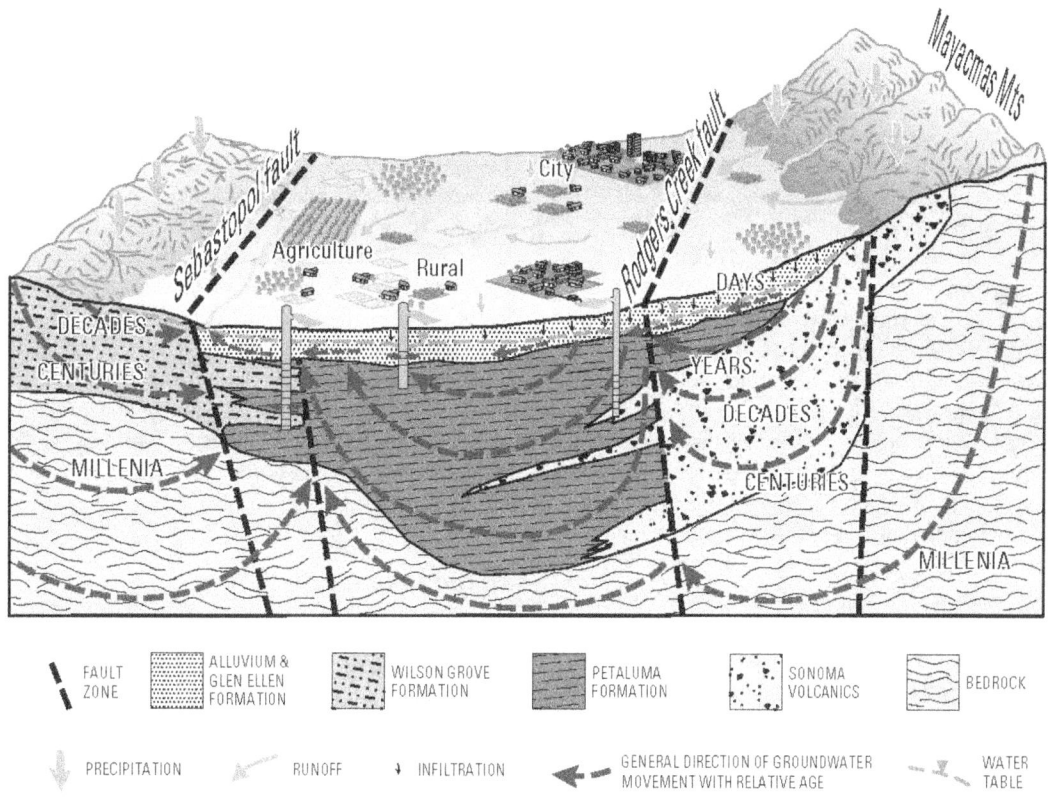

Figure 3.   Conceptual model of the groundwater-flow system, Santa Rosa Plain watershed, Sonoma County, California.

Groundwater Flow

In general, groundwater flows from the mountains in the UPL storage unit through the VAL storage unit and into the WB and CB storage units to the west, and from the Wilson Grove highlands in the WG storage unit on the west toward the Laguna de Santa Rosa on the western edge of the CB storage unit (fig. 3).

Samples from springs in the Mayacmas Mountains are a mixed cation-bicarbonate type water that have dissolved-solids concentrations less than about 100 mg/L—the lowest dissolved-solids concentrations in the SRPW. As groundwater migrates through the UPL and VAL storage units, the dissolved-solids concentration of the groundwater increases as a result of water-rock interactions and anthropogenic inputs, such as irrigation return flows and septic-tank discharge. The median dissolved-solids concentrations of well samples in the UPL and VAL storage units were 330 and 392 mg/L, respectively. The isotopic values of well samples in the UPL and VAL storage units were similar to the spring 2009 isotopic values from surface-water sites in the Mayacmas Mountains (see *chapter C*, figs. 6*B* and 6*D*), indicating that precipitation and runoff from the Mayacmas Mountains is the probable

source of recharge to the wells sampled in these storage units. Only three of the nine (about 33 percent) samples from the UPL and VAL storage units had detectable tritium concentrations, indicative of modern water (see *chapter C*, fig. 7*A*). The absence of modern water in many of these samples probably reflects the low vertical permeability of the basement rocks and Sonoma Volcanics that compose the UPL and VAL storage units. Samples from wells perforated in the Sonoma Volcanics, in the VAL storage unit, had uncorrected ^{14}C ages of 10,000 years before present, indicating that groundwater movement through the Sonoma Volcanics is relatively slow.

Groundwater-level and water-quality data indicate that the Rodgers Creek fault zone is a barrier to groundwater flow between the VAL storage unit and the WB and CB storage units, which compose the SRP. Groundwater-level contours east of the Rodgers Creek fault zone indicated that groundwater is discharging to the stream channel, which probably is the result of the low permeability of the fault zone limiting the lateral movement of groundwater across the fault zone, so that it is forced into the stream channel (fig. 3).

A deep well sample in the Rogers Creek fault zone, near Santa Rosa, had a dissolved-solids concentration of 1,510 mg/L, which is more than three times greater than the dissolved-solids concentrations analyzed in almost all other wells. The sample was a sodium-chloride type water with concentrations of boron in excess of 2,000 mg/L, which is hypothesized to be from deeply circulating groundwater rising along the fault zone. The uncorrected ^{14}C age of the sample was 27,000 years before present. The very old age of the water supported the hypothesized source of water to this well.

Comparison of data from the Rodgers Creek fault zone to available data from downgradient (west) of the fault zone indicated that the high dissolved-solids concentration of water in the fault zone did not significantly affect the water quality of wells in the WB and CB storage units (see *chapter C*, fig. 3C). Almost all well samples west of the Rodgers Creek fault zone had lower dissolved-solids concentrations than both the shallow and deep samples from the Rodgers Creek fault zone. The lower dissolved solids concentrations in samples from wells west of the Rodgers Creek fault zone indicated that groundwater underflow across the fault zone is minimal. The Rodgers Creek fault zone probably restricts the lateral movement of water from the UPL and VAL storage units to the WB and CB storage units, forcing groundwater to discharge toward the stream channels east of the fault zone, as indicated by the groundwater-level contours (see *chapter B*, fig. 29A). The lower dissolved-solids concentration sampled from wells downgradient of the Rodgers Creek fault zone probably is the result of recharge west (downgradient) of the fault zone.

In the SRP, samples from most shallow wells perforated or completed in the Glen Ellen Formation are a mixed cation-bicarbonate type water in which magnesium is the most abundant cation. Samples from wells perforated or completed in the Petaluma Formation, or in both the Glen Ellen and Petaluma Formations in the SRP, are a mixed cation-bicarbonate type water in which sodium is the most abundant cation or are a sodium-bicarbonate water. In general, the dissolved-solids concentrations of samples from the shallow wells were higher than dissolved-solids concentrations of samples from the deep wells, which could be the result of anthropogenic factors such as irrigation-return flows. The samples from wells perforated in the Petaluma Formation, or in both the Glen Ellen and Petaluma Formations in the CB storage unit, had higher dissolved-solids concentrations and a higher proportion of sodium than samples from wells perforated in the Wilson Grove Formation on the western end of the CB storage unit (see *chapter C*, fig. 3C). The difference in dissolved-solids concentration and water type indicated that the unnamed fault, east of the Sebastopol fault, is at least a partial barrier to groundwater flow.

The stable isotopes of water (oxygen-18 and deuterium) indicated a mixture of water sources in the SRP (see *chapter C*, fig. 6D). Most of the samples from wells in the SRP were isotopically heavier (less negative) than streamflow and well samples from the UPL and VAL storage units, indicating that streamflow and groundwater underflow from these storage units cannot be the sole source of recharge to the SRP. The heavier isotopic values indicated that at least some of the recharge to the SRP originates as precipitation falling directly on the lower elevations of the SRP.

The dissolved-solids concentrations in samples from wells perforated almost exclusively in the Wilson Grove Formation were lower than in samples from wells perforated in the Glen Ellen and Petaluma Formations (see *chapter C*, fig. 3C). The samples from wells with perforations in the Glen Ellen and Petaluma Formations immediately east of the WG storage unit are sodium-bicarbonate type water with dissolved-solids concentrations in excess of 420 mg/L (see *chapter C*, fig. 3C). The difference in water type and dissolved-solids concentrations indicated that there is little groundwater interaction between the Wilson Grove Formation and the neighboring Glenn Ellen and Petaluma Formations. Apparently, the Sebastopol fault limits the lateral groundwater movement from the WG storage unit to the CB storage unit, and low permeability clay layers in the Petaluma Formation limit the vertical groundwater movement between the Wilson Grove Formation and overlying Petaluma Formation.

Groundwater Age

Tritium (^{3}H) concentrations were analyzed in 35 samples collected from 30 wells in the SRPW and ranged from less than 0.3 (detection limit of samples collected in 2004) to 2.4 tritium units (TU). About 43 percent of the samples had detectable tritium concentrations, indicating that these samples contained some modern water (water recharged since 1952). As expected, modern recharge was more prevalent in the shallow well samples. The vertical migration of recharge in the SRP probably is retarded by the presence of low permeability clay deposits in the Glen Ellen and Petaluma Formations.

Measured ^{14}C activities for the 16 well samples analyzed ranged from 1.3 to 84.7 percent modern carbon. These ^{14}C activities represent uncorrected ages of 1,000–34,000 years before present. The deep-well samples all had uncorrected ^{14}C ages of 4,000 years or older, and five of the deep well samples were 10,000 years or older. The relatively old age of the deep-well samples supported the observation made from the tritium data that the vertical migration of modern recharge probably is restricted by low-permeability clay deposits in the Glen Ellen and Petaluma Formations. The low-permeability clay deposits also confine the deeper aquifer systems, which explains the rapid response of the deeper aquifer systems to pumping and cessation of pumping. The oldest water sampled was from a well near the Laguna de Santa Rosa, which is near the end of a long groundwater-flow-line through the SRP groundwater basin (fig. 3).

Data Gaps

Analyses of hydrologic, hydrogeologic, and geochemical data collected for this study indicated gaps in the data available for the SRPW. These gaps included a lack of depth-dependent data, additional water-quality data, and agricultural and domestic pumpage data.

There is only one nested piezometer site (wells W33–35) in the SRPW, and most wells are perforated over multiple aquifers. This lack of depth-dependent water-level and water-quality data makes it difficult to calibrate a groundwater-flow model by aquifer or layer.

The variability observed in the water-quality data (for example, chloride and specific-conductance data) cannot be explained by using the available dataset. Additional data collection (for example, isotopes, boron, and age dates) is necessary.

Groundwater pumping is a primary sink from the saturated zone, and the maximum total annual pumpage could be as much as 53,400 acre-ft. Of this total, only municipal pumping, which could make up only 15 percent of the total, is reported. The balance is composed of agricultural and domestic pumping, which are unreported. In order to develop and calibrate a numerical model of the SRPW, improved estimates of the unreported pumpage must be made, and the locations of the wells identified.

References Cited

California Department of Water Resource, 1994, California water plan update: Bulletin 160-93, October 1994, v. 2, 315 p.

California Irrigation Management Information (CIMIS), 2005, Department of Water Resources and University of California, Davis: data accessed on March 31, 2010, at URL *http://www.cimis.water.ca.gov/*.

Cardwell, G.T., 1958, Geology and ground water in the Santa Rosa and Petaluma areas, Sonoma County, California: U.S. Geological Survey Water-Supply Paper 1427, 273 p. and 5 plates.

Daly, Christopher, Gibson, W.P., Doggett, Matthew, Smith, Joseph, and Taylor, George, 2004, Up-to-date monthly climate maps for the conterminous United States: Proceedings of the 14th American Meteorological Society Conference on Applied Climatology, 84th AMS Annual Meeting Combined Preprints, American Meteorological Society, Seattle, Washington, January 13–16, 2004, Paper P5.1, CD-ROM.

Herbst, C.M., Jacinto, D.M., and McGuire, R.A., 1982, Evaluation of ground water resources, Sonoma County, volume 2: Santa Rosa Plain: California Department of Water Resources Bulletin 118-4, 107 p., 1 plate.

Hevesi, J.A., Woolfenden, L.R., Niswonger, R.G., Regan, R.S., and Nishikawa, Tracy, 2011, Decoupled application of the integrated hydrologic model, GSFLOW, to estimate agricultural irrigation in the Santa Rosa Plain, California, *in* Maxwell, R.M., Poeter, E.P., Hill, M.C., and Zheng, Chunmiao, eds., MODFLOW and More 2011: Integrated hydrologic modeling–Conference proceedings, June 5–8, 2011: Golden, Colorado, International Groundwater Modeling Center, p. 115–119.

Hillel, Daniel, 1982, Introduction to soil physics: San Diego, CA, Academic Press, Inc., 364 p.

Linsley, R.K., Kohler, M.A., and Paulus, J.L.H., 1982, Hydrology for engineers (3d. ed.): New York, NY, McGraw-Hill, Inc., 508 p.

Maidment, D.R., 1993, Handbook of hydrology: New York, NY, McGraw-Hill, Inc., 750 p.

Markstrom, S.L., Niswonger, R.G., Regan, R.S., Prudic, D.E., and Barlow, P.M., 2008, GSFLOW—Coupled groundwater and surface-water flow model based on the integration of the Precipitation-Runoff Modeling System (PRMS) and the Modular Ground-Water Flow Model (MODFLOW-2005): U.S. Geological Survey Techniques and Methods 6-D1, 240 p.

Simley, J.D. and Carswell Jr., W.J., 2009, The National Map – Hydrography: U.S. Geological Survey Fact Sheet 2009-3054, 4 p.

U.S. Department of Agriculture, Natural Resources Conservation Service, 2007, National Engineering Handbook, Part 630 Hydrology, Chapter 7 Hydrologic Soil Groups: Washington, DC, 16 p.

Woolfenden, L.R., Hevesi, J.A., Niswonger, R.G., Regan, R.S., and Nishikawa, Tracy, 2011, Modeling a complex hydrologic system with an integrated hydrologic model *in* Maxwell, R.M., Poeter, E.P., Hill, M.C., and Zheng, Chunmiao, eds., MODFLOW and More 2011: Integrated hydrologic modeling–Conference proceedings, June 5–8, 2011: Golden, Colorado, International Groundwater Modeling Center, p. 134–138.

Appendix A.   Selected chemical and physical properties and inorganic constituents in samples from selected springs, and wells, Santa Rosa Plain Watershed, Sonoma County, California, 1947–2010.

Appendix A provided separately as a Microsoft Excel® file.

www.ingramcontent.com/pod-product-compliance
Lightning Source LLC
Chambersburg PA
CBHW081443170526
45166CB00008B/2301